Lecture Notes in Computer Science

T0238385

Commenced Publication in 1973
Founding and Former Series Editors:
Gerhard Goos, Juris Hartmanis, and Jan van Leeuwen

Thomas Erlebach   Christos Kaklamanis (Eds.)

Approximation and Online Algorithms

4th International Workshop, WAOA 2006
Zurich, Switzerland, September 14-15, 2006
Revised Papers

 Springer

Volume Editors

Thomas Erlebach
University of Leicester
Department of Computer Science
University Road, Leicester, LE1 7RH, UK
E-mail: t.erlebach@mcs.le.ac.uk

Christos Kaklamanis
University of Patras
Department of Computer Engineering and Informatics
26500, Rio, Patras, Greece
E-mail: kakl@ceid.upatras.gr

Library of Congress Control Number: 2006939787

CR Subject Classification (1998): F.2.2, G.2.1-2, G.1.2, G.1.6, I.3.5, E.1

LNCS Sublibrary: SL 1 – Theoretical Computer Science and General Issues

ISSN   0302-9743
ISBN-10   3-540-69513-3 Springer Berlin Heidelberg New York
ISBN-13   978-3-540-69513-4 Springer Berlin Heidelberg New York

Springer is a part of Springer Science+Business Media

springer.com

© Springer-Verlag Berlin Heidelberg 2006
Printed in Germany

Typesetting: Camera-ready by author, data conversion by Scientific Publishing Services, Chennai, India
Printed on acid-free paper   SPIN: 11970125   06/3142   5 4 3 2 1 0

Preface

The 4th Workshop on Approximation and Online Algorithms (WAOA 2006) focused on the design and analysis of algorithms for online and computationally hard problems. Both kinds of problems have a large number of applications from a variety of fields. WAOA 2006 took place at ETH Zurich in Zurich, Switzerland, during September 14–15, 2006. The workshop was part of the ALGO 2006 event that also hosted ESA, WABI, IWPEC, and ATMOS. The three previous WAOA workshops were held in Budapest (2003), Rome (2004), and Palma de Mallorca (2005). The proceedings of these previous WAOA workshops have appeared as LNCS volumes 2909, 3351 and 3879, respectively.

Topics of interest for WAOA 2006 were: algorithmic game theory, approximation classes, coloring and partitioning, competitive analysis, computational finance, cuts and connectivity, geometric problems, inapproximability results, mechanism design, network design, packing and covering, paradigms for design and analysis of approximation and online algorithms, randomization techniques, real-world applications, and scheduling problems. In response to the call for papers, we received 62 submissions. Each submission was reviewed by at least three referees, and the vast majority by at least four referees. The submissions were mainly judged on originality, technical quality, and relevance to the topics of the conference. Based on the reviews, the Program Committee selected 26 papers.

We are grateful to Andrei Voronkov for providing the EasyChair conference system, which was used to manage the electronic submissions, the review process, and the electronic PC meeting. It made our task much easier.

We would also like to thank all the authors who submitted papers to WAOA 2006 as well as the local organizers of ALGO 2006.

November 2006   Thomas Erlebach
                                                  Christos Kaklamanis

Organization

Program Co-chairs

Thomas Erlebach   University of Leicester
Christos Kaklamanis   University of Patras

Program Committee

Evripidis Bampis	University of Evry
Reuven Bar-Yehuda	Technion Haifa
Leah Epstein	University of Haifa
Thomas Erlebach	University of Leicester
Klaus Jansen	Universität Kiel
Christos Kaklamanis	University of Patras
Jochen Könemann	University of Waterloo
Danny Krizanc	Wesleyan University
Madhav Marathe	Virginia Tech
Seffi Naor	Microsoft Research and Technion, Israel
Alessandro Panconesi	University of Rome "La Sapienza"
Pino Persiano	Università di Salerno
Martin Skutella	Universität Dortmund
Roberto Solis-Oba	University of Western Ontario
Rob van Stee	Universität Karlsruhe

Additional Referees

Amjad Aboud	Sergio De Agostino	Laurent Gourvès
Ernst Althaus	Gianluca De Marco	Gregory Gutin
Eric Angel	Florian Diedrich	M.T. Hajiaghayi
Spyros Angelopoulos	György Dósa	Alex Hall
Vincenzo Auletta	Christoph Dürr	Han Hoogeveen
Nikhil Bansal	Pierre-Francois Dutot	Csanád Imreh
Gill Barequet	Alon Efrat	Yuval Ishai
Cristina Bazgan	Ran El-Yaniv	Liran Katzir
Eli Ben-Sasson	Roee Engelberg	Rohit Khandekar
Jit Bose	Guy Even	Samir Khuller
Niv Buchbinder	Lene M. Favrholdt	Stavros Kolliopoulos
Alberto Caprara	Dimitris Fotakis	Goran Konjevod
Deepti Chafekar	Martin Fürer	Alexander Kononov
JiangZhuo Chen	Stefan Funke	Guy Kortsarz

Sven O. Krumke
V.S. Anil Kumar
Christian Laforest
Asaf Levin
Matthew Macauley
Ionnis Milis
Jérôme Monnot
Shlomo Moran
Pat Morin
Petra Mutzel
Lata Narayanan
Tom O'Connell
Ojas Parekh
Marco Pellegrini
Kirk Pruhs
Dror Rawitz

Joachim Reichel
Yossi Richter
Guido Schäfer
Heiko Schilling
Roy Schwartz
Ulrich M. Schwarz
Danny Segev
Hadas Shachnai
Sunil Shende
Gennady Shmonin
Mohit Singh
René Sitters
Alexander Souza
Mauro Sozio
S. S. Ravi
Nicolas Stier

Tami Tamir
Orestis A. Telelis
Nicolas Thibault
Shripad Thite
Ralf Thöle
Alessandro Tiberi
Eric Torng
Denis Trystram
Carmine Ventre
Tjark Vredeveld
Oren Weimann
Prudence Wong
Michal Ziv-Ukelson
Vadim Zverovich

Table of Contents

Approximation Algorithms for Scheduling Problems with Exact Delays*

Alexander A. Ageev and Alexander V. Kononov

Sobolev Institute of Mathematics, pr. Koptyuga 4, Novosibirsk, Russia
{ageev,alvenko}@math.nsc.ru

Abstract. We give first constant-factor approximations for various cases of the coupled-task single machine and two-machine flow shop scheduling problems with exact delays and makespan as the objective function.
In particular, we design 3.5- and 3-approximation algorithms for the
general cases of the single-machine and the two-machine problems, respectively. We also prove that the existence of a $(2 - \varepsilon)$-approximation
algorithm for the single-machine problem as well as the existence of a
$(1.5 - \varepsilon)$-approximation algorithm for the two-machine problem implies
P=NP. The inapproximability results are valid for the cases when the
operations of each job have equal processing times and for these cases
the approximation ratios achieved by our algorithms are very close to
best possible: we prove that the single machine problem is approximable
within a factor of 2.5 and the two-machine problem is approximable
within a factor of 2.

1   Introduction

In this paper we consider two scheduling problems with exact delays. In both
problems a set $J = \{1, \ldots, n\}$ of independent jobs is given. Each job $j \in J$
consists of two operations with processing times a_j and b_j separated by a given
intermediate delay l_j, which means that the second operation of job j must start
processing exactly l_j time units after the first operation of job j has been completed. In the single machine problem all operations are executed by a single
machine. In the two-machine (flow shop) problem the first operations are executed by the first machine and the second ones by the second one. It is assumed
that at any time no machine can process more than one operation and no pre-
emptions are allowed in processing of any operation. The objective is to minimize
the makespan (the schedule length). Extending the standard three-field notation
scheme introduced by Graham et al. [6] we denote the single machine problem
by $1 \mid \text{exact } l_j \mid C_{\max}$ and the two-machine problem by $F2 \mid \text{exact } l_j \mid C_{\max}$.

The problems with exact delays arise in command-and-control applications in
which a centralized commander distributes a set of orders (associated with the
first operations) and must wait to receive responses (corresponding to the second

---

* Research supported by the Russian Foundation for Basic Research, projects 05-01-
00960, 06-01-00255.

T. Erlebach and C. Kaklamanis (Eds.): WAOA 2006, LNCS 4368, pp. 1–14, 2006.

operations) that do not conflict with any other (for more extensive discussion on the subject, see [4,8]). Research papers on problem $1 \mid \text{exact } l_j \mid C_{\max}$ are mainly motivated by applications in pulsed radar systems, where the machine is a multifunctional radar whose purpose is to simultaneously track various targets by emitting a pulse and receiving its reflection some time later [2,5,7,4,8]. Coupled-task scheduling problems with exact delays also arise in chemistry manufacturing where there often may be an exact technological delay between the completion time of some operation and the initial time of the next operation.

1.1   Related Work

Farina and Neri [2] present a greedy heuristic for a special case of problem $1 \mid \text{exact } l_j \mid C_{\max}$. Izquierdo-Fuente and Casar-Corredera [5] develop a Hopfield neural network for the problem. Elshafei et al. [4] present a Lagrange relaxation algorithm based on a discretization of the time horizon. Orman and Potts [7] establish that the problem is strongly NP-hard even in some special cases. In particular, they prove it for $1 \mid \text{exact } l_j, a_j = b_j = l_j \mid C_{\max}$. Yu [9], [10] proves that the two machine problem $F2 \mid \text{exact } l_j \mid C_{\max}$ is strongly NP-hard even in the case of unit processing times, which implies that the single machine problem is strongly NP-hard in the case of unit processing times as well (see [10]). Ageev and Baburin [1] present non-trivial constant-factor approximation algorithms for both the single and two machine problems under the assumption of unit processing times. More specifically, in [1] it is shown that problem $1 \mid \text{exact } l_j, a_j = b_j = 1 \mid C_{\max}$ is approximable within a factor of 7/4 and problem $F2 \mid \text{exact } l_j \ a_j = b_j = 1 \mid C_{\max}$, within a factor of 3/2.

1.2   Our Results

In this paper we present first constant-factor approximation algorithms for the general cases of $1 \mid \text{exact } l_j \mid C_{\max}$ and $F2 \mid \text{exact } l_j \mid C_{\max}$. We construct a 3.5-approximation algorithm for solving the single machine problem and 3-approximation algorithms for its special cases when $a_j \leq b_j$, or $a_j \geq b_j$ for all $j \in J$. We also show that the latter algorithms provide a 2.5-approximation for the case when $a_j = b_j$ for all $j \in J$. Moreover, we prove that problem $1 \mid \text{exact } l_j \mid C_{\max}$ is not $(2 - \varepsilon)$-approximable unless P=NP even in the case of $a_j = b_j$ for all $j \in J$. Addressing problem $F2 \mid \text{exact } l_j \mid C_{\max}$ we present a 3-approximation algorithm for the general case and show that it provides a 2-approximation for the cases when $a_j \leq b_j$, or $a_j \geq b_j$ for all $j \in J$. Furthermore, we prove that the problem is not $(1.5 - \varepsilon)$-approximable unless P=NP even in the case of $a_j = b_j$ for all $j \in J$. All designed algorithms can be implemented in $O(n \log n)$ time. The inapproximability results show that the approximation ratios achieved by our algorithms in the cases when $a_j = b_j$ for all $j \in J$ are very close to best possible: the single machine problem is approximable within a factor of 2.5 and not approximable within a factor of $(2 - \varepsilon)$; the two machine problem is approximable within a factor of 2 and not approximable within a factor of $(1.5 - \varepsilon)$. Approximability results established to date for the problems are summarized in Table 1.

Table 1. A summary of the approximability results

problem	appr. factor	inappr. bound	ref.
$1 \mid$ exact $l_j, \mid C_{\max}$	3.5	$2 - \varepsilon$	this paper
$1 \mid$ exact $l_j, a_j \leq b_j \mid C_{\max}$	3	$2 - \varepsilon$	this paper
$1 \mid$ exact $l_j, a_j \geq b_j \mid C_{\max}$	3	$2 - \varepsilon$	this paper
$1 \mid$ exact $l_j, a_j = b_j \mid C_{\max}$	2.5	$2 - \varepsilon$	this paper
$1 \mid$ exact $l_j, a_j = b_j = 1 \mid C_{\max}$	1.75		[1]
$F2 \mid$ exact $l_j \mid C_{\max}$	3	$1.5 - \varepsilon$	this paper
$F2 \mid$ exact $l_j, a_j \leq b_j \mid C_{\max}$	2	$1.5 - \varepsilon$	this paper
$F2 \mid$ exact $l_j, a_j \geq b_j \mid C_{\max}$	2	$1.5 - \varepsilon$	this paper
$F2 \mid$ exact $l_j, a_j = b_j = 1 \mid C_{\max}$	1.5		[1]

1.3 Basic Notation

For both problems an instance will be represented as a collection of triples $\{(a_j, l_j, b_j) : j \in J\}$ where $J = \{1, \ldots, n\}$ is the set of jobs, a_j and b_j are the lengths of the first and the second operations of job j, respectively and l_j is the given delay between these operations. As usual, we assume that all input numbers are nonnegative integers. For a schedule σ and any $j \in J$, denote by $\sigma(j)$ the starting time of the first operation of job j. As the starting times of the first operations uniquely determine the starting times of the second operations, any feasible schedule is uniquely specified by the collection of starting times of the first operations $\{\sigma(1), \ldots, \sigma(n)\}$. For a schedule σ and any $j \in J$, denote by $C_j(\sigma)$ the completion time of job j in σ; note that $C_j(\sigma) = \sigma(j) + l_j + a_j + b_j$ for all $j \in J$. The length of a schedule σ is denoted by $C_{\max}(\sigma)$ and thus $C_{\max}(\sigma) = \max_{j \in J} C_j(\sigma)$. The length of a shortest schedule is denoted by $C_{\max}^*$.

The remainder of the paper is organized as follows. In Section 2 and 3 we describe and analyze the algorithms for the single machine and two-machine problems, respectively. Section 4 contains the inapproximability results.

2 Algorithms for the Single Machine Problem

In this section we describe and analyze approximation algorithms for the general and some special cases of the single machine problem.

2.1 Algorithms for Special Cases

We begin with presenting algorithm 1M$\leq$ for the case when $a_j \leq b_j$ for all $j \in J$.

Informally, the algorithm sorts the jobs in nonincreasing order of delays and then successively constructs segments of the output schedule, which we call blocks. An s-th block is the maximum possible subsequence of jobs $\{j_s, \ldots, j_{s+1} - 1\}$ that admits a feasible schedule in which the machine continuously processes the second operations of $j_1, \ldots, j_{s+1} - 1$. The performance analysis is based on the remarkable observation that the total idle time within each block except the first one can be evaluated via the processing times of the previously scheduled jobs.

ALGORITHM 1M$\leq$.

PHASE I *(jobs ordering)*. Number the jobs in the following way:

$$a_1 + l_1 \geq a_2 + l_2 \geq \ldots \geq a_n + l_n \ . \tag{1}$$

PHASE II *(constructing indices j_s)*. By examining the set of jobs in the order $j = 1, \ldots, n$ compute the indices $j_1 < j_2 < \ldots < j_r \leq n$ in the following way.

Step 1. Set $j_1 = 1$. If $\sum_{s=1}^{t-1} b_s \leq l_t$ for all $t = 2, \ldots, n$, then set $r = 1$, otherwise go to Step 2.

Step $k(k \geq 2)$. Set j_k to be equal to the minimum index among indices $t > j_{k-1}$ such that $\sum_{s=j_{k-1}}^{t-1} b_s > l_t$. If $j_k = n$ or $\sum_{s=j_k}^{t-1} b_s \leq l_t$ for all $t = j_k + 1, \ldots, n$, then set $r = k$, otherwise go to Step $k + 1$.

PHASE III *(constructing the schedule)*. Set $\sigma(j_1) = \sigma(1) = 0$. If $r > 1$, then for $s = 2, \ldots, r$ set

$$\sigma(j_s) = \sigma(j_{s-1}) + a_{j_{s-1}} + l_{j_{s-1}} + \sum_{k=j_{s-1}}^{j_s-1} b_k \ . \tag{2}$$

For every $j \in J \setminus \{j_1, \ldots, j_r\}$, set

$$\sigma(j) = \sigma(j_s) + a_{j_s} + l_{j_s} - a_j - l_j + \sum_{k=j_s}^{j-1} b_k \tag{3}$$

where s is the maximum index such that $j_s < j$.

Example. Consider the following instance of problem $1 \mid$ exact $l_j, a_j \leq b_j \mid C_{\max}$ (the jobs are ordered according to Phase I):

$$\{(1, 6, 2), (2, 4, 3), (1, 5, 4), (1, 3, 2), (1, 3, 1), (1, 2, 3)\}.$$

Phase II finds that $i_1 = 1$, $i_2 = 4$, $i_3 = 6$, i. e., we have three blocks: $B_1 = \{1, 2, 3\}$, $B_2 = \{4, 5\}$, $B_3 = \{6\}$. Finally, Phase III computes the schedule σ: $\sigma(1) = 0$, $\sigma(2) = 3$, $\sigma(3) = 6$, $\sigma(4) = 16$, $\sigma(5) = 18$, $\sigma(6) = 23$ (see Fig. 1).

Correctness and running time. For convenience, set $j_{r+1} = n + 1$. Note that the set of jobs J splits into r disjoint subsets $B_s = \{j_s, \ldots, j_{s+1} - 1\}$, $s = 1, \ldots, r$ (we will further refer to them as *blocks*). The following lemma shows that algorithm 1M$\leq$ constructs a feasible schedule and describes its structure.

Lemma 1. *Let $1 \leq s \leq r$.*

(i) *For any two jobs $j', j'' \in B_s$ such that $j' < j''$, $\sigma(j'') \geq \sigma(j') + a_{j'}$.*
(ii) *For any job $j \in B_s$, the first operation of j completes before starting the second operation of job j_s.*
(iii) *The completion time of job $j_{s+1} - 1$ coincides with the starting time of the first operation of job j_{s+1}.*

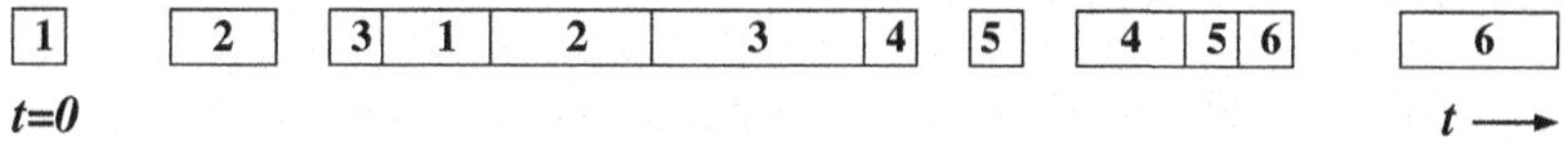

Fig. 1. The schedule constructed by algorithm 1M$\leq$

(iv) *Within the time interval*

$$[\sigma(j_s), \sigma(j_s) + a_{j_s} + l_{j_s} + \sum_{k=j_s}^{j_{s+1}-1} b_k]$$

the machine executes both operations of each job in B_s and only these operations.

(v) *Within the time interval $[\sigma(j_s), \sigma(j_s) + a_{j_s} + l_{j_s}]$ the machine processes the first operations of all jobs in B_s in the order $j_s, \ldots, j_{s+1} - 1$ and only these operations (with possible idle times).*

(vi) *Within the time interval*

$$[\sigma(j_s) + a_{j_s} + l_{j_s}, \sigma(j_s) + a_{j_s} + l_{j_s} + \sum_{k=j_s}^{j_{s+1}-1} b_k]$$

the machine without idle times processes the second operations of all jobs in B_s in the order $j_s, \ldots, j_{s+1} - 1$.

Proof. First observe that (i), (iii), and (vi) imply (iv); (i), (ii), (iii) yield (v) as well.

Now let $j', j'' \in B_s$ and $j'' > j'$. By (3) we obtain that

$$\sigma(j'') - \sigma(j') = \sigma(j_s) + a_{j_s} + l_{j_s} - a_{j''} - l_{j''} + \sum_{k=j_s}^{j''-1} b_k$$

$$- \left(\sigma(j_s) + a_{j_s} + l_{j_s} - a_{j'} - l_{j'} + \sum_{k=j_s}^{j'-1} b_k \right)$$

$$= a_{j'} + l_{j'} - a_{j''} - l_{j''} + \sum_{k=j'}^{j''-1} b_k \; . \tag{4}$$

By using $b_{j'} \geq a_{j'}$, (4), and (1) we obtain

$$\sigma(j'') - \sigma(j') - a_{j'} \geq b_{j'} + l_{j'} - a_{j''} - l_{j''} \geq a_{j'} + l_{j'} - a_{j''} - l_{j''} \geq 0 \; ,$$

which proves (i). Let $j \in B_s$. Then by the construction of j_s, $\sum_{k=j_s}^{j-1} b_k \leq l_j$. By (3) it follows that

$$\sigma(j) + a_j = \sigma(j_s) + a_{j_s} + l_{j_s} - l_j + \sum_{k=j_s}^{j-1} b_k \leq \sigma(j_s) + a_{j_s} + l_{j_s} \; ,$$

which yields (ii). Next we have

$$
\begin{aligned}
C_{j_{s+1}-1}(\sigma) &= \sigma(j_{s+1}-1) + a_{j_{s+1}-1} + l_{j_{s+1}-1} + b_{j_{s+1}-1} \\
\text{(by (3)} &= \sigma(j_s) + a_{j_s} + l_{j_s} - a_{j_{s+1}-1} - l_{j_{s+1}-1} \\
&\quad + \sum_{k=j_s}^{j_{s+1}-2} b_k + a_{j_{s+1}-1} + l_{j_{s+1}-1} + b_{j_{s+1}-1} \\
&= \sigma(j_s) + a_{j_s} + l_{j_s} + \sum_{k=j_s}^{j_{s+1}-1} b_k \\
\text{(by (2))} &= \sigma(j_{s+1}) \ ,
\end{aligned}
$$

which establishes (iii). By (4) we have that

$$
\sigma(j'') + a_{j''} + l_{j''} = \sigma(j') + a_{j'} + l_{j'} + \sum_{k=j'}^{j''-1} b_k \geq \sigma(j') + a_{j'} + l_{j'} + b_{j'} \ , \quad (5)
$$

which means that the second operation of job j'' starts after the completion of job j'. Moreover, if $j'' = j'+1$, then the inequality in (5) holds with equality, which means that the second operation of job j'' starts exactly after the completion of job j' and thereby (vi) is verified. □

It is easy to see that the most time consuming part of the algorithm is the sorting on Phase I and so its running time is $O(n \log n)$.

Approximation ratio. First, note that $C^*_{\max}$ is at least the load of the machine and the maximum length of a job, i. e.,

$$
C^*_{\max} \geq \max\{\sum_{j=1}^{n}(a_j + b_j), \max_{j \in J}(a_j + b_j + l_j)\} \ . \quad (6)
$$

For $s = 1, \ldots, r$, set $H_s = a_{j_s} + l_{j_s} + \sum_{k=j_s}^{j_{s+1}-1} b_k$. By (iv) and (vi) of Lemma 1

$$
C_{\max}(\sigma) = \sum_{s=1}^{r} H_s = \sum_{s=1}^{r}(a_{j_s} + l_{j_s}) + \sum_{j=1}^{n} b_j \ . \quad (7)
$$

Recall that by the construction of j_s for each $s \geq 2$, $\sum_{k=j_{s-1}}^{j_s-1} b_k > l_{j_s}$. Hence (7) implies that

$$
\begin{aligned}
C_{\max}(\sigma) &\leq \sum_{s=1}^{r} a_{j_s} + l_1 + \sum_{s=2}^{r}\sum_{k=j_{s-1}}^{j_s-1} b_k + \sum_{j=1}^{n} b_j \\
&\leq \left(\sum_{s=1}^{r} a_{j_s} + \sum_{j=1}^{n} b_j\right) + \sum_{j=1}^{n} b_j + l_1 \ . \quad (8)
\end{aligned}
$$

Thus by (6) for the case when $a_j \leq b_j$ for all $j \in J$, we have $C_{\max}(\sigma) \leq 3 \cdot C_{\max}^*$. In the case when $a_j = b_j$ for all $j \in J$, (6) implies that $\sum_{j=1}^{n} b_j \leq \frac{1}{2} C_{\max}^*$, which together with (8) yields $C_{\max}(\sigma) \leq \frac{5}{2} \cdot C_{\max}^*$. Summing up we obtain the following

Theorem 1

  (i) *Algorithm 1M≤ finds a schedule of length at most thrice the length of a shortest schedule.*
 (ii) *When applied to problem* $1 \mid$ *exact* $l_j, a_j = b_j \mid C_{\max}$ *algorithm 1M≤ finds a schedule of length at most 2.5 times the length of a shortest schedule.* □

Observe that problem $1 \mid$ exact $l_j, a_j \geq b_j \mid C_{\max}$ reduces to problem $1 \mid$ exact l_j, $a_j \leq b_j \mid C_{\max}$ by the standard inverse of the time axis. Thus $1 \mid$ exact $l_j, a_j \geq b_j \mid C_{\max}$ can be solved within a factor of 3 of the length of an optimal schedule as well.

2.2   Algorithm for the General Case

Let $I = \{(a_j, l_j, b_j) : j \in J\}$ be an instance of $1 \mid$ exact $l_j \mid C_{\max}$.

ALGORITHM 1M.

  1. If $\sum_{j=1}^{n} a_j > \sum_{j=1}^{n} b_j$, replace $I = \{(a_j, l_j, b_j) : j \in J\}$ by the symmetrical instance $\{(b_j, l_j, a_j) : j \in J\}$ (which is equivalent to the inverse of the time axis).
  2. Form the new instance $I^* = \{(a_j, l_j, \bar{b}_j) : j \in J\}$ where $\bar{b}_j = \max\{a_j, b_j\}$ (note that I^* is an instance of $1 \mid$ exact $l_j, a_j \leq b_j \mid C_{\max}$.)
  3. By applying Algorithm 1M≤ to I^* find a schedule σ.
  4. If $\sum_{j=1}^{n} a_j \leq \sum_{j=1}^{n} b_j$ output σ; otherwise output the inverse of σ.

Running time. It is clear that the running time of Algorithm 1M is of the same order as that of Algorithm 1M≤, i. e., the algorithm runs in time $O(n \log n)$.

Approximation ratio. Clearly, we may assume that

$$\sum_{j=1}^{n} a_j \leq \sum_{j=1}^{n} b_j \ . \tag{9}$$

By (8) and the construction of $\bar{b}_j$, we have

$$C_{\max}(\sigma) \leq \left(\sum_{s=1}^{r} a_{j_s} + \sum_{j=1}^{n} \bar{b}_j \right) + \sum_{j=1}^{n} \bar{b}_j + l_1$$

$$= \left(\sum_{s=1}^{r} a_{j_s} + \sum_{j=1}^{n} \max\{a_j, b_j\} \right) + \sum_{j=1}^{n} \max\{a_j, b_j\} + l_1$$

$$\leq \sum_{s=1}^{r} a_{j_s} + 2 \sum_{j=1}^{n} (a_j + b_j) + l_1 \ .$$

Since by (9), $\sum_{s=1}^{r} a_{j_s} \leq \sum_{j=1}^{n} a_j \leq \frac{1}{2} \sum_{j=1}^{n}(a_j + b_j)$, it follows that

$$C_{\max}(\sigma) \leq \frac{5}{2} \sum_{j=1}^{n}(a_j + b_j) + l_1$$

$$\text{(by (6))} \leq \frac{7}{2} C_{\max}^* .$$

Thus we arrive at the following

Theorem 2. *Algorithm 1M finds a schedule of length at most 3.5 times the length of a shortest schedule.*   □

Tightness. We now present an example demonstrating that the approximation bound established for Algorithm $1M \leq$ cannot be improved on with respect to the lower bound (6). Let x be a positive integer and $k = 2x - 1$. Consider the instance of problem $1 \mid$ exact $l_j, a_j \leq b_j \mid C_{\max}$ consisting of one job $(1, k(x + 1), x)$ and k identical jobs $(1, x - 1, x)$. So we have $n = k + 1$. It is easy to see that algorithm $1M\leq$ outputs a schedule σ consisting of n blocks and thus

$$C_{\max}(\sigma) = 1 + k(x+1) + x + k(1 + x - 1 + x) = 3kx + k + 1 + x = 3kx + 3x .$$

On the other hand, the lower bound (6) is

$$LB = \max\{1 + x + k + kx, 1 + x + k(x+1)\} = (k+1)(x+1) .$$

Thus

$$\frac{C_{\max}(\sigma)}{LB} = \frac{3kx + 3x}{(k+1)(x+1)} = \frac{3(2x-1)x + 3x}{2x(x+1)} = \frac{6x^2}{2x^2 + 2x} ,$$

which tends to 3 as $x \to \infty$.

A similar construction shows that an approximation factor of 2.5 is tight with respect to the lower bound (6) for the case when $a_j = b_j$ for all $j \in J$.

3   Algorithm for the Two-Machine Problem

In the section we present a constant-factor approximation algorithm for the two-machine problem. We analyze its performance in general and some special cases of the problem. Theorem 5 shows that the approximation ratio of this rather simple algorithm is surprisingly close to best possible.

ALGORITHM 2M.

PHASE I *(jobs ordering)*. Number the jobs in the following way:

$$a_1 + l_1 \leq a_2 + l_2 \leq \ldots \leq a_n + l_n . \tag{10}$$

PHASE II *(constructing the schedule)*. Set $\sigma(1) = 0$. For $j = 2, \ldots, n$, set

$$\sigma(j) = \max\{\sigma(j-1) + a_{j-1}, \sigma(j-1) + b_{j-1} + a_{j-1} + l_{j-1} - a_j - l_j\} . \tag{11}$$

Example. Consider the following instance of problem $F2 \mid \text{exact } l_j \mid C_{\max}$ (the jobs are ordered according to Phase I):

$$\{(1,2,3),(3,1,1),(1,3,4),(2,3,2)\} .$$

Phase II computes the schedule σ with $\sigma(1) = 0$, $\sigma(2) = 2$, $\sigma(3) = 5$, $\sigma(4) = 8$ (see Fig. 2).

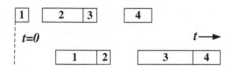

Fig. 2. The schedule constructed by algorithm 2M

Correctness and running time. By (11) for any $j = 2, \ldots, n$, $\sigma(j) \geq \sigma(j - 1) + a_{j-1}$, which guarantees that the first operations of different jobs do not overlap. Moreover, by (11) for any $j = 2, \ldots, n$,

$$C_j(\sigma) - b_j = \sigma(j) + a_j + l_j \geq \sigma(j-1) + a_{j-1} + b_{j-1} + l_{j-1} = C_{j-1}(\sigma) , \quad (12)$$

which means that the second operation of job j starts after the completion of the second operation of the previous job $j - 1$, $j = 2, \ldots, n$. Therefore the second operations of different jobs do not overlap as well. Thus σ is a feasible schedule.

It is clear that the algorithm can be implemented in $O(n \log n)$ time.

Approximation ratio

Theorem 3

(i) *Algorithm 2M finds a schedule of length at most thrice the length of a shortest schedule.*

(ii) *When applied to the special cases where $a_j \leq b_j$, or $a_j \geq b_j$ for all $j \in J$, algorithm 2M finds a schedule of length at most twice the length of a shortest schedule.*

Proof. Observe that $C_{\max}^*$ is at least the maximum load of machines and the maximum job length, i. e.,

$$C_{\max}^* \geq \max\{\max_{j \in J}(a_j + b_j + l_j), \sum_{j \in J} a_j, \sum_{j \in J} b_j\} . \quad (13)$$

By (12) job n is the last job processed on machine 2, which means that the length of σ coincides with the completion time of this job, i. e.,

$$C_{\max}(\sigma) = \sigma(n) + a_n + b_n + l_n . \quad (14)$$

Since $\sigma(1) = 0$, we have

$$\sigma(n) = \sum_{j=2}^{n} \left(\sigma(j) - \sigma(j-1) \right) . \tag{15}$$

By (10) and (11), for $j = 2, \ldots, n$,

$$\sigma(j) \leq \max\{\sigma(j-1) + a_{j-1}, \sigma(j-1) + b_{j-1}\}$$
$$= \sigma(j-1) + \max\{a_{j-1}, b_{j-1}\} .$$

By (15), it follows that $\sigma(n) \leq \sum_{j=2}^{n} \max\{a_{j-1}, b_{j-1}\}$ and by (14),

$$C_{\max}(\sigma) \leq \sum_{j=2}^{n} \max\{a_{j-1}, b_{j-1}\} + a_n + b_n + l_n . \tag{16}$$

By (13) it follows that in the case of arbitrary a_j and b_j, $C_{\max}(\sigma) \leq 3C_{\max}^*$ and in the case when $a_j \geq b_j$ or $b_j \geq a_j$ for all $j \in J$, $C_{\max}(\sigma) \leq 2C_{\max}^*$. $\square$

Tightness. As above we present an example validating that an approximation factor of 3 cannot be improved on with respect to the lower bound (13). Let k be a positive integer. Consider the instance of $F2 \mid$ exact $l_j \mid C_{\max}$ consisting of $k+1$ identical jobs $(1, k^2, k)$ and k identical jobs $(k, k^2 - k + 1, 1)$, i. e., the total number of jobs $n = 2k + 1$. As $a_j + l_j = 1 + k^2$ for all $j \in J$, Phase I may index the jobs in the following alternating order:

$$(1, k^2, k), (k, k^2 - k + 1, 1), (1, k^2, k), (k, k^2 - k + 1, 1), \ldots, (1, k^2, k) .$$

It is easy to see that using this order Phase II computes the schedule σ with $\sigma(2s + 1) = 2ks$ and $\sigma(2s) = k + 2ks$ for $s = 1, \ldots, k$ (the case of $k = 3$ is depicted in Fig. 3). Therefore the completion time of the last operation on the first machine is $2k^2 + 1$ and so

$$C_{\max}(\sigma) = 2k^2 + 1 + k^2 + k = 3k^2 + k + 1 .$$

On the other hand, the lower bound $LB = k^2 + 2k$. Thus

$$\frac{C_{\max}(\sigma)}{LB} = \frac{3k^2 + k + 1}{k^2 + 2k} ,$$

which tends to 3 as $k \to \infty$.

The tightness of an approximation factor of 2 for the case of $a_j = b_j$ for all $j \in J$ is an easy exercise.

4   Inapproximability Lower Bounds

In this section we establish inapproximability lower bounds for both problems. To this end we construct specific polynomial-time reductions from the following well-known NP-complete problem [3]:

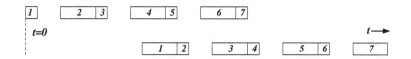

Fig. 3. The case of $k = 3$

PARTITION

Instance: Nonnegative numbers $w_1, \ldots, w_m$.

Question: Does there exist a subset $X \subseteq \{1, \ldots, m\}$ such that $\sum_{k \in X} w_k = S$ where $S = \frac{1}{2} \sum_{k=1}^{m} w_k$?

4.1   Problem $1 \mid$ exact $l_j, \ a_j = b_j \mid C_{\max}$

Let I be an instance of PARTITION. Define an instance $I^* = \{(a_j, l_j, b_j) : j \in J\}$ of problem $1 \mid$ exact $l_j, \ a_j = b_j \mid C_{\max}$. Let $J = \{1, \ldots, m+3\}$ and

$$a_j = b_j = w_j, \quad l_j = (2q+3)S - w_j \quad \text{for } j = 1, \ldots, m \ ,$$
$$a_j = b_j = S, \quad l_j = (2q+4)S \quad \text{for } j = m+1, m+2 \ ,$$
$$a_{m+3} = b_{m+3} = qS, \quad l_{m+3} = 0$$

where q is a positive integer.

Lemma 2

(i) *If $\sum_{k \in X} w_k = S$ for some subset $X \subseteq \{1, \ldots, m\}$, then $C^*_{\max} = (2q+8)S$.*
(ii) *If $\sum_{k \in X} w_k \neq S$ for all $X \subseteq \{1, \ldots, m\}$, then $C^*_{\max} \geq (4q+3)S$.*

Proof.

(i). First, observe that since $(2q+8)S$ is the load of the machine, $C^*_{\max} \geq (2q+8)S$. Now we present a schedule σ with $C_{\max}(\sigma) = (2q+8)S$. Set $\sigma(m+1) = 0$, $\sigma(m+2) = 2S$, $\sigma(m+3) = 4S$, and in an arbitrary order put the first operations of $j \in X$ within the interval $[S, 2S]$ and the first operations of $j \in \{1, \ldots, m\} \setminus X$ within the interval $[3S, 4S]$ (see Fig. 4). Then the second operations of jobs in X and in $\{1, \ldots, m\} \setminus X$ will be processed in the same order within the vacant time intervals $[2qS + 4S, 2qS + 5S]$ and $[2qS + 6S, 2qS + 7S]$, respectively (see Fig. 4). Thus $C_{\max}(\sigma) = (2q+8)S$, as required.

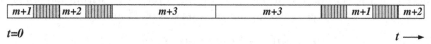

Fig. 4. The jobs in $\{1, \ldots, m\}$ are executed within the shaded intervals

(ii). Assume to the contrary that $C_{\max}(\sigma) < (4q+3)S$ for some feasible schedule σ of I^*. We first claim that both operations of job $m+3$ are processed between the first and second operations of each of the remaining jobs. Indeed, if job $m+3$ is not processed between the operations of some job in $\{m+1, m+2\}$, then

$$C_{\max}(\sigma) \geq 2qS + (2q+4)S = 4qS + 4S$$

and if job $m+3$ is not processed between the operations of some job in $\{1, \ldots, m\}$, then

$$C_{\max}(\sigma) \geq 2qS + (2q+3)S = 4qS + 3S \ .$$

Since the jobs $m+1$ and $m+2$ are identical, we may assume that $\sigma(m+1) < \sigma(m+2) < \sigma(m+3)$. Then by the claim and the construction of I^*,

$$\sigma(m+3) < C_{m+3}(\sigma) < C_{m+1}(\sigma) < C_{m+2}(\sigma) \ .$$

Since the first operations of all jobs in $\{1, \ldots, m+2\}$ are processed before time $\sigma(m+3)$,

$$\sigma(m+3) \geq 4S \ . \tag{17}$$

By the above claim, $\sigma(j) < \sigma(m+3) < C_{m+3}(\sigma) < C_j(\sigma)$ for all $j \in \{1, \ldots, m\}$. Assume that the first executable job is $j \in \{1, \ldots, m\}$. Hence $0 = \sigma(j) \leq \sigma(m+1)$. Then by the definition of job j,

$$C_{m+3}(\sigma) \leq C_j(\sigma) - w_j = \sigma(j) + 2w_j + 2qS + 3S - 2w_j = 2qS + 3S \ .$$

Since $C_{m+3}(\sigma) = \sigma(m+3) + 2qS$, it follows that $\sigma(m+3) \leq 3S$, contradicting the fact that the first operations of all jobs in $\{1, \ldots, m+2\}$ are processed before time $\sigma(m+3)$. Thus $m+1$ is the first executable job, i. e., $\sigma(m+1) = 0$. By a similar way it can be shown (this also follows from the time axis symmetry) that job $m+2$ is the last executable job. It follows that the second operations of all jobs in $\{1, \ldots, m\}$ are processed within the interval $[C_{m+3}(\sigma), C_{m+2}(\sigma) - S]$ and thus we have

$$C_{m+2}(\sigma) - C_{m+3}(\sigma) \geq 4S \ . \tag{18}$$

Let $T' = [\sigma(m+3) - 3S, \sigma(m+3)]$, $T'' = [C_{m+3}(\sigma), C_{m+3}(\sigma) + 3S]$. Then by the construction of I^*, for any $j \in \{1, \ldots, m\}$, $\sigma(j) \in T'$ and $C_j(\sigma) \in T''$. Next, by (17) we have that

$$C_{m+3}(\sigma) = \sigma(m+3) + 2qS \geq 2qS + 4S \ .$$

On the other hand, (18) implies that

$$\sigma(m+3) \leq \sigma(m+2) + 2S \ . \tag{19}$$

Thus the first operation of job $m+2$ is processed within interval T' while (17) implies that the second operation of job $m+1$ is processed within interval T''. Therefore we may define the following subintervals of T' and T'':

$$\begin{aligned}
T_1 &= [\sigma(m+3) - 3S, \sigma(m+2)] \ , \\
T_2 &= [\sigma(m+2) + S, \sigma(m+3)] \ , \\
T_3 &= [C_{m+3}(\sigma), C_{m+1}(\sigma) - S] \ , \\
T_4 &= [C_{m+1}(\sigma), C_{m+3}(\sigma) + 3S] \ .
\end{aligned}$$

As $|T_1| + |T_2| = |T_3| + |T_4| = 2S$ and the operations of all jobs in $\{1, \ldots, m\}$ are processed within $\bigcup_{k=1}^{4} T_k$, we have that $\sum_{j:\sigma_j \in T_k} w_j = |T_k|$ for $k = 1, 2, 3, 4$, where $|T_k|$ stands for the length of T_k. In particular, it follows that within interval T_1 the machine without idle times executes the first operations of some jobs in $\{1, \ldots, m\}$. Then the second operations of these jobs are executed within interval $[C_{m+3}(\sigma), C_{m+3}(\sigma) + |T_1|]$, which therefore cannot contain the second operation of job $m + 1$. This implies that

$$C_{m+3}(\sigma) + |T_1| = C_{m+3}(\sigma) - \sigma(m + 3) + 3S + \sigma(m + 2)$$
$$= 2qS + 3S + \sigma(m + 2)$$
$$\leq C_{m+1}(\sigma) - S$$
$$= 2qS + 6S - S = 2qS + 5S$$

or, equivalently, $\sigma(m + 2) \leq 2S$. By (19) it follows that $\sigma(m + 3) \leq 4S$, which together with (17) yields $\sigma(m + 3) = 4S$. Then

$$|T_3| = C_{m+1}(\sigma) - S - C_{m+3}(\sigma) = 2qS + 6S - S - 4S - 2qS = S$$

and thus $\sum_{j:C_j(\sigma) \in T_3} w_j = |T_3| = S$, which contradicts the assumption of (ii).

$\square$

The following is a straightforward corollary of the lemma.

Theorem 4. *The existence of a $(2 - \varepsilon)$-approximation algorithm for problem $1 \mid exact\ l_j,\ a_j = b_j \mid C_{\max}$ implies P=NP.* $\square$

4.2   Problem $F2 \mid$ exact l_j, $a_j = b_j \mid C_{\max}$

Let I be an instance of PARTITION. Define an instance $I^* = \{(a_j, l_j, b_j) : j \in J\}$ of $F2 \mid$ exact l_j, $a_j = b_j \mid C_{\max}$ in the same way as in the above subsection, i. e., set $J = \{1, \ldots, m + 3\}$ and

$$a_j = b_j = w_j, \quad l_j = (2q + 3)S - w_j \quad \text{for } j = 1, \ldots, m \ ,$$
$$a_j = b_j = S, \quad l_j = (2q + 4)S \quad \text{for } j = m + 1, m + 2 \ ,$$
$$a_{m+3} = b_{m+3} = qS, \quad l_{m+3} = 0$$

where q is a positive integer. The proof of the next lemma is quite similar to that of Lemma 2.

Lemma 3

(i) *If $\sum_{k \in X} w_k = S$ for some subset $X \subset \{1, \ldots, m\}$, then $C_{\max}^* \leq (2q + 8)S$ (a schedule of length $(2q + 8)S$ is depicted in Fig. 5).*
(ii) *If $\sum_{k \in X} w_k \neq S$ for all $X \subset \{1, \ldots, m\}$, then $C_{\max}^* \geq (3q + 3)S$.* $\square$

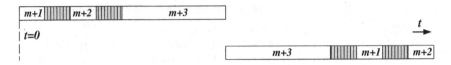

Fig. 5. The jobs in $\{1, \ldots, m\}$ are executed within the shaded intervals

As above we arrive at the following corollary:

Theorem 5. *The existence of a* $(1.5 - \varepsilon)$-*approximation algorithm for problem* $F2 \mid exact\ l_j,\ a_j = b_j \mid C_{\max}$ *implies P=NP.* □

References

1. A. A. Ageev and A. E. Baburin, Approximation Algorithms for the Single and Two-Machine Scheduling Problems with Exact Delays, to appear in Operations Research Letters.
2. A. Farina and P. Neri, Multitarget interleaved tracking for phased array radar, IEEE Proc. Part F: Comm. Radar Signal Process. 127 (1980) (4), 312–318.
3. M.R. Garey and D.S. Johnson, Computers and intractability: A guide to the theory of NP-completeness, Freeman, San Francisco, CA, 1979.
4. M. Elshafei, H. D. Sherali, and J.C. Smith, Radar pulse interleaving for multi-target tracking, Naval Res. Logist. 51 (2004), 79–94.
5. A. Izquierdo-Fuente and J. R. Casar-Corredera, Optimal radar pulse scheduling using neural networks, in: IEEE International Conference on Neural Networks, vol. 7, 1994, 4588-4591.
6. R. L.Graham, E.L. Lawler, J.K. Lenstra, and A.H.G. Rinnooy Kan, Optimization and approximation in deterministic sequencing and scheduling: a survey. Annals of Discrete Mathematics 5 (1979), 287-326.
7. A. J. Orman and C. N. Potts, On the complexity of coupled-task scheduling, Discrete Appl. Math. 72 (1997), 141–154.
8. H. D. Sherali and J. C. Smith, Interleaving two-phased jobs on a single machine, Discrete Optimization 2 (2005), 348–361.
9. W. Yu, The two-machine shop problem with delays and the one-machine total tardiness problem, Ph.D. thesis, Technische Universiteit Eindhoven, 1996.
10. W. Yu, H. Hoogeveen, and J. K. Lenstra, Minimizing makespan in a two-machine flow shop with delays and unit-time operations is NP-hard. J. Sched. 7 (2004), no. 5, 333–348.

Bidding to the Top: VCG and Equilibria of Position-Based Auctions

Gagan Aggarwal, Jon Feldman, and S. Muthukrishnan

Google, Inc.
76 Ninth Avenue, 4th Floor, New York, NY, 10011
1600 Amphitheatre Pkwy, Mountain View, CA, 94043
{gagana,jonfeld,muthu}@google.com

Abstract. Many popular search engines run an auction to determine the placement of advertisements next to search results. Current auctions at Google and Yahoo! let advertisers specify a single amount as their bid in the auction. This bid is interpreted as the maximum amount the advertiser is willing to pay per click on its ad. When search queries arrive, the bids are used to rank the ads linearly on the search result page. Advertisers seek to be high on the list, as this attracts more attention and more clicks. The advertisers pay for each user who clicks on their ad, and the amount charged depends on the bids of all the advertisers participating in the auction.

We study the problem of ranking ads and associated pricing mechanisms when the advertisers not only specify a bid, but additionally express their preference for positions in the list of ads. In particular, we study *prefix position auctions* where advertiser i can specify that she is interested only in the top κ_i positions.

We present a simple allocation and pricing mechanism that generalizes the desirable properties of current auctions that do not have position constraints. In addition, we show that our auction has an *envy-free* [1] or *symmetric* [2] Nash equilibrium with the same outcome in allocation and pricing as the well-known truthful Vickrey-Clarke-Groves (VCG) auction. Furthermore, we show that this equilibrium is the best such equilibrium for the advertisers in terms of the profit made by each advertiser. We also discuss other position-based auctions.

1 Introduction

In the sponsored search market on the web, advertisers bid on keywords that their target audience might be using in search queries. When a search query is made, an online (near-real time!) auction is conducted among those advertisers with matching keywords, and the outcome determines where the ads are placed and how much the advertisers pay. We will first review the existing auction model before describing the new model we study (a description can be found in Chapter 6 of [3]).

T. Erlebach and C. Kaklamanis (Eds.): WAOA 2006, LNCS 4368, pp. 15–28, 2006.
© Springer-Verlag Berlin Heidelberg 2006

Current Auctions. Consider a specific query consisting of one or more *keywords*. When a user issues that search query, the search engine not only displays the results of the web search, but also a set of "sponsored links." In the case of Google, Yahoo, and MSN, these ads appear on a portion of the page near the right border, and are linearly ordered in a series of slots from top to bottom. (On Ask.com, they are ordered linearly on the top and bottom of the page).

Formally, for each search query, we have a set of n advertisers interested in advertising. This set is usually derived by taking a union over the sets of advertisers interested in the individual keywords that form the query. Advertiser i bids b_i, which is the maximum amount the advertiser is willing to pay for a click. There are $k < n$ positions available for advertisements. When a query for that keyword occurs, an online auction determines the set of advertisements, their placement in the positions, and the price per click each has to pay.

The most common auction mechanism in use today is the *generalized second-price* (GSP) auction (sometimes also referred to as the *next-price auction*). Here the ads are ranked in decreasing order of bid, and priced according to the bid of the next advertiser in the ranking. In other words, suppose wlog that $b_1 \geq b_2 \geq \ldots \geq b_n$; then the first k ads are placed in the k positions, and for all $i \in [1, k]$, bidder i gets placed in position i and pays b_{i+1} per click.[1]

We note two properties ensured by this mechanism:

1. *(Ordering Property)* The ads that appear on the page are ranked in decreasing order of b_i.
2. *(Minimum Pay Property)* If a user clicks on the ad at position i, the advertiser pays the minimum amount she would have needed to *bid* in order to be assigned the position she occupies.

Search engine companies have made a highly successful business out of these auctions. In part, the properties above have dissuaded advertisers from trying to game the auction. In particular, the minimum-pay property ensures that an advertiser has no incentive to lower a winning bid by a small amount in order to pay a lower price for the same position. Still, the GSP auction is not truth-revealing, that is, an advertiser may be incentivized to bid differently than her true value under certain conditions [4].

Only recently have we obtained a detailed formal understanding of the properties of this auction. Authors in [1,2,4] have analyzed the auction in terms of its equilibria. They show that when the click-through rates are *separable*, i.e. the click-through rate of an ad at a given position is the product of an ad-specific factor and a position-specific factor, the GSP has a Nash equilibrium whose outcome is equivalent to the famous Vickrey-Clarke-Groves (VCG) mechanism [5,6,7] which is known to be truthful. [1,2] go on to show that this equilibrium is *envy-free*, that is, each advertiser prefers the current outcome (as it applies to her) to being placed in another position and paying the price-per-click being paid by the current occupant of the position. Further, among

---

[1] The Google auction actually ranks according to $w_i b_i$, for some weight w_i related to the quality of the ad, and then sets the price for bidder i to $w_{i+1}b_{i+1}/w_i$. All our results generalize to this "weighted" bid case as well.

all the envy-free equilibria, the VCG equilibrium is bidder-optimal; that is, for each advertiser, her price-per-click is minimum under this equilibrium. We note that when the click-through rates are separable, the outcome produced by the VCG mechanism has the ordering property. Authors in [4] also show that even when the click-through rates are arbitrary, there is a pricing method with the ordering property that is truthful. (This pricing method reduces to the VCG pricing method when the click-through rates are separable.) Furthermore, they show that the GSP has a Nash equilibrium that has the same outcome as their mechanism. Together, these results provide some understanding of the current auctions. That in turn provides confidence in the rules of the auction, and helps support the vast market for search keyword advertising.

Emerging Position-Based Auctions. As this market matures, advertisers are becoming increasingly sophisticated. For example they are interested in the relative performance of their ads and keywords, and so the search engines provide tools to track statistics. As advertisers learn when and where their ads are most effective, they need more control over their campaigns than is provided by simply stating a keyword and a bid.

One of the most important parameters affecting the performance of an advertisement is its position on the page. Indeed, the reason the auction places the ads in descending order on the page is that the higher ads tend to get clicked on more often than the lower ones. In fact, having an ad place higher on the page not only increases the chances of a click, it also has value as a branding tool, regardless of whether the ad gets clicked. Indeed, a recent empirical study by the Interactive Advertising Bureau and Nielsen//NetRatings concluded that higher ad positions in paid search have a significant brand awareness effect [8]. Because of this, advertisers would like direct control over the position of their ad, beyond just increasing the bid. Ideally, the search engine would conduct a more general auction that would take such position preferences into account; we refer to this as a *position-based auction*.

Our Results. In this paper, we initiate the study of position-based auctions where advertisers can impose position constraints. In particular, we study the most relevant case of *prefix* position constraints, inspired by the branding advertiser: advertiser i specifies a position κ_i and a bid b_i, which says that the advertiser would like to appear only in the top κ_i positions (or not at all) and is willing to pay at most b_i per click. Upon receiving bids from a set of n such advertisers, the search engine must conduct an auction and place ads into k positions while respecting the prefix constraints.

Our main results are as follows. We present a simple auction mechanism that has both the ordering and the minimum pay property, just like the current auctions. The mechanism is highly efficient to implement, taking near-linear time. Further, we provide a characterization of its equilibria. We prove that this auction has a Nash equilibrium whose outcome is equivalent in allocation and pricing to that of VCG. Additionally, we prove that this equilibrium is *envy-free* and that among all envy-free equilibria, this particular one is bidder-optimal.

Our results generalize those in [1,2], which proved the same thing for the GSP without position constraints. The main difficulty in generalizing these results lies in the fact that once you allow position constraints, the allocation function of VCG no longer obeys the ordering property, thus making it challenging to engineer an appropriate equilibrium. Our principal technical contributions are new structural properties of the VCG allocation that allow us to relate the VCG allocation with an auction that preserves the ordering property.

In the future, advertisers may want even more control over ad position. We discuss more general position-based auctions at the end of the paper.

2   Prefix Position Auction Mechanisms

Formally, the prefix position auction problem is as follows. There are n advertisers for a search keyword. They submit bids $b_1, \ldots, b_n$ respectively. There are k positions for advertisements numbered $1, \ldots, k$, top to bottom. Each advertiser $i \in \{1, \ldots, n\}$ also submits a cutoff position $\kappa_i \leq k$, and requires that their advertisements should not appear below position κ_i.

An auction mechanism consists of two functions:

- an *allocation* function that maps bids to a matching of advertisers to positions, as well as
- a *pricing* function that assigns a price per click ppc_j to each position won by an advertiser. We restrict our attention to mechanisms where the prices always respect the bids; i.e., we have $\text{ppc}_j \leq b_i$ if i is assigned to j.

A natural allocation strategy that retains the ordering property is as follows: rank the advertisers in decreasing order of b_i as in GSP. Now, go through the ranking one position at a time, starting at the top; if you encounter an advertiser that appears below her bottom position κ_i, remove her from the ranking and move everyone below that position up one position, and continue checking down the rest of the list.

Two natural pricing strategies immediately come to mind here: (1) Set prices according to the subsequent advertiser in the ranking *before* any advertiser is removed, or (2) set prices according to the subsequent advertiser in the ranking *after* all the advertisers are removed (more precisely, the ones that appear below her position). It turns out that neither of these options achieves the minimum pay property as shown by the following examples. Assume for the sake of these examples that $0.05 is the amount used to charge for the last position.

Example 1. Suppose we set prices before removing out-of-position advertisers. Now suppose we have the following ranges and bids where the number in parentheses is the position constraint κ_i:

A: (5) $5   B: (5) $4   C: (5) $3   D: (2) $2   E: (5) $1

We run the auction, and the order is (A, B, C, D, E). If we price now, the prices are ($4, $3, $2, $1, $0.05). Bidder D gets removed and so we end up with (A, B, C, E), and we charge ($4, $3, $2, $0.05). However, if bidder C had bid $1.50, which is below what she was charged, the auction would still have ended up as (A, B, C, E). Thus, the minimum pay property is violated by charging too much.

For a more intuitive reason why this is a bad mechanism, it would allow a form of "ad spam". Suppose a bidder sets her bottom cutoff to (2), but then bids an amount that would never win position one or two. In this case, she drives up the price for those that are later in the auction (e.g., competitors), at no risk and no cost to herself.

Example 2. Now suppose we set the prices after removing out-of-position advertisers, and we have the following bids and prefix constraints:

A: (5) $5   B: (5) $4   C: (2) $3   D: (5) $2

We run the auction, and the order is (A, B, C, D). Now we remove C and we get the order (A, B, D). We price according to this order and so the prices are ($4, $2, $0.05). Bidder B bid $4 and paid $2; however, if B changed her bid to $2.50, then bidder C would have gotten second position. Thus the minimum pay property is violated, but this time because we are charging too little.

As for intuition, this option opens up a possible "race for the bottom" situation. Suppose we have a group of bidders only interested in positions 1-4 (perhaps because those appear on the page without scrolling). The winners of the top three positions pay according to the competitive price for those top positions, but the winner of position 4 pays according to the winner of position 5, who could be bidding a much lower amount. Thus, these top bidders have an incentive to lower their prices so that they can take advantage of this bargain.

But now consider a third alternative, which will turn out to be the one that achieves the minimum-pay property: *For each advertiser that is allocated a particular position j, set the price according to the first advertiser that appears later in the original ranking that included j in her range.* For an example of this pricing method, consider the situations from the examples above:

In Example 1, the advertisers would be ranked (A, B, C, D, E), and then (A, B, C, E) after removal. The price for A is set to $4, since B had position 1 in its range. Similarly, the price for B is set to $3 since C had position 2 in its range. The price for C is set to $1, however, since D did not include position 3 in its range. The price for C is set to $0.05.

In Example 2, the advertisers would be ranked (A, B, C, D) and after removal we get (A, B, D). The price for A is $4, but the price for B is now $3; even though C did not win any position, it was still a participant in the auction, and was bidding for position 2. The price for D is $0.05.

Top-down Auction. We now define an auction mechanism for prefix position constraints that is equivalent to the final proposal above, and is easily seen to have the minimum-pay property. Furthermore, this mechanism is exceedingly easy to implement, taking time $O(n \log n)$.

Definition 1. *The* top-down auction mechanism *works as follows: For each position in order from the top, iteratively run a simple second-price auction (with one winner) among those advertisers whose prefix range includes the position being considered. By a "simple second-price auction," we mean that the highest bidder in the auction is allocated the position, and pays a price-per-click equal to the second-highest bid. This winner is then removed from the pool of advertisers for subsequent auctions and the iteration proceeds.*

3   Analysis of the Top-Down Prefix Auction

We have found a natural generalization of GSP to use with prefix position constraints, and now we would like to know what properties this auction has. Since GSP is a special case, we already know that the auction is not truthful [4]. But from [2,1,4] we at least know something about the equilibria of GSP. It is natural to ask whether or not these results hold true in our more general setting.

In this section, we answer this in the affirmative, and prove that the top-down prefix auction has an "envy-free" Nash equilibrium whose outcome (in terms of allocation and pricing) is equivalent to that of VCG. ("Envy-freeness" is a stronger condition than is imposed by the Nash equilibrium, dictating that no bidder envies the allocation and price of any other bidder.) We go on to prove that this equilibrium is the bidder-optimal envy-free Nash equilibrium in the sense that it maximizes the "utility" (or profit) made by each advertiser.

Definitions. Each position j has an associated *click-through rate* $c_j > 0$ which is the probability that a user will click on an ad in that position. Using the idea that higher positions receive more clicks, we may assume $c_1 > c_2 > \ldots > c_k$. To make the discussion easier, we will abuse this notation and say that an ad in position j "receives c_j clicks," even allowing $c_j > 1$ for some examples.

Each advertiser has a valuation v_i it places on a click, as long as that click comes from one of its desired positions. Using the "branding" motivation, we assume a valuation of $-\infty$ if an ad even *appears* at a position below its bottom cutoff κ_i. Since $c_j > 0$ for all positions j, we can (equivalently) think of this as a valuation of $-\infty$ on a *click* below position κ_j. So, given some total price p (for all the clicks) for a position j, the *utility* of bidder i is defined as $u_i = c_j v_i - p$ if $j \leq \kappa_i$, and $-\infty$ otherwise.[2]

The Vickrey-Clarke-Groves (VCG) Auction. The VCG auction mechanism [5,6,7] is a very general technique that can be applied in a wide range of settings. Here we give its application to our problem. For a more general treatment, we refer the reader to Chapter 23 of [9].

---

[2] Note that we are making the assumption that click-through rates are dependent only on the position and not on the ad itself. Our results hold as long as the click-through rates are *separable*, i.e. the click-through rate of an ad at a given position is the product of a per-position factor and a per-advertiser factor. More general forms of click-through rate would require further investigation.

Let Θ represent the allocation of bidders to positions that maximizes the *total valuation* on the page; i.e., Θ is a matching M of advertisers i to positions j that respects the position constraints ($j \leq \kappa_i$), and maximizes $\sum_{(i,j)\in M} v_i c_j$. Note that this assignment could also have empty slots, but they must be contiguous at the bottom end. The Θ allocation is the most "efficient" allocation, but an allocation function in an auction mechanism has access to the bids b_i not the valuations v_i. So instead, the VCG allocation M^* is the matching M that maximizes $\sum_{(i,j)\in M} b_i c_j$.

Intuitively, the VCG price for a particular bidder is the total difference in others' valuation caused by that bidder's presence. To define this pricing function formally, we need another definition: Let M^*_{-x} be the VCG allocation that would result if bidder x did not exist. More formally, this allocation is the matching M that does not include bidder x and maximizes $\sum_{(i,j)\in M} b_i c_j$.

The VCG price for bidder i in position j is then $p_j = M^*_{-i} - M^* + c_j b_i$. (Here we are abusing notation and using M^* and M^*_{-i} to denote the total valuation of the allocation as well as the allocation itself.) Note that p_j is a *total* price for all clicks at that position, not a per-click price. Only in the case that $b_i = v_i$ does the VCG mechanism actually successfully compute Θ. However, it is well-known (see [9] for example) that the pricing method of VCG ensures that each bidder is incentivized to actually reveal their true valuation and set $b_i = v_i$. This holds *regardless of the actions of the other bidders,* a very strong property referred to as "dominant-strategy truthfulness." Thus in equilibrium, we get $b_i = v_i$, $M^* = \Theta$, and $M^*_{-i} = \Theta_{-i}$, where Θ_{-i} is the Θ allocation that would result if bidder i did not exist.

For convenience, for the remainder of paper we rename the bidders by the slots to which they were assigned in Θ, even when we are talking about the top-down prefix auction. The unassigned bidders are renamed to $(k+1, \ldots, n)$ arbitrarily. We will use $p_i = \Theta_{-i} - \Theta + c_i v_i$ to denote the VCG price (at equilibrium) for position (and bidder) i.

Envy-Free Nash Equilibria and the GSP Auction. The VCG mechanism is desirable because it has an equilibrium that results in the most efficient allocation according to the true valuations of the bidders. Furthermore this equilibrium occurs when each bidder actually reveals their true valuations. The GSP auction (without position constraints) does not have this second property, but in fact it does have the first: namely that it has an equilibrium whose allocation is the most efficient one (this was proved in [1,2,4]). Furthermore, this equilibrium also results in the same *prices* that result from VCG. This validates the GSP from an incentive-compatibility point of view, and shows that the ordering property does not preclude efficiency. This equilibrium also has the following property:

Definition 2. *An allocation and pricing is an* envy-free equilibrium *if each bidder prefers the current outcome (as it applies to her) to being placed in another position and paying the price-per-click being paid by the current occupant of the position.*

Moreover, among all envy-free Nash equilibria, this particular one is *bidder-optimal*, in the sense that it results in the lowest possible price for each particular advertiser. Note that in GSP, for a particular bidder, the only position for which envy-freeness is *not* implied by Nash is the position directly above.

3.1   Equilibrium in the Top-Down Auction

It is natural to ask if all these properties also hold true in the presence of position constraints. One of the difficulties in proving this comes from the fact that the VCG allocation no longer preserves the ordering property, as shown by the following simple example. Suppose advertiser A has bottom cutoff (2) and a bid of $2, advertiser B has cutoff (1) and a bid of $1, and we have $c_1 = 101$ and $c_2 = 100$. The VCG allocation gives position 1 to B and position 2 to A, for a total revenue of $\approx$ $300. The top-down auction will give position 1 to A and position 2 will be unfilled. The revenue is equal to $\approx$ $200.

Despite this, it turns out that there is an equilibrium of the top-down auction where bidders end up in the optimal allocation, which we prove in our main theorem:

Theorem 1. *In the top-down prefix auction, there exists a set of bids and stated position cutoffs such that*

(a) each bidder is allocated to the same slot as she would be in the dominant-strategy equilibrium of VCG,

(b) the winner of each slot pays the same total price as she would have in the dominant-strategy equilibrium of VCG, and

(c) the bidders are in an envy-free Nash equilibrium.

Furthermore (d), for each advertiser, her utility under VCG outcomes is the maximum utility she can make under any envy-free equilibrium. In other words, a VCG outcome is a bidder-optimal envy-free equilibrium of the top-down auction.

The remainder of this section is devoted to proving this theorem. The bids that satisfy this theorem are in fact quite simple: we set $b_i = p_{i-1}/c_{i-1}$ for all bidders i assigned in Θ. Thus, if we show that $b_1 > b_2 > \ldots > b_k$, we would get that the top-down auction assigns the bidders exactly like Θ and sets the same prices (modulo some technical details). This would prove (a) and (b) above.

The chain. To show that the bids are indeed decreasing, and to show (c), it turns out that we need to prove some technical lemmas about the difference between Θ and Θ_{-i} for some arbitrary bidder i. In Θ_{-i}, some bidder i' takes the place of i (unless i is in the last slot, in which case perhaps no bidder takes this slot). In turn, some bidder i'' takes the slot vacated by i', etc., until either the vacated slot is the bottom slot k, or some previously unassigned bidder is introduced into the solution. We call this sequence of bidder movements ending at slot i the "chain" of moves of Θ_{-i}. Note that the chain has the property that it begins either with an unassigned bidder, or with the bidder from the last slot and ends at slot i. If we consider the slots not on the chain, we claim that (wlog) the assignment does not change on these slots when we go from Θ to Θ_{-i}. This is easily seen by substituting a purported better assignment on these slots back into Θ. Note that this implies that Θ_{-i} has at most one new bidder (that wasn't in Θ), and that no bidder besides i that was assigned in Θ has dropped out. The chain is said to have *minimum length* if there is no shorter chain that achieves the same valuation as Θ_{-i}. A *link* in this chain refers to the movement of a bidder i from slot i to some slot i'. We say that this is a *downward* link if $i' > i$; otherwise it is an upward link.

Lemma 1. *The minimum length chain for Θ_{-i} does not contain a downward link followed by an upward link.*

Proof. Suppose it does contain such a sequence. Then, some bidder i_1 moved from slot i_1 to slot $i_2 > i_1$, and bidder i_2 moved from slot i_2 to a slot $i_3 < i_2$. An alternate solution, and thus a candidate solution for Θ_{-i} is to have bidder i_1 move from slot i_1 to slot i_3, have bidder i_2 remain in slot i_2, and keep everything else the same. (Bidder i_1 can move to slot i_3 since $i_3 < i_2$ and i_2 is in range for bidder i_1 (by the fact that i_1 moved to i_2 in Θ_{-i}).)

The difference between the two solutions is $c_{i_2}(v_{i_2} - v_{i_1}) + c_{i_3}(v_{i_1} - v_{i_2}) = (c_{i_3} - c_{i_2})(v_{i_1} - v_{i_2})$. We know $c_{i_3} > c_{i_2}$ since $i_3 < i_2$. We also know $v_{i_1} \geq v_{i_2}$ since otherwise Θ could switch bidders i_1 and i_2 (note again that bidder i_1 can move to slot i_2, since it did so in Θ_{-i}). Thus the difference is non-negative, and so this alternate solution to Θ_{-i} has either greater valuation or a shorter chain. $\square$

Lemma 2. *Let x and y be arbitrary bidders assigned to slots x and y in Θ, where $x < y$. Then, (i) if slot y is in the range of bidder x, we have $\Theta_{-y} \geq \Theta_{-x} + c_y(v_x - v_y)$, and (ii) $\Theta_{-x} \geq \Theta_{-y} + c_x(v_y - v_x)$.*

Proof. **(i)** Consider the assignment of bidder y in Θ_{-x}. Recall that for any i, all bidders besides i present in Θ are also present in Θ_{-i}. Thus y is present somewhere in Θ_{-x}. Note also that the minimum-length chain for Θ_{-x} ends at slot x, and so if y is present in this chain, it cannot follow a downward link; otherwise the chain would contradict Lemma 1, since x is above y. Thus we may conclude that y ends up in position $y' \leq y$. Since slot y is in range for bidder x by assumption, we also have that y' is in range for bidder x; thus we can construct a candidate solution for Θ_{-y} by replacing (in Θ_{-x}) bidder y with bidder x. We may conclude that $\Theta_{-y} \geq \Theta_{-x} + c_{y'}(v_x - v_y) \geq \Theta_{-x} + c_y(v_x - v_y)$.

(ii) This time we need to consider the assignment of x in Θ_{-y}. By the same logic as above, bidder x is present somewhere, and if x either stayed in the same place of moved up, we can replace x with y (in Θ_{-y}) to get a candidate for Θ_{-x}, and we are done. The only remaining case is when x moves down in Θ_{-y} and this is a bit more involved.

Consider the section of the chain of Θ_{-y} from bidder x to the end at bidder y (who is below x). Since x is on a downward link, and downward links cannot be followed by an upward link (Lemma 1), it must be the case that this section of the chain is entirely downward links. Let $x \to x_1 \to x_2 \to \ldots \to x_\ell \to y$ be this chain, and so we have $x < x_1 < x_2 \ldots < x_\ell < y$.

We write the assignment of Θ to these $\ell + 2$ places using the notation $[x, x_1, x_2, \ldots, x_\ell, y]$, and consider other assignments to these slots using the same notation. The solution Θ_{-y} assigns these slots as $[w, x, x_1, \ldots, x_\ell]$, where w is the bidder before x in the chain. For notational purposes define $x_{\ell+1} = y$.

Consider the following alternate solution constructed from Θ_{-y}: change only the assignments to these special $\ell+2$ slots to $[w, x_1, \ldots, x_\ell, y]$. This is a candidate for Θ_{-x} and so by calculating the difference in valuation between this candidate solution and Θ_{-y} we get

$$\Theta_{-x} \geq \Theta_{-y} + \left(\sum_i^\ell v_{x_i}(c_{x_i} - c_{x_{i+1}}) \right) + c_y v_y - c_{x_1} v_x \qquad (1)$$

Putting this aside for now, consider the following alternate solution for Θ. Take the assignment in Θ and change the assignment to only those $\ell + 2$ positions to $[y, x, x_1, \ldots, x_\ell]$. This is feasible since y moves up, and the remaining changes are identical to Θ_{-y}. Since this solution must have valuation at most that of Θ,

$$c_x v_y + c_{x_1} v_x + \sum_1^\ell v_{x_i} c_{x_{i+1}} \leq c_x v_x + \left(\sum_1^\ell v_{x_i} c_{x_i} \right) + c_y v_y$$

$$\Longleftrightarrow c_x(v_y - v_x) \leq \left(\sum_1^\ell v_{x_i}(c_{x_i} - c_{x_{i+1}}) \right) + c_y v_y - c_{x_1} v_x$$

This, combined with (1), implies (ii).   □

Now we are ready to prove the first part of our main theorem: that our bids give the same outcome as VCG, and are indeed an envy-free equilibrium.

Proof of Theorem 1(a-c). The bids of the equilibrium are defined as follows. For all bidders $i > 1$ assigned in Θ, we set $b_i = p_{i-1}/c_{i-1}$. We set b_1 to any number greater than b_2. For all bidders assigned in Θ, we set their stated cutoff to their true cutoff κ_i. If there are more than k bidders, then for some bidder α that was not assigned in Θ, we set $b_\alpha = p_k/c_k$, and set the stated cutoff of bidder j to the bottom slot k. For all other bidders not in Θ, we set their bid to zero, and their cutoff to their true cutoff.

Consider two arbitrary bidders x and y assigned in Θ, where $x < y$. Using Lemma 2(ii), we get $\Theta_{-x} \geq \Theta_{-y} + c_x(v_y - v_x)$. Substituting for Θ_{-x} and Θ_{-y} using the definitions of p_x and p_y, respectively, we get:

$$p_x - c_x v_x \geq p_y - c_y v_y + c_x(v_y - v_x) \quad \Longleftrightarrow \quad \left(v_y - \frac{p_y}{c_y} \right) c_y \geq \left(v_y - \frac{p_x}{c_x} \right) c_x$$

Since $c_y < c_x$, we get $\frac{p_x}{c_x} > \frac{p_y}{c_y}$.

Since we chose x and y arbitrarily, we have just showed that $b_2 > \ldots > b_k$, and $b_k > b_\alpha$ if bidder α exists. We have $b_1 > b_2$ by definition, and all other bids are equal to zero. Thus the bids are decreasing in the VCG order, and so the top-down auction will choose the same allocation as VCG. By construction, the top-down auction will also have the same prices as VCG.

It remains to show that this allocation and pricing is an envy-free equilibrium. Consider again two bidders x and y assigned in Θ with $x < y$. The utilities of x and y are $u_x = c_x v_x - p_x$ and $u_y = c_y v_y - p_y$. We must show that x does not envy y, and that y does not envy x.

If y is out of range of bidder x, then certainly x does not envy y. If x is in range of bidder y, then by Lemma 2(i), we get $\Theta_{-y} \geq \Theta_{-x} + c_y(v_x - v_y)$. Substituting for Θ_{-y} and Θ_{-x} using the definitions of p_y and p_x, we get

$$\Theta + p_y - c_y v_y \geq \Theta + p_x - c_x v_x + c_y(v_x - v_y)$$
$$\Longleftrightarrow c_y v_x - p_y \leq c_x v_x - p_x = u_x.$$

Thus x does not envy y. Similarly, Lemma 2(ii) shows that y does not envy x.

Now consider some bidder z not assigned in Θ. We must show that bidder z does not envy any bidder that is assigned a slot in the desired range of z. Consider some such bidder y; replacing y with z creates a candidate for Θ_{-y}. Thus we have $\Theta_{-y} \geq \Theta + c_y(v_z - v_y)$, which becomes $p_y = \Theta_{-y} - \Theta + c_y v_y \geq c_y v_z$. This implies that z does not envy y. $\square$

Now it remains to show the second part of Theorem 1, namely that among all envy-free equilibria, the one we define is optimal for each bidder. First we give a lemma showing that envy-freeness in the top-down auction implies that the allocation is the same as VCG. Then we use this to compare our equilibrium with an arbitrary envy-free equilibrium.

Lemma 3. *Any envy-free equilibrium of the top-down auction has an allocation with optimal valuation.*

Proof. For the purposes of this proof, we will extend any allocation of bidders to slots to place all n bidders into "slots". For this, we will introduce dummy slots indexed by integers greater than k, with click-through rate $c_i = 0$. We index the bidders according to their (extended) allocation in Θ.

For the purposes of deriving a contradiction, let E, p be the allocation and pricing for an envy-free equilibrium of the top-down auction such that the valuation of E is less than Θ. Thus, p_i refers to the price of slot i in this envy-free equilibrium. Define a graph on n nodes, one for each slot. For each bidder i, make an edge from i to j, where j is the slot in which bidder i is placed in E; i.e., bidder i is in slot i in Θ and in slot j in E. Note that this graph is a collection of cycles (a self-loop is possible, and is defined as a cycle).

Define the weight of an edge (i, j) to be the change in valuation caused by bidder i moving from slot i in Θ to slot j in E. So, we have that the weight of (i, j) is equal to $v_i(c_j - c_i)$. Since the total change in valuation from Θ to E is negative by definition, the sum of the weights of the edges is negative. This implies that there is a negative-weight cycle Y in the graph, and so we have

$$\sum_{(i,j)\in Y} v_i(c_j - c_i) < 0. \tag{2}$$

By the fact that E is envy-free, for each edge (i,j), we also have that bidder i would rather be in slot j than in slot i (under the prices p imposed by the envy-free equilibrium). In other words, $v_i c_j - p_j \geq v_i c_i - p_i$. Rearranging and summing over the edges in Y, we get

$$\sum_{(i,j)\in Y} v_i(c_j - c_i) \geq \sum_{(i,j)\in Y} p_j - p_i = 0. \tag{3}$$

(The sum on the right-hand side equals zero from the fact that Y is a cycle.) Equations (2) and (3) together give us a contradiction.   □

Note that the profit of an advertiser i is the same under all VCG outcomes, and is equal to the difference in valuation between Θ and Θ_{-i}.

Proof of Theorem 1(d). Consider some envy-free equilibrium E of the top-down auction. This equilibrium must have an allocation with optimal valuation (by Lemma 3). We will call this allocation Θ. Let $\{p_i^E\}_i$ be the price of slot i in this equilibrium; We will rename the bidders such that bidder i is assigned to slot i by allocation Θ. Consider one such bidder x assigned to slot x. Consider the chain $x_\ell \to x_{\ell-1} \to \ldots \to x_0 = x$ for Θ_{-x}. (Here bidder x_j moves from slot x_j in Θ to slot x_{j-1} in Θ_{-x}.) By the fact that E is envy-free, for all $j \in [0, \ell-1]$ we have

$$v_{x_{j+1}} c_{x_{j+1}} - p_{x_{j+1}}^E \geq v_{x_{j+1}} c_{x_j} - p_{x_j}^E$$
$$\iff p_{x_j}^E \geq v_{x_{j+1}}(c_{x_j} - c_{x_{j+1}}) + p_{x_{j+1}}^E.$$

(Each move is this chain is feasible, since it was made by Θ_{-x}.) Composing these equations for $j = 0, \ldots, \ell-1$, we get

$$p_x^E = p_{x_0}^E \geq v_{x_1}(c_{x_0} - c_{x_1}) + v_{x_2}(c_{x_1} - c_{x_2}) + \ldots + v_{x_\ell}(c_{x_{\ell-1}} - c_{x_\ell})$$

But note that each term of the right-hand side of this inequality represents the difference in valuation for a bidder on the chain of Θ_{-x}. Thus the sum of these terms is exactly the VCG price p_x, and we have $p_x^E \geq p_x$. Hence, the profit of advertiser x under equilibrium E is no less than her profit under VCG.   □

4   Concluding Remarks

The generalized second-price auction has worked extraordinarily well for search engine advertising. We believe that the essential properties of this auction that make it a success are that it preserves the ranking order inherent in the positions, and that it is stable in the sense that no bidder has an incentive to change her bid by a small amount for a small advantage. We have given a simple new prefix position auction mechanism that preserves these properties and has the same equilibrium properties as the regular GSP.

A natural question arises if advertisers will have preference for positions that go beyond the top κ_i's. It is possible that there are other considerations that make lower positions more desirable. For example, the last position may be preferable to the next-last. Also, appearing consistently at the same position may be desirable for some. Some of the advertisers may not seek the topmost positions in order to weed out clickers who do not persist through the topmost advertisements to choose the most appropriate one. Thus, there are a variety of factors that govern the position preference of an advertiser. In the future, this may lead to more general position auctions than the prefix auctions we have studied here. We briefly comment on two variants.

Arbitrary Ranges. If we allow top cutoffs (i.e., bidder i can set a valuation α_i and never appear above position α_i), we can consider running essentially the same top-down auction: For each position in order from the top, run a simple Vickrey auction (with one winner) among those advertisers whose range includes the position being considered; the winner is allocated the position, pays according to the next-ranked advertiser, and is removed from the pool of advertisers for subsequent auctions.

The difference here is that we can encounter a position j where there are not advertisers willing to take position j, but there are still advertisers willing to take positions lower than j. (This cannot occur with prefix ranges.) On a typical search page, the search engine must fill in something at position j, or else the subsequent positions do not really make sense. Practically speaking, one could fill in this position with some sort of "filler" ad. Given some sort of resolution of this issue, the top-down auction maintains the minimum pay property for general ranges, by essentially the same argument as the prefix case in this paper.

However, the property that there is an equilibrium that matches the VCG outcome is no longer true, as shown by the following example:

Example 3. Suppose we have three bidders, and their ranges and valuations are given as follows: A (1,1) \$3; B (2,3) \$2; C (1,3) \$1. We also have three positions, and we get 100, 99 and 98 clicks in them, respectively. The VCG outcome is an allocation of [A,B,C], and prices [\$2, \$1, \$0] (for *all* clicks). To achieve this outcome in the top-down range auction, we must have A with the highest bid, and it is ∞ wlog. Since C is the only other bidder competing for the first slot, the price of A (which must be \$2) is determined by the bid of C, and thus $\$2 = p_1 = b_C c_1 = 100 b_C$. Therefore we have that $b_C = 2/100$. Since bidder B wins the second slot, we must have bidder B outbidding bidder C, and the price of B is also determined by the bid of C; so we get $p_2 = b_C c_2 = 99(2/100)$. This is inconsistent with the VCG price of \$1.

General Position Bids. One is tempted to generalize the position-based auction so that instead of enforcing a ranking, each advertiser submits separate bids for each position and the market decides which positions are better. Suppose we allowed such bids, and let $b_{i,j}$ denote the bid of advertiser i for position j.

In this setting, the ordering property no longer makes sense, but it still might be interesting to consider the minimum-pay property. We do need to clarify our definition of this property; because the advertiser has control of more than just the bid that gave her the victory; we need to make sure that altering the *other* bids cannot give the advertiser a bargain for this particular position.

A natural mechanism and pricing scheme is as follows: Given the bids $b_{i,j}$, compute the maximum matching of bids to positions (i.e., the VCG allocation). Now for each winning bid b_j^i, do the following. delete all other edges from i, and lower this bid until the max matching no longer assigns i to j. Set the price per click ppc_j to the bid where this happens.

Note that this price has the property that if the winner of a position bids between the bid and the price, then they either get the same position at the same price, or perhaps one of their other bids causes them to get a different position. But, we still have the property that the winner cannot get the position she won for a lower price.

It turns out that this is exactly the VCG mechanism, as seen by the following argument. In the following, let M be the valuation of the maximum matching, and for some i let M_{-i} be the valuation of the maximum matching that does not include bidder i. Note also that if bidder i is assigned position j, the VCG price is $M_{-i} - (M - b_{i,j}c_j)$.

In the suggested auction, when setting the price for bidder i, consider the moment when the bid is lowered to ppc_j. The total valuation of the matching at this point is $M - (b_{i,j} - ppc_j)c_j$. But the valuation of the matching at this point also is equal to M_{-i} since lowering the bid below ppc_j makes the matching no longer assign i to j (and all other edges are deleted, so i is not assigned anywhere else). So we get $M - (b_{i,j} - ppc_j)c_j = M_{-i}$, and therefore $ppc_jc_j = M_{-i} - (M - b_{i,j}c_j)$, which is the VCG price.

References

1. Edelman, B., Ostrovsky, M., Schwarz, M.: Internet advertising and the generalized second price auction: Selling billions of dollars worth of keywords. In: Second Workshop on Sponsored Search Auctions. (2006)
2. Varian, H.: Position auctions (2006) Working Paper, available at http://www.sims.berkeley.edu/~hal/Papers/2006/position.pdf.
3. Aggarwal, G.: Privacy Protection and Advertising in a Networked World. PhD thesis, Stanford University (2005)
4. Aggarwal, G., Goel, A., Motwani, R.: Truthful auctions for pricing search keywords. In: ACM Conference on Electronic Commerce (EC06). (2006)
5. Vickrey, W.: Counterspeculation, auctions and competitive sealed tenders. Journal of Finance **16** (1961) 8–37
6. Clarke, E.: Multipart pricing of public goods. Public Choice **11** (1971) 17–33
7. Groves, T.: Incentives in teams. Econometrica **41** (1973) 617–631
8. Nielsen//NetRatings: Interactive advertising bureau (IAB) search branding study (2004) Commissioned by the IAB Search Engine Committee. Available at http://www.iab.net/resources/iab_searchbrand.asp.
9. Mas-Collel, A., Whinston, M., Green, J.: Microeconomic Theory. Oxford University Press (1995)

Coping with Interference: From Maximum Coverage to Planning Cellular Networks

David Amzallag, Joseph (Seffi) Naor, and Danny Raz

Computer Science Department
Technion - Israel Institute of Technology
Haifa 32000, Israel
{amzallag,naor,danny}@cs.technion.ac.il

Abstract. Cell planning includes planning a network of base stations providing a coverage of the service area with respect to current and future traffic requirements, available capacities, interference, and the desired quality-of-service. This paper studies cell planning under budget constraints through a very close-to-practice model. This problem generalizes several problems such as budgeted maximum coverage, budgeted unique coverage, and the budgeted version of the facility location problem.

We present the first study of the budgeted cell planning problem. Our model contains capacities, non-uniform demands, and interference that are modeled by a penalty-based mechanism that may reduce the contribution of a base station to a client as a result of simultaneously covering this client by other base stations. We show that this very general problem is **NP**-hard to approximate and thus we define a restrictive version of the problem that covers all interesting practical scenarios. We show that although this variant remains **NP**-hard, it can be approximated within a factor of $\frac{e-1}{2e-1}$ of the optimum.

1 Introduction

Consider a set $I = \{1, 2, \ldots, m\}$ of possible configurations of base stations and a set $J = \{1, 2, \ldots, n\}$ of clients. Each base station $i \in I$ has capacity w_i, opening cost c_i, and every client $j \in J$ has a demand d_j. The demand is allowed to be simultaneously satisfied by more than one base station. Each base station i has a *coverage area* represented by a set $S_i \subseteq J$ of clients admissible to be covered (or satisfied) by it; this base station can satisfy at most w_i demand units of the clients in S_i.

When a client is belong to the coverage area of more than one base station, interference between the servicing stations may occur. These interference are modeled by a penalty-based mechanism and may reduce the contribution of a base station to a client. Let P be an $m \times m \times n$ matrix of *interference*, where $p(i_1, i_2, j) \in [0, 1]$ represents the fraction of i_1's service which client j loses as a result of interference with i_2 (defining $p(i, i, j) = 0$ for every $i \in I$, $j \in J$,

T. Erlebach and C. Kaklamanis (Eds.): WAOA 2006, LNCS 4368, pp. 29–42, 2006.
© Springer-Verlag Berlin Heidelberg 2006

and $p(i, i', j) = 0$ for every $j \notin S_{i'}$)[1]. This means that the interference caused as a result of a coverage of a client by more than one base station depends on the geographical position of the related "client". Followed by the above setting, we denote by $Q(i, j)$ the net contribution of base station i to client j, for every $j \in J$, $i \in I$, after incorporating the interference. A detailed description of $Q(i, j)$ is given later in this section.

The *budgeted cell planning problem* (BCPP) asks for a subset of base stations $I' \subseteq I$ whose cost does not exceed a given budget B, such that the total number of *fully* satisfied clients is maximized. That is, a solution to BCPP needs to maximize the number of clients for which $\sum_{i \in I} Q(i, j) \geq d_j$.

This problem generalizes several problems such as budgeted maximum coverage [10], budgeted unique coverage [5], and the budgeted version of the facility location problem (analyzed in Section 2.4). So far, these problems were studied (in the sense of approximation algorithms) without considering capacities or non-uniform demands. Coping with interference in covering problems is a great algorithmic challenge; unlike problems where there are no interference, during the time the solution is established, adding a new candidate (e.g., set, bin, item) to the cover may *decrease* the value of the solution. Furthermore, this problem involves full coverage (also known as *all-or-nothing* coverage) which usually makes the approximation task more complex (see [4] for example).

Cell planning is one of the most significant steps in the planning and management of cellular networks and it is among the most fundamental problems in the field of optimization of cellular networks. Cell planning includes planning a network of base stations that provides a (full or partial) coverage of the service area with respect to current and future traffic requirements, available capacities, interference, and the desired QoS. Under these constraints, the objective is, in general, to minimize the operator's total system cost. Cell planning is employed not only when new networks are built or when modifications to a current networks are made, but also (and mainly) when there are changes in the traffic demands, even within a small local area (e.g., building a new mall in the neighborhood or opening new highways). Planning cellular networks under budget limitations is practically the most important optimization problem in the planning stage. Since budget restrictions may lead to failure in achieving the required coverage, the objective, in this case, is hence to maximize the number of covered clients. This paper studies cell planning under budget constraints where the goal is to have a theoretical model that can be used in practical setting.

Computing $Q(i, j)$. Our technique for solving BCPP is independent in the structure of $Q(i, j)$. We describe here two general models for computing $Q(i, j)$.

Let x_{ij} be the fraction of the capacity w_i of a base station i that is supplied to client j. Recall that $I' \subseteq I$ is the set of base stations selected for opening, the contribution of base station i to client j is, in general is defined by

---

[1] For simplicity, we do not consider here interference of higher order. These can be further derived and extended from our model.

$$Q(i,j) = w_i x_{ij} \cdot \prod_{i' \in I'} \left(1 - p(i,i',j)\right). \tag{1}$$

This means that the net contribution of base station i to client j depends on all other base stations i' that contains j in their coverage areas. Each of these base stations "interferes" base station i to service j and reduces the contribution of $w_i x_{ij}$ by a factor of $p(i,i',j)$.

Since (1) is a high-order expression we use the following first-order approximation[2]

$$\prod_{i' \in I'} \left(1 - p(i,i',j)\right) = \left(1 - p(i,i'_1,j)\right)\left(1 - p(i,i'_2,j)\right) \ldots \approx 1 - \sum_{i' \in I'} p(i,i',j) \tag{2}$$

Combining (1) and (2) we get

$$Q(i,j) \approx \begin{cases} w_i x_{ij}\left(1 - \sum_{i' \in I'} p(i,i',j)\right), & \sum_{i' \in I'} p(i,i',j) < 1 \\ 0, & \text{otherwise.} \end{cases} \tag{3}$$

Consider, for example, a client j belonging to the coverage areas of two base stations i_1 and i_2, and assume that just one of these base stations, say i_1, is actually participating in j's satisfaction (i.e., $x_{i_1 j} > 0$ but $x_{i_2 j} = 0$). According to the above model, the mutual interference of i_2 on i_1's contribution ($w_1 x_{i_1 j}$) should be considered, although i_2 is not involved in the coverage of client j.

In most cellular wireless technologies, this is the usual behavior of interference. However, in some cases a base station can affect the coverage of a client *if and only if* it is participating in its demand satisfaction. The contribution of base station i to client j in this case is defined by

$$Q(i,j) \approx \begin{cases} w_i x_{ij}\left(1 - \sum_{i' \neq i \in I_j} p(i,i')\right), & \sum_{i' \neq i \in I_j} p(i,i') < 1 \\ 0, & \text{otherwise.} \end{cases} \tag{4}$$

where I_j is the set of base stations that participates in the coverage of client j, i.e., $I_j = \{i \in I : x_{ij} > 0\}$. Notice that in this model the interference function does not depend on the geographic position of the clients.

Our contributions. In this paper we present the first study of the budgeted cell planning problem. To the best of our knowledge, despite the extensive research of non-budgeted cell planning problems (i.e., minimum-cost cell planning, as descried in Section 2.1), there is no explicit study in the literature of the BCPP (in both theoretical and, surprisingly, also in practical settings). We survey, in Section 2, some previous work related to BCPP. Budgeted maximum coverage, budgeted unique coverage, budgeted facility location, and maximizing submodular set functions are among the reviewed problems. In Section 3 we show that approximating BCPP is **NP**-hard. Then we define a restrictive version of BCPP,

---

[2] Notice that, in this context, one can precisely estimate the "cost" of such approximation using [11,9]. However, for simplicity we do not include these works in this practical model.

the $k4k$-budgeted cell planning, by making additional assumptions that are motivated by practical considerations. The additional property is that every set of k-opened base stations can fully satisfy at least k clients, for every integral value of k. In Section 4 we show that this problem remains **NP**-hard and present an $\frac{e-1}{2e-1}$ (≈ 0.3873) factor approximation algorithm for this problem.

2   Related Work

2.1   The Minimum-Cost Cell Planning Problem

The *minimum-cost cell planning problem* asks for a minimum-cost subset $I' \subseteq I$ that satisfies the demands of *all* the clients. This important problem is one of the most studied on the area of cellular network optimization. Previous work dealt with a wide variety of special cases (e.g., cell planning without interference, frequency planning, uncapacitated models, antenna-type limitations, and topological assumptions regarding coverage). These works range from meta-heuristics (e.g., genetic algorithms, simulated annealing, etc.) and greedy approaches, through exponential-time algorithms that compute an optimal solution, to approximation algorithms for special cases of the problem. A comprehensive survey of various works on minimum-cost cell planning problems appears in [3]. An $O(\log W)$-approximation algorithm for the non-interference version of the minimum-cost cell planning problem is presented in [2], where W is the largest given capacity of a base station.

2.2   Base Stations Positioning Under Geometric Restrictions

A PTAS for the uncapacitated BCPP with unit demands (i.e., $w_i = \infty$ and $d_j = 1$ for all $i \in I, j \in J$) and without interference is given in [7]. In this case the problem is studied under geometric restrictions of disks of constant radius D (i.e., S_i is the set of clients located within a distance no greater than D from the geometric location of i, for every $i \in I$), a minimal distance between different base stations that have to be kept, and clients as well as base stations are associated with points in the Euclidean plane.

2.3   Budgeted Maximum Coverage and Budgeted Unique Coverage

BCPP is closely related to the budgeted maximum coverage and the budgeted unique coverage version of set cover. Given a collection of subsets S of a universe U, where each element in U has a specified weight and each subset has a specified cost, and a budget B. The *budgeted maximum coverage problem* asks for a subcollection $S' \subseteq S$ of sets, whose total cost is at most B, such that the total weight of elements covered by S' is maximized. The *budgeted unique coverage problem* is a similar problem where elements in the universe are uniquely covered, i.e., appears in exactly one set of S'. Both problems are special cases of BCPP in which elements are clients with unit demands, every set $i \in I$ corresponds to a base station i containing all clients in its coverage area $S_i \subseteq J$, and $w_i \geq |S_i|$ for all base stations in I. In this setting, budgeted maximum coverage is the case (in the sense that a solution for BCPP is optimal if and only if it is optimal for the budgeted maximum coverage) when there are no interference (i.e., P is the zero matrix), while budgeted unique coverage is when the interference is taking to be the highest (i.e., $p(i', i'', j) = 1$ for every $i' \neq i''$, and $p(i', i'', j) = 0$ otherwise).

For the budgeted maximum coverage problem, there is a $(1-\frac{1}{e})$-approximation algorithm [10,1], and this is the best approximation ratio possible unless **NP=P** [6]. For the budgeted unique coverage problem, there is an $\Omega(1/\log n)$-approximation algorithm [5] and, up to a constant exponent depending on ϵ, $O(1/\log n)$ is the best possible ratio assuming **NP** $\not\subseteq$ **BPTIME** $(2^{n^{\epsilon}})$ for some $\epsilon > 0$. Interestingly enough, we will show in the next section that our generalization for both of these problems is hard to approximate.

2.4   Budgeted Facility Location

The budgeted version of the (uncapacitated) facility location problem is also closely related to BCPP. In the traditional (uncapacitated) facility location problem we wish to find optimal locations in which to build facilities, from a given set I, to serve a given set J of clients, where building a facility in location i incurs a cost of f_i. Each client j must be assigned to one facility, thereby incurring a cost of c_{ij} (without assuming the triangle inequality). The objective is to find a solution of minimum total cost. The *budgeted facility location problem* is to find a subset $I' \subseteq I$ such that the total cost of opening facilities and connecting clients to open facilities does not exceed a given budget B, and the total number of connected clients is maximized.

Given an instance of the budgeted (uncapacitated) facility location problem, we show in the following that this problem is a special case of the budgeted maximum coverage problem. By a *star* we mean a pair (i, Q) with $i \in I$ and $Q \subseteq J$. The cost of a star (i, Q) is $c(i, Q) = f_i + \sum_{j \in Q} c_{ij}$, and its *effectiveness* is $\frac{|Q|}{c(i,Q)}$. Then the budgeted (uncapacitated) facility location problem is a special case of the budgeted maximum coverage problem: set J is the set of elements that need to be covered, and let $S = 2^J$, where $c(Q)$ is the minimum-cost of a star (i, Q) (we take the same budget for both instances).

However, the resulting budgeted maximum coverage instance has exponential size, and therefore this reduction cannot be used directly. Nevertheless, we can apply the algorithm of [10] without generating the instance explicitly, as proposed by Hochbaum [8]: In each step, we have to find a most effective star, open its facility and henceforth disregard all clients in this star. Although there are exponentially many stars, it is easy to find the most effective one as it suffices to consider stars (i, Q_k^i), for $i \in I$ and $k \in \{1, 2, \ldots, |J|\}$. Here Q_k^i denotes the first k clients in a linear order with nondecreasing c_{ij}. Clearly, other stars cannot be more effective. Hence, we get the same approximation ratio as the budgeted maximum coverage problem. Moreover, since the budgeted maximum coverage can be described, by a simple reduction, as a special case of the budgeted facility location, the best we can hope for the budgeted facility location problem is the same approximation factor as the budgeted maximum coverage problem.

2.5   Maximizing Submodular Set Functions

Let $U = \{1, \ldots, n\}$, let $c_u, u \in U$ be a set of nonnegative weights, and let B be a nonnegative budget. The problem of *maximizing nondecreasing submodular set function with budget constraint* is

$$\max_{S \subseteq U} \left\{ f(S) : \sum_{u \in S} c_u \leq B \right\},$$

where $f(S)$ is a nonnegative nondecreasing submodular polynomially computable set function (a set function is submodular if $f(S) + f(T) \geq f(S \cup T) + f(S \cap T)$ for all $S, T \subseteq U$ and nondecreasing if $f(S) \leq f(T)$ for all $S \subseteq T$). For this problem, there is a $(1 - \frac{1}{e})$-approximation algorithm [12], and as this problem is a generalization of the budgeted maximum coverage ($c_u = 1$, for all $u \in U$, and $f(S)$ denotes the maximum weight that can be covered by the set S), this ratio is the best achievable.

Although this problem seems, at least from a natural perspective, to be closely related to BCPP, observe that set (covering) functions are not submodular, in general, when interference are involved. Consider, for example, an instance of BCPP in which $I = \{1, 2, 3\}$ with $w_1 = w_2 = 1$ and $w_3 = 1/4$, a single client with $d = 2$ that can be satisfied by all base stations, and symmetric penalties $p(1, 3) = p(2, 3) = 1/2$, while $p(1, 2) = 0$. Taking $S = \{1\} \cup \{3\}$ and $T = \{2\} \cup \{3\}$ we have $f(S) + f(T) \ngeq f(S \cup T) + f(S \cap T)$, where $f(S)$ is defined to be the maximum number of fully satisfied clients that can be covered by the set S of base stations.

3   Inapproximability

As mentioned earlier, the budgeted maximum coverage as well as budgeted unique coverage can be seen as special cases of BCPP. In both cases the approximation algorithms are based on the greedy technique of Khuller, Moss, and Naor [10]. This means picking at each step the most effective set until either no element is left to be covered or the budget limitation is exceeded. Combining this method with the enumeration technique yields the $(1 - \frac{1}{e})$-approximation algorithm of [10].

Unfortunately, a natural attempt to adapt the ideas from [10] to the setting of BCPP fails, as stated by the next theorem.

Theorem 1. *It is **NP**-hard to find a feasible solution to the budgeted cell planning problem.*

Proof. The proof is via a reduction from the subset sum problem. Given an instance of the subset sum problem, i.e., a set of natural numbers $A = \{a_1, a_2, \ldots, a_n\}$ and an additional natural number $T = \frac{1}{2} \sum_{i=1}^{n} a_i$. We build an instance of the budgeted BCPP with $I = \{1, 2, \ldots, n\}, |J| = 1$ and $w_i = c_i = a_i$ for every $i \in I$; the budget and the single client's demand are $B = d = T$ and no interference are assumed.

It is easy to see that the client is satisfied if and only if there exists $S \subseteq A$ with $\sum_{i \in S} a_i = T$. Since there is only a single client, any polynomial-time approximation algorithm must produce a full coverage, solving the subset sum problem in polynomial time.   ∎

4   The $k4k$-Budgeted Cell Planning Problem

In light of the above inapproximability result, we turn to define a restrictive version of BCPP which is general enough to cover all interesting practical cases. In order to do that, we use the fact that in general, the number of base stations in cellular networks is much smaller than the number of clients. Notice that when planning cellular networks, the notion of "clients" sometimes means mobile-clients and sometimes it represents the total traffic demand created by many mobile-clients at a given location. Our models support both forms of representations. Moreover, when there is a relatively large cluster of antennas in a given location, this cluster is usually addressed to meet the traffic requirements of a high-density area of clients. Thus for both interpretations of "clients" the number of satisfied clients is always much bigger than the number of base stations. Followed by the above discussion, we define the $k4k$-*budgeted cell planning problem* ($k4k$-BCPP) to be BCPP with the additional property that every set of k base stations can fully satisfy at least k clients, for every integer k (and we refer to this property as "$k4k$ property").

In this section we show that this problem can be approximated within a factor of $\frac{e-1}{2e-1}$ of the optimum. First, we show that this problem remains hard.

Theorem 2. *The $k4k$-budgeted cell planning problem is* **NP**-*hard.*

Proof. Via a reduction from the budgeted maximum coverage problem. Consider an instance of the budgeted maximum coverage problem, that is, a collection of subsets $\mathcal{S} = \{S_1, \ldots, S_m\}$ with associated costs $\{c_i\}_{i=1}^m$ over a domain of elements $X = \{x_1, \ldots, x_n\}$, and a budget L.

We can construct an instance of $k4k$-BCPP such that an optimal solution to this problem gives an optimal solution to the budgeted maximum coverage problem. First, we construct a bipartite graph of elements vs. sets, derived from the budgeted maximum coverage instance: there is an edge (x_i, S_j) if and only if element x_i belongs to set S_j. The instance of $k4k$-BCPP is as follows: the set of clients is $\{x_1, \ldots, x_n\} \cup \{y_1, \ldots, y_m\}$, where each of the x_j's is of unit demand and each of the y_r's is of zero demand, the set of potential base stations is $\{S_1, \ldots, S_m\}$, each of opening cost c_i, a capacity $w_i = |S_i|$, and a set of admissible clients for covering $S_i \cup \{y_1, \ldots, y_m\}$, for every $i = 1, \ldots, m$, and $j = 1, \ldots, n$, and a budget $B = L$, while no interference are assumed.

Clearly, a solution to $k4k$-BCPP is optimal if and only if the corresponding solution of the budgeted maximum coverage instance is optimal.   ∎

4.1 The Structure of BCPP Solutions

Our algorithm is based on a combinatorial characterization of the solution set to BCPP[3] (and in particular to $k4k$-BCPP). The following lemma is a key component in the analysis of our approximation algorithm.

Lemma 1. *Every solution to the $k4k$-budgeted cell planning problem can be transformed to a solution in which the number of clients that are covered by more than one base station is at most the number of opened base stations. Moreover, this transformation leaves the number of fully satisfied clients as well as the solution cost unchanged.*

Proof. Consider a solution $\Delta = \{I', J', \mathbf{x}\}$ to the $k4k$-BCPP, where $I' \subseteq I$ is the set of base stations selected for opening, $J' \subseteq J$ is the set of fully satisfied clients, x_{ij}'s are the base station-client coverage rates, and $J'' \subseteq J'$ is the set of clients that are satisfied by more than one base station. Without loss of generality we may assume that every client has a demand greater than zero, since there is no need for "covering" clients with zero demand. We associate the weighted bipartite graph $G_\Delta = (I' \cup J', E)$ with every such solution. In this graph, $(i, j) \in E$ has weight $w(i, j) = w_i x_{ij}$ if and only if $x_{ij} > 0$, and $w(i, j) = 0$, otherwise. Two cases need to be considered:

1. If G_Δ is acyclic then we are done (i.e., no transformation is needed); in this case $|J''| < |I'|$. To see this, let T be a forest obtained from G_Δ by fixing an arbitrary base station vertex as the root (in each of the connected components of G_Δ) and trimming all client leaves. These leaves correspond to clients who are covered, in the solution, by a single base station. Since the height of the tree is even, the number of internal client-vertices is at most the number of base station-vertices, hence $|J''| < |I'|$.

2. Otherwise, we transform $G_\Delta = (I' \cup J', E)$ into an acyclic bipartite graph $G_{\Delta'} = (I' \cup J', E')$ using a cycle canceling algorithm. For simplicity, we first describe the following algorithm for the case that interference do not exist.

   Algorithm 1 [CYCLE CANCELING WITHOUT INTERFERENCE]. As long as there are cycles in G_Δ, pick a cycle C and let γ be the weight of a minimum-weight edge on this cycle. Take a minimum-weight edge on C and, starting from this edge, alternately, in clockwise order along the cycle, decrease and increase the weight of every edge by γ.

   It is easy to verify that at the end of the algorithm every client receives, and every base station supplies, the same amount of demand units as before. Moreover, the only changes here are the values of the x_{ij}'s. Hence, Algorithm 1 preserves the number as well as the identity of the satisfied clients. Since at each iteration at least one edge is removed, $G_{\Delta'}$ is acyclic, thus yielding $|J''| < |I'|$ as in the former case.

---

[3] Results in this section can be applied also for non-$k4k$ versions of the BCPP. For simplicity, we concentrate here on the $k4k$ versions of the problem.

When interferences exist but $Q(i,j)$ is independent on the x_{ij}'s, we can still use Algorithm 1 to preserves the number as well as the identity of the satisfied clients. In this case change in the x_{ij}'s does not affect the $Q(i,j)$ of any client. However, when $Q(i,j)$ is a function of the x_{ij}'s this algorithm can no longer guarantee this. This is true because of the way the fully satisfied clients "use" base stations not on the cycle depends on the interference, and thus the modifications of the edge weights on the cycle are not enough. To overcome this problem we generalize the method of cycle canceling. Consider a cycle $C = (v_1, \ldots, v_k = v_1)$ in G_Δ, such that odd vertices correspond to base stations. Let v_i be any client-vertex in C. Now suppose the base station which corresponds to v_{i-1} increases its supply to v_i by α units of demand. The basic idea of the generalization is to compute the exact number of demand units the base station which corresponds to v_{i+1} *must* subtract from its coverage, in order to preserve the satisfaction of that client, taking into account all the demand (with its interference) supplied by base station vertices which are outside the cycle.

Notice that increasing a certain $w(v_i, v_{i+1})$ *does not necessary* increase the supply to client v_i. When interferences are considered, it could actually happen that increasing $w(v_i, v_{i+1})$ *decreases* the supply to v_i (if the new interference penalties outweigh the increased supply). Similarly, decreasing some $w(v_i, v_{i+1})$ could actually increase the supply to v_i. However, one can assume for optimal solutions that these cases do not occur (as the solution could be transformed into an equivalent solution where such edges have $w(v_i, v_{i+1}) = 0$).

To demonstrate the idea of canceling cycles when interferences exist let us assume, for simplicity, that there is only a single base station which is not on the cycle, denoted by v_o, which participates in the coverage of client v_i, and the interference model is assumed to be the one in (4). Then, the total contribution of base stations v_{i-1}, v_{i+1}, and v_o to the coverage of client v_i is, by (1),

$$\delta(v_i) = Q(v_0, v_i) + Q(v_{i+1}, v_i) + Q(v_{i-1}, v_i).$$

Given that the supply of base station v_{i-1} to client v_i is increased by α units of demand (i.e., $w'(v_{i-1}, v_i) = w(v_{i-1}, v_i) + \alpha$, where w' is the updated weight function of the edges), base station v_{i+1} must decrease its supply to this client by β units of demand (i.e., $w'(v_{i+1}, v_i) = w(v_{i+1}, v_i) - \beta$) in order to preserve the satisfaction of client v_i (assuming v_o's supply remains the same). Then, the value of β can be computed via a solution to the following equation (in variable β),

$$\delta'(v_i) = Q'(v_0, v_i) + Q'(v_{i+1}, v_i) + Q'(v_{i-1}, v_i) = \delta(v_i).$$

Notice that our cycle canceling algorithms are used for the proof of existence and such computations are not necessary for the execution of our approximation algorithm.

Algorithm 2 [CYCLE CANCELING WITH INTERFERENCE]. As long as there are cycles in G_Δ, pick a cycle $C = (v_1, \ldots, v_k = v_1)$ where odd vertices represent base stations. As before, every edge $e_i = (v_i, v_{i+1})$ on the cycle has a weight $w(v_i, v_{i+1})$ associated with it, representing the amount of demand supplied by the base-station-vertex in e_i to the client-vertex in e_i. For simplicity, let d'_i denote this value.

We recursively define a sequence of weights $\{y_i\}_{i=1}^{k-1}$, with alternating signs which represent a shift in the demand supply of base stations to clients along the cycle. Start by setting $y_1 = \epsilon$, representing an increase of ϵ to the demand supplied by the base-station-vertex v_1 to the client-vertex v_2. This increase of supply may not all be credited to vertex v_2 due to interference, some of which are possibly due to base stations which are outside the cycle, which contribute to v_2's satisfaction. Set y_2 to be the maximal decrease in the demand supplied by base-station-vertex v_3 to client-vertex v_2, such that v_2 remains satisfied. This amount of demand is now "available" to base-station-vertex v_3, hence we allow v_3 to supply this surplus to client-vertex v_4. We continue in this manner along the cycle.

If, by repeating this procedure, we end up with $|y_{k-1}| \geq \epsilon$, then we say the cycle is ϵ-*adjustable*. Otherwise, redefine the values of $\{y_i\}_{i=1}^{k-1}$ in a similar manner, but in reverse order, i.e., starting from y_{k-1} and ending with y_1. However, it is easy to verify that at least one direction, the cycle is ϵ-adjustable, for some value of ϵ.

Let $\epsilon_{\max}$ be the largest value for which the cycle is adjustable, and consider its corresponding values of y_i, $i = 1, \ldots, k-1$. Note that the y_i's have alternating signs, and for any client-vertex v_i, $y_i = -y_{i-1}$. Define the quotients $z_i = d'_i/y_i$ for every $i = 1, \ldots, k-1$, and let $z_{\min} = \min_{z_i < 0} |z_i|$. Now increase the amount of demand supplied on every edge on the cycle to be $w'(v_i, v_{i+1}) = y_i \cdot z_{\min}$, where w' is the updated weight function of the edges, as before.

Two important invariants are maintained throughout our cycle-canceling procedure. The first is that $w'(i, j) \geq 0$ for every edge (i, j) of the cycle. The second is that there exists at least one edge $e = (i, j)$ on the cycle for which $w'(i, j) = 0$. Therefore Algorithm 2 preserves the number and the identity of the satisfied clients and $G_{\Delta'}$ is also a solution. Since at each iteration at least one edge is removed, $G_{\Delta'}$ is acyclic and $|J''| < |I'|$ as before. ∎

4.2   An $\frac{e-1}{2e-1}$ -Approximation Algorithm

We are now ready to present a $\frac{e-1}{2e-1}$-approximation algorithm for $k4k$-BCPP. We combine ideas from [10] together with our characterization of the optimal solution set of $k4k$-BCPP. Throughout this section we use the following notation. Let N_i be the maximum number of clients that can be covered by a single base station i (i.e., without allowing simultaneously covering of a client), $i = 1, 2, \ldots, m$. Let $N(I')$ denote the total number of clients that can be covered by I' in such a way that each client is covered by a single base station, and let

Algorithm 3. $k4k$-BUDGETED CELL PLANNING
1: $J' \leftarrow \emptyset$; $H_3 \leftarrow 0$;
2: $H_1 \leftarrow$ maximum number of base stations having a total opening cost less than or equal to B
3: $H_2 \leftarrow \text{argmax}\{N(S), \text{ such that } S \subseteq I, |S| < \ell, \text{ and } c(S) \leq B\}$
4: **for all** $S \subseteq I$, such that $|S| = \ell$ and $c(S) \leq B$ **do**
5:     $\mathcal{I} \leftarrow I \setminus S$
6:     **repeat**
7:         select $i \in \mathcal{I}$ that maximizes $\frac{N'_i}{c_i}$
8:         **if** $c(S) + c_i \leq B$ **then**
9:             $S \leftarrow S \cup \{i\}$
10:            $J' \leftarrow J' \cup J_i$
11:            update w_i by the demand units supplied to J_i
12:            $c(S) \leftarrow c(S) + c_i$
13:        **end if**
14:        $\mathcal{I} \leftarrow \mathcal{I} \setminus \{i\}$
15:    **until** $\mathcal{I} = \emptyset$
16:    **if** $N(S) > H_3$ **then** $H_3 \leftarrow N(S)$
17: **end for**
18: Output the solution having the largest value from $\{H_1, H_2, H_3\}$

N'_i denote the maximum number of clients that can be covered by a single base station i, but not covered by any other base stations in I'. Finally, we denote by J_i the set of clients that are fully satisfied by base station i. Without loss of generality we may assume that the opening cost of any base station does not exceed B, since base stations of cost greater than B do not belong to any feasible solution.

We first observe that the greedy algorithm that opens at each step a base station maximizing the ratio $\frac{N'_i}{c_i}$ has an unbounded approximation factor. Consider, for example, two base stations and $M + 2$ clients $J = \{1, \ldots, M, M + 1, M + 2\}$ having unit demands. Let $w_1 = 2, c_1 = 1, S_1 = \{M + 1, M + 2\}$, where $w_2 = c_2 = M, S_2 = \{1, \ldots, M\}$. The overall budget in this example is taken to be M. The optimal solution opens the second base station satisfying exactly M clients, while the solution obtained by the greedy algorithm opens the first base station satisfying exactly 2 clients. The approximation ratio for this instance is $M/2$, and is therefore unbounded.

Our algorithm comprises of two phases. In the first phase, the algorithm computes the maximum number of base stations having a total opening cost less than or equal to B. Since our instances are "$k4k$", this is a lower bound on the optimal solution of $k4k$-BCPP. Furthermore, it can be computed in linear-time by picking base stations in non-decreasing order of their opening cost. Another set of candidate solutions concentrates, in the second phase, on the number of clients that can be fully satisfied by a single base station. This is also a lower bound on the optimal solution and it is computed as the best of two possible candidates (in a similar way to [10]). For a fixed integer $\ell \geq 3$, the first candidate consists of all subsets of I of cardinality less than ℓ which have cost at most B, while the second one enumerates all feasible solutions of cardinality ℓ having cost at most B, and then completes each subset to a candidate solution using the greedy algorithm. Based on both phases the algorithm outputs the candidate solution having the maximum number of satisfied clients.

The problem of computing the optimal value of $N(S)$, for a given set of base stations, S, is **NP**-hard. In fact, this problem is a generalization of the budgeted maximum coverage problem containing capacities as well as non-uniform demands. Fortunately, a straightforward extension of [10] gives a $(1-\frac{1}{e})$-approximation algorithm for this generalization.

Theorem 3. *Algorithm 3 is a $\frac{e-1}{2e-1}$-approximation algorithm for the k4k-budgeted cell planning problem.*

Proof. Let $\tilde{n}$ be the solution obtained by Algorithm 3, and let n^* be the maximum number of satisfied clients as obtained by the optimal solution. In the latter, n_1^* denotes the number of clients that are satisfied by a single base station, and n_2^* is the number of clients satisfied by more than one base station. Finally, we denotes I^* to be the set of base stations opened (by the optimal solution) for satisfying these $n^* = n_1^* + n_2^*$ clients.

Now, if $\hat{n}_1$ denotes the maximum number of clients that can be satisfied by a single base station then $\hat{n}_1 \geq n_1^*$. Since computing $\hat{n}_1$ is done using the extension of the algorithm of [10], we have

$$\tilde{n} \geq \left(1 - \frac{1}{e}\right)\hat{n}_1. \tag{5}$$

Combining the above discussion, gives

$$\left(2 - \frac{1}{e}\right)\tilde{n} = \tilde{n} + \left(1 - \frac{1}{e}\right)\tilde{n} \tag{6}$$

$$\geq \tilde{n} + \left(1 - \frac{1}{e}\right)|I^*| \tag{7}$$

$$\geq \left(1 - \frac{1}{e}\right)\hat{n}_1 + \left(1 - \frac{1}{e}\right)|I^*| \tag{8}$$

$$\geq \left(1 - \frac{1}{e}\right)n_1^* + \left(1 - \frac{1}{e}\right)n_2^* \tag{9}$$

$$\geq \left(1 - \frac{1}{e}\right)n^* \tag{10}$$

where inequality (7) follows from the fact that every set of k opened base stations can satisfy at least k clients (as used by the first candidate of our algorithm), inequality (8) is based on (5), and inequality (9) follows from Lemma 1. ∎

Finding x_{ij}'s values. Algorithm 3 outputs a set $J' \subseteq J$ of fully satisfied clients and a set $I' \subseteq I$ of base stations providing this coverage. However, the values of the x_{ij}'s are not inclusively outcome from the algorithm. Since in this setting we are given a set of *already opened* base stations, these values can be efficiently determined by any feasible solution of the following linear program (**LP**). Notice that the objective function in this linear program is not important and any feasible point will do.

$$\max \sum_{i \in I'} \sum_{j \in J'} x_{ij} \qquad \qquad \text{(LP)}$$

$$\text{s.t.} \sum_{i \in I'} Q(i,j) \geq d_j \qquad \forall j \in J' \qquad (11)$$

$$\sum_{j \in J'} x_{ij} \leq 1 \qquad \forall i \in I' \qquad (12)$$

$$0 \leq x_{ij} \leq 1 \qquad \forall i \in I', j \in S_i$$

$$x_{ij} = 0 \qquad \forall i \in I', j \notin S_i$$

In this linear program constraints (11) ensures that every client will be fully satisfied (notice that $Q(i,j)$ is the same as in (3) without the need to open base stations), while constraints (12) maintains the capacity bounds for the base stations.

5 Conclusions and Future Work

In this paper we present a theoretical study of the budgeted cell planning, a central complex optimization problem in planning of cellular networks. As far as we know, no performance guarantee was given so far to this problem. We show that although this problem is **NP**-hard to approximate, we can still cover all practical scenarios by adopting a very practical assumption, called the $k4k$-property, satisfied by every real cellular network, and we give a fully combinatorial $\frac{e-1}{2e-1}$-approximation algorithm for this problem. We believe that taking capacities, non-uniform demands, and interference into considerations makes a significant step towards making approximation algorithms a key ingredient in practical solutions to many planning and covering problems in cellular networks.

An interesting open problem that is closely related to BCPP is the all-or-nothing demand maximization problem. In this problem we are given a set $I = \{1, 2, \ldots, m\}$ of base stations that are already opened, a set $J = \{1, 2, \ldots, n\}$ of clients. Each base station $i \in I$ has capacity w_i, and every client $j \in J$ has a profit p_j and a demand d_j which is allowed to be simultaneously satisfied by more than one base station. Each base station i has a *coverage area* represented by a set $S_i \subseteq J$ of clients admissible to be covered (or satisfied) by it. Let P be an $m \times m \times n$ matrix of *interference* for satisfying a client by several base stations, as in BCPP. The *all-or-nothing demands maximization problem* asks for a maximum-profit subset $J' \subseteq J$ of clients that can be fully satisfied by I. As one can noticed, this problem is a special case of BCPP by taking $c_i = 0$, and $p_j = 1$, for every $i \in I, j \in J$.

Acknowledgments

We would like to thank Gabi Scalosub for his comments for an earlier version of this paper. This research was supported by REMON - Israel 4G Mobile Consortium, sponsored by Magnet Program of the Chief Scientist Office in the Ministry of Industry and Trade of Israel.

References

1. A. Ageev and M. Sviridenko. Approximation algorithms for maximum coverage and max cut with given sizes of parts. In *Proceedings of the Conference on Integer Programming and Combinatorial Optimization*, volume 1610 of *Lecture Notes in Computer Science*, pages 17–30. Springer-Verlag, Berlin, 1999.
2. D. Amzallag, R. Engelberg, J. Naor, and D. Raz. Approximation algorithms for cell planning problems. Manuscript, 2006.
3. D. Amzallag, M. Livschitz, J. Naor, and D. Raz. Cell planning of 4G cellular networks: Algorithmic techniques, and results. In *Proceedings of the 6th IEE International Conference on 3G & Beyond (3G'2005)*, pages 501–506, 2005.
4. C. Chekuri, S. Khanna, and F. B. Shepherd. The all-or-nothing multicommodity flow problem. In *Proceedings of the 36th Annual ACM Symposium on Theory of Computing*, pages 156–165, 2004.
5. E. D. Demaine, U. Feige, M. Hajiaghayi, and M. R. Salavatipour. Combination can be hard: Approximability of the unique coverage problem. In *Proceedings of the 17th Annual ACM-SIAM Symposium on Discrete Algorithms*, pages 162–171, 2006.
6. U. Feige. A threshold of $\ln n$ for approximating set cover. *J. ACM*, 45:634–652, 1998.
7. C. Glaßer, S. Reith, and H. Vollmer. The complexity of base station positiong in cellular networks. In *Workshop on Approximation and Randomized Algorithms in Communication Networks*, pages 167–177, 2000.
8. D. Hochbaum. Heuristics for the fixed cost median problem. *Mathematical Programming*, 22(2):148–162, 1982.
9. J. Kahn, N. Linial, and A. Samorodnitsky. Inclusion-exclusion : exact and approximate. *Combinatorica*, 16:465–477, 1996.
10. S. Khuller, A. Moss, and J. Naor. The budgeted maximum coverage problem. *Information Processing Letters*, 70:39–45, 1999.
11. N. Linial and N. Nisan. Approximate inclusion-exclusion. *Combinatorica*, 10:349–365, 1990.
12. M. Sviridenko. A note on maximizing a submodular set function subject to knapsack constraint. *Operations Research Letters*, 32:41–43, 2004.

Online Dynamic Programming Speedups*

Amotz Bar-Noy[1], Mordecai J. Golin[2], and Yan Zhang[2]

[1] Brooklyn College, 2900 Bedford Avenue Brooklyn, NY 11210
amotz@sci.brooklyn.cuny.edu
[2] Hong Kong University of Science and Technology, Kowloon, Hong Kong
{golin,cszy}@cse.ust.hk

Abstract. Consider the Dynamic Program $h(n) = \min_{1 \leq j \leq n} a(n,j)$ for $n = 1, 2, \ldots, N$. For arbitrary values of $a(n,j)$, calculating all the $h(n)$ requires $\Theta(N^2)$ time. It is well known that, if the $a(n,j)$ satisfy the *Monge property*, then there are techniques to reduce the time down to $O(N)$. This speedup is inherently static, i.e., it requires N to be known in advance.

In this paper we show that if the $a(n,j)$ satisfy a stronger condition, then it is possible, without knowing N in advance, to compute the values of $h(n)$ in the order of $n = 1, 2, \ldots, N$, in $O(1)$ amortized time per $h(n)$. This *maintains the DP speedup online*, in the sense that the time to compute all $h(n)$ is $O(N)$. A slight modification of our algorithm restricts the worst case time to be $O(\log N)$ per $h(n)$, while maintaining the amortized time bound. For $a(n,j)$ that satisfy our stronger condition, our algorithm is also simpler to implement than the standard Monge speedup.

We illustrate the use of our algorithm on two examples from the literature. The first shows how to make the speedup of the D-median on a line problem in an online settings. The second shows how to improve the running time for a DP used to reduce the amount of bandwidth needed when paging mobile wireless users.

1 Introduction

Consider the class of problems defined by

$$h(n) = \min_{1 \leq j \leq n} a(n,j), \quad \forall\, 1 \leq n \leq N \tag{1}$$

where the goal is to compute $h(n)$ for $1 \leq n \leq N$. In many applications, (1) is a Dynamic Program (DP), in the sense that the values of $a(n,j)$ depend upon $h(i)$, for some $1 \leq i < n$. In this paper, we always assume any particular $a(n,j)$ can be computed in $O(1)$ time, provided that the values of $h(i)$ it depends on are known. For a generally defined function $a(n,j)$, it requires $\Theta(N^2)$ time to compute all the $h(n)$. It is well known, though [1], that if the values of $a(n,j)$ satisfy the *Monge property* (see Section 1.1), then the SMAWK algorithm [2] can compute all the $h(n)$, for $1 \leq n \leq N$, in $O(N)$ time. To be precise, if

* The research of the second and third authors was partially supported by Hong Kong RGC CERG grant HKUST6312/04E.

T. Erlebach and C. Kaklamanis (Eds.): WAOA 2006, LNCS 4368, pp. 43–54, 2006.
© Springer-Verlag Berlin Heidelberg 2006

1. the value of N is known in advance;
2. and for any $1 \leq j \leq n \leq N$, the value of $a(n, j)$ can be computed in $O(1)$ time, i.e., $a(n, j)$ does not depend on $h(i)$;
3. and the values of $a(n, j)$ satisfy the Monge property defined by (4),

then the SMAWK algorithm [2] can compute all of the $h(n)$ for $1 \leq n \leq N$ in $O(N)$ time.

The main purpose of this paper is to consider the DP formula (1) in *online* settings. By this we mean that the values of $h(n)$ are computed in the order $n = 1, 2, \ldots, N$ *without* knowing the parameter N in advance, and the values of $a(n, j)$ are allowed to depend on *all* previously-computed values of $h(i)$ for $1 \leq i < n$. To be precise, our main result is

Theorem 1. *Consider the DP defined by (1). If*

1. $\forall\, 1 \leq j \leq n \leq N$, the value of $a(n, j)$ can be computed in $O(1)$ time, provided that the values of $h(i)$ for $1 \leq i < n$ are known;
2. and $\forall\, 1 \leq j < n \leq N$,

$$a(n, j) - a(n - 1, j) = c_n + \delta_j \beta_n \qquad (2)$$

where c_n, β_n and δ_j are constants satisfying
(a) $\forall\, 1 < n \leq N$, $\beta_n \geq 0$;
(b) and $\delta_1 > \delta_2 > \cdots > \delta_{N-1}$,

then, there is an algorithm that computes the values of $h(n)$ in the order $n = 1, 2, \ldots, N$ in $O(1)$ amortized and $O(\log N)$ worst-case time per $h(n)$. The algorithm does not know the value of N until $h(N)$ has been computed.

We call the Condition 2 in Theorem 1 (including Conditions (a) and (b)) the *online Monge property*. As we will see in Section 1.1, the online Monge property is a stronger Monge property. The SMAWK algorithm is a $\Theta(N)$ speedup of the computation of (1) when $a(n, j)$ satisfy the Monge property. Theorem 1 says that if $a(n, j)$ satisfy the online Monge property, then the same speedup can be maintained online, in the sense that the time to compute all $h(n)$ is still $O(N)$. Section 2 will give the main algorithm, which achieves the $O(1)$ amortized bound. In Section 2.3, we modify the algorithm a little bit to achieve the worst case $O(\log N)$ bound. Section 3 shows two applications of this technique.

Note that the online Monge property only says that c_n, β_n and δ_j exist. It does not say that c_n, β_n and δ_j are given. However, if δ_j is given, then the algorithm will be easier to understand. So, throughout this paper we will assume we have an extra condition:

– The values of δ_j can be computed in $O(1)$ time, provided that the values of $h(i)$ for $1 \leq i < j$ are known.

This condition is not really necessary. In Appendix A, we will show how it is implied by other conditions in Theorem 1.

As a final note we point out that there is a body of literature already discussing "online" problems of (1), e.g., [3,4,5,6,7]. We should clarify that the "online" in those papers actually had a different meaning than the one used here. More specifically, the result they have is that if

1. the value of N is known in advance;
2. and for any $1 \le j \le n \le N$, the value of $a(n, j)$ can be computed in $O(1)$ time, provided that the values of $h(i)$ for $1 \le i < j$ are known;
3. and the values of $a(n, j)$ satisfy the Monge property defined by (4),

then both the Galil-Park algorithm [6] and the Larmore-Schieber algorithm [7] can compute all of the $h(n)$ for $1 \le n \le N$ in $O(N)$ time. As we can see, their definition of "online" is only that the $a(n, j)$ can depend upon part of the previously-computed values of $h(i)$, i.e., for $1 \le i < j$. It does not mean that $h(n)$ can be computed without knowing the problem size N in advance.

1.1 Relations to Monge

In this section, we briefly introduce the definition of Monge property. See the survey [1] for more details. Consider an $N \times N$ matrix A. Denote by $R(n)$ the *index* of the rightmost minimum of row n of A, i.e.,

$$R(n) = \max\{j : A_{n,j} = \min_{1 \le i \le N} A_{n,i}\}.$$

A matrix A is *monotone* if $R(1) \le R(2) \le \cdots \le R(N)$, A is *totally monotone* if all submatrices[1] of A are monotone. The SMAWK algorithm [2] says that if A is totally monotone, then it can compute all of the $R(n)$ for $1 \le n \le N$ in $O(N)$ time.

For our problem, if we set

$$A_{n,j} = \begin{cases} a(n, j) & 1 \le j \le n \le N \\ \infty & \text{otherwise} \end{cases} \tag{3}$$

then $h(n) = a(n, R(n))$. Hence, if we can show the matrix A defined by (3) is totally monotone, then the SMAWK algorithm can solve our problem (offline version) in $O(N)$ time. Totally monotone properties are usually established by showing a slightly stronger property, the Monge Property (also known as the *quadrangle inequality*). A matrix A is Monge if $\forall\, 1 \le n < N$ and $\forall\, 1 \le j < N$,

$$A_{n,j} + A_{n+1,j+1} \le A_{n+1,j} + A_{n,j+1}.$$

It is easy to show that A is totally monotone if it is Monge. So, for the offline version of our problem, we only need to show that the matrix A defined by (3) is Monge, i.e., $\forall\, 1 \le j < n < N$,

$$a(n, j) + a(n + 1, j + 1) \le a(n + 1, j) + a(n, j + 1). \tag{4}$$

---

[1] In this paper, submatrices can take non-consecutive rows and columns from the original matrix, and are not necessarily square matrices.

By the conditions in Theorem 1,

$$a(n+1,j) + a(n,j+1) - a(n,j) - a(n+1,j+1) = (\delta_j - \delta_{j+1})\beta_{n+1} \geq 0.$$

So, the matrix A defined by (3) is Monge, and the SMAWK algorithm solves the offline problem.

Our problem is a special case of Monge. But how special a case? Referring to Section 2.2 of [1] for more details, we see that if we only consider the finite entries, then a matrix A is Monge if and only if $\forall A_{n,j} \neq \infty$,

$$A_{n,j} = P_n + Q_j + \sum_{k=n}^{N} \sum_{i=1}^{j} F_{ki} \tag{5}$$

where P and Q are vectors, and F is an $N \times N$ matrix, called the *distribution matrix*, whose entries are all nonnegative. For our problem, let $\delta_0 = \delta_1$. Then

$$a(n,j) = a(N,j) - \sum_{k=n+1}^{N} c_k - \delta_j \sum_{k=n+1}^{N} \beta_k$$

$$= a(N,j) - \sum_{k=n+1}^{N} c_k - \delta_0 \sum_{k=n+1}^{N} \beta_k + (\delta_0 - \delta_j) \sum_{k=n+1}^{N} \beta_k$$

So, in our problem,

$$P_n = - \sum_{k=n+1}^{N} (c_k + \delta_0 \beta_k), \qquad Q_j = a(N,j), \qquad F_{ki} = (\delta_{i-1} - \delta_i)\beta_{k+1},$$

where we define $\beta_{N+1} = 0$. This shows that our problem is a special case of the Monge property where the distribution matrix has rank 1.

Conversely, if the distribution matrix F has rank 1, then the values of $a(n,j)$ satisfy the conditions of Theorem 1. So, Theorem 1 is really showing that the row minima of any Monge matrix defined by a rank 1 distribution matrix can be found online.

2   The Algorithm

In this section, we show the main algorithm that achieves the $O(1)$ amortized bound in Theorem 1. We will show the algorithm at step n, where the values of $h(i)$ have been computed for $1 \leq i < n$, and we want to compute the value of $h(n)$. By the conditions in Theorem 1 and the extra condition, all the values $a(n,j)$ and δ_j for $1 \leq j \leq n \leq N$ are known.

The key concept of the algorithm is a set of straight lines defined as follows.

Definition 2. $\forall 1 \leq j \leq n \leq N$, we define

$$L_j^n(x) = a(n,j) + \delta_j \cdot x \tag{6}$$

So, $h(n) = \min_{1 \leq j \leq n} L_j^n(0)$. To compute $\min_{1 \leq j \leq n} L_j^n(x)$ at $x = 0$ efficiently, the algorithm maintains $\min_{1 \leq j \leq n} L_j^n(x)$ for the entire range $x \geq 0$, i.e., at step n, the algorithm maintains the *lower envelope* of the set of lines $\{L_j^n(x) : 1 \leq j \leq n\}$ in the range $x \in [0, \infty)$.

2.1   The Data Structure

The only data structure used is an array, called the *active-indices array*, $Z = (z_1, \ldots, z_t)$ for some length t. It will be used to represent the lower envelope. It stores, from left to right, the indices of the lines that appear on the lower envelope in the range $x \in [0, \infty)$. That is, at step n, if we walk along the lower envelope from $x = 0$ to the right, then we will sequentially encounter the lines $L_{z_1}^n(x), L_{z_2}^n(x), \ldots, L_{z_t}^n(x)$. Since $\delta_1 > \delta_2 > \cdots > \delta_n$, and by the properties of lower envelopes, we have $z_1 < z_2 < \cdots < z_t = n$, and no line can appear more than once in the active-indices array.

Once we have the active-indices array, computing $h(n)$ becomes easy as $h(n) = a(n, z_1)$. So, the problem is how to obtain the active-indices array. Inductively, when the algorithm enters step n from step $n - 1$, it maintains an active-indices array for step $n - 1$, which represents the lower envelope of the lines $\{L_j^{n-1}(x) : 1 \leq j \leq n - 1\}$. So, the main part of the algorithm is to *update* the old active-indices array to the new active-indices array for $\{L_j^n(x) : 1 \leq j \leq n\}$.

Before introducing the algorithm, we introduce another concept, the *break-point* array, $X = (x_0, \ldots, x_t)$, where $x_0 = 0$, $x_t = \infty$ and x_i ($1 \leq i < t$) is the x-coordinate of the intersection point of lines $L_{z_i}^n(x)$ and $L_{z_{i+1}}^n(x)$. The break-point array is *not* stored explicitly, since for any i, the value of x_i can be computed in $O(1)$ time, given the active-indices array.

2.2   The Main Algorithm

In step n, we need to consider n lines $\{L_j^n(x) : 1 \leq j \leq n\}$. The algorithm will first deal with the $n - 1$ lines $\{L_j^n(x) : 1 \leq j \leq n - 1\}$, and then add the last line $L_n^n(x)$. Figure 1 illustrates the update process by an example. Figure 1(a) shows what we have from step $n - 1$, Figure 1(b) shows the considerations for the first $n - 1$ lines, and Figure 1(c) shows the adding of the last line.

Deal with the first $n-1$ lines. For the first $n-1$ lines $\{L_j^n(x) : 1 \leq j \leq n-1\}$, the key observation is the following lemma.

Lemma 3. $\forall\, 1 < n \leq N$ and $\forall\, x$,

$$L_j^n(x) = L_j^{n-1}(x + \beta_n) + c_n, \qquad \forall\, 1 \leq j \leq n - 1.$$

Proof. By (2) and (6),

$$\begin{aligned}
L_j^n(x) &= [a(n, j) - \delta_j \beta_n] + \delta_j (x + \beta_n) \\
&= [a(n - 1, j) + c_n] + \delta_j (x + \beta_n) \\
&= L_j^{n-1}(x + \beta_n) + c_n.
\end{aligned}$$

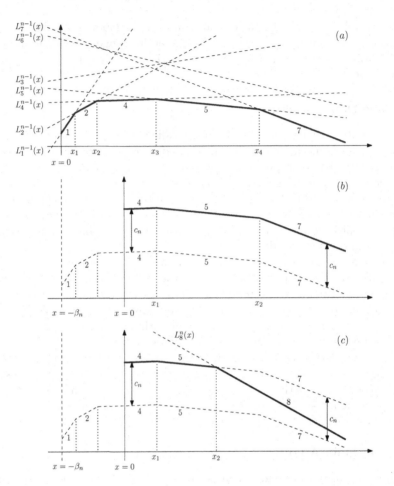

Fig. 1. The update of the active-indices array from Step $n-1$ to Step n, where $n = 8$. The thick solid chains are the lower envelopes. Figure (a) shows the lower envelope for the lines $\{L_j^{n-1}(x) : 1 \leq j \leq n-1\}$, Figure (b) shows the lower envelope for the lines $\{L_j^n(x) : 1 \leq j \leq n-1\}$, and Figure (c) shows the lower envelope for the lines $\{L_j^n(x) : 1 \leq j \leq n\}$. The numbers beside the line segments are the indices of the lines. The active-indices array changes from (a)(1, 2, 4, 5, 7), to (b)(4, 5, 7), then to (c)(4, 5, 8).

Lemma 3 says that if we translate the line $L_j^{n-1}(x)$ to the left by β_n and upward by c_n, then we obtain the line $L_j^n(x)$. The translation is independent of j, for $1 \leq j \leq n-1$. So,

Corollary 4. *The lower envelope of the lines $\{L_j^n(x) : 1 \leq j \leq n-1\}$ is the translation of the lower envelope of $\{L_j^{n-1}(x) : 1 \leq j \leq n-1\}$ to the left by β_n and upward by c_n.*

As an example, see Figure 1, (a) and (b). From Figure 1(a) to 1(b), the entire lower envelope translates to the left by β_n and upward by c_n.

We call an active-index z_i *negative* if the part of $L_{z_i}^n(x)$ that appears on the lower envelope is completely contained in the range $x \in (-\infty, 0]$. By Corollary 4, to obtain the active-indices array for $\{L_j^n(x) : 1 \le j \le n-1\}$ from the old active-indices array, we only need to delete those active-indices who becomes negative due to the translation. This can be done by a simple sequential scan. We scan the old active-indices array from left to right, check each active-index whether it becomes negative. If it is, we delete it. As soon as we find the first active-index that is nonnegative, we can stop the scan, since the rest of the indices are all nonnegative.

To be precise, we scan the old active-indices array from z_1 to z_t. For each z_i, we compute x_i, the right break-point of the segment z_i. If $x_i < 0$, then z_i is negative. Let $z_{\min}$ be the first active-index that is nonnegative, then the active-indices array for $\{L_j^n(x) : 1 \le j \le n - 1\}$ is $(z_{\min}, \ldots, z_t)$.

Adding the last line. We now add the line $L_n^n(x)$. Recall Condition (a) in Theorem 1. Since $L_n^n(x)$ has the smallest slope over all lines, it must be the rightmost segment on the lower envelope. And since no line can appear on the lower envelope more than once, we only need to find the intersection point between $L_n^n(x)$ and the lower envelope of $\{L_j^n(x) : 1 \le j \le n - 1\}$. Assume they intersect on segment $z_{\max}$, then the new lower envelope should be $(z_{\min}, \ldots, z_{\max}, n)$. See Figure 1(c), in the example, $z_{\max} = 5$.

To find $z_{\max}$, we also use a sequential scan, but from right to left. We scan the active-indices array from z_t to $z_{\min}$. For each z_i, we compute x_{i-1}, the left break-point of segment z_i, and compare the values of $L_n^n(x_{i-1})$ and $L_{z_i}^n(x_{i-1})$. If $L_n^n(x_{i-1})$ is smaller, then z_i is deleted from the active-indices array. Otherwise, we find $z_{\max}$.

The running time. The two sequential scans use amortized $O(1)$ time per step, since each line can be added to or deleted from the active-indices array at most once.

2.3   The Worst-Case Bound

To achieve the worst-case bound, we can use binary search to find $z_{\min}$ and $z_{\max}$. Since for a given index z and value x the function $L_z^n(x)$ can be computed in $O(1)$ time, the binary search takes $O(\log N)$ time worst case.

To keep both the $O(1)$ amortized time and the $O(\log N)$ worst-case time, we run both the sequential search and the binary search in parallel, interleaving their steps, stopping when the first one of the two searches completes.

3   Applications

We will now see two applications. Both will require *multiple* applications of our technique, and both will be in the form

$$H(d,n) = \min_{d-1 \leq j \leq n-1} \left(H(d-1,j) + W_{n,j}^{(d)} \right),$$ (7)

where the value of $W_{n,j}^{(d)}$ can be computed in $O(1)$ time, and the values of $H(d,n)$ for $d = 0$ or $n = d$ are given. The goal is to compute $H(D, N)$. Setting

$$a^{(d)}(n,j) = H(d-1,j) + W_{n,j}^{(d)},$$

for each fixed d $(1 \leq d \leq D)$, the values of $a^{(d)}(n,j)$ satisfy the online Monge property in Theorem 1, i.e.,

$$a^{(d)}(n,j) - a^{(d)}(n-1,j) = W_{n,j}^{(d)} - W_{n-1,j}^{(d)} = c_n^{(d)} + \delta_j^{(d)} \beta_n^{(d)}.$$ (8)

where $\delta_j^{(d)}$ decreases as j increases, and $\beta_n^{(d)} \geq 0$.

As before, we want to compute $H(d,n)$ in online fashion, i.e., as n increases from 1 to N, at step n, we want to compute the set $\mathcal{H}_n = \{H(d,n) \mid 1 \leq d \leq D\}$. By Theorem 1, this can be done in $O(D)$ amortized time per step. This gives a total of $O(DN)$ time to compute $H(D, N)$, while the naive algorithm requires $O(DN^2)$ time.

3.1   D-Medians on a Directed Line

The first application comes from [8]. It is the classic D-median problem when the underlying graph is restricted to a directed line. In this problem we have N points (users) $v_1 < v_2 < \cdots < v_N$, where we also denote by v_i the x-coordinate of the point. Each user v_i has a *weight*, denoted by w_i, representing the amount of requests. We want to choose a subset $S \subseteq V$ as servers (medians) to provide service to the users' requests. The line is *directed*, in the sense that the requests from a user can only be serviced by a server to its left. So, v_1 must be a server. Denote by $\ell(v_i, S)$ the distance from v_i to the nearest server to its left, i.e., $\ell(v_i, S) = \min\{v_i - v_l \mid v_l \in S, v_l \leq v_i\}$. The objective is to choose D servers (not counting v_1) to minimize the *cost*, which is $\sum_{i=1}^{N} w_i \ell(v_i, S)$.

The problem can be solved by the following DP. Let $H(d,n)$ be the minimum cost of servicing $v_1, v_2, \ldots, v_n$ using exactly d servers (not counting v_1). Let $W_{n,j} = \sum_{l=j+1}^{n} w_l(v_l - v_{j+1})$ be the cost of servicing $v_{j+1}, \ldots, v_n$ by server v_{j+1}. Then

$$H(d,n) = \begin{cases} 0 & n = d \\ W_{n,0} & d = 0, \ n \geq 1 \\ \min_{d-1 \leq j \leq n-1} (H(d-1,j) + W_{n,j}), & 1 \leq d < n \end{cases}$$

The optimal cost we are looking for is $H(D, N)$.

To see the online Monge property, since

$$W_{n,j} - W_{n-1,j} = w_n(v_n - v_{j+1}),$$

we have $c_n = w_n v_n$, $\delta_j = -v_{j+1}$ and $\beta_n = w_n$, satisfying (8). So, Theorem 1 will solve the online problem in $O(D)$ amortized time per step. Hence, the total time to compute $H(D, N)$ is $O(DN)$.

[8] also gives an $O(DN)$ time algorithm, by observing the standard Monge property and applying the SMAWK algorithm. The algorithm in this paper has smaller constant factor in the $O(\cdot)$ notation, and hence is more efficient in practice. Further more, the online problem makes sense in this situation. It is known as the *one-sided* online problem. In this problem, a new user is added from right in each step. When a new user comes, our algorithm recomputes the optimal solution in $O(D)$ time amortized and $O(D \log N)$ time worst case.

We note that the corresponding online problem for solving the D-median on an *undirected* line was treated in [9], where a problem-specific solution was developed. The technique in this paper is a generalization of that one.

3.2   Wireless Mobile Paging

The second application comes from wireless networking [10]. In this problem, we are given N regions, called the *cells*, and there is a *user* somewhere. We want to find which cell contains the user. To do this, we can only query a cell whether the user is in or not, and the cell will answer yes or no. For each cell i, we know in advance the probability that it contains the user, denote it by p_i. We assume $p_1 \geq p_2 \geq \cdots \geq p_N$. We also approximate the real situation by assuming the cells are *disjoint*, so p_i is the probability that cell i contains the user *and* no other cell does.

There is a tradeoff issue between the delay and the bandwidth requirement. For example, consider the following two strategies. The first strategy queries all cells simultaneously, while the second strategy consists of N rounds, querying the cells one by one from p_1 to p_N, and stops as soon as the user is found. The fist strategy has the minimum delay, which is only one round, but has the maximum bandwidth requirement since it queries all N cells. The second strategy has the maximum worst case delay of N rounds, but the expected bandwidth requirement is the minimum possible, which is $\sum_{i=1}^{N} i p_i$ queries. In the tradeoff, we are given a parameter D, which is the worst case delay that can be tolerated, and we are going to find an optimal strategy that minimize the expected number of queries.

It is obvious that a cell with larger p_i should be queried no later than one with smaller p_i. So, the optimal strategy actually breaks the sequence $p_1, p_2, \ldots, p_N$ into D contiguous subsequences, and queries one subsequence in each round. Let $0 = r_0 < r_1 < \cdots < r_D = N$, and assume in round i, we query the cells from $p_{r_{i-1}+1}$ to p_{r_i}. Recall that the cells are disjoint. The expected number of queries, defined as the *cost*, is

$$\sum_{i=1}^{D} r_i \left(\sum_{l=r_{i-1}+1}^{r_i} p_l \right). \tag{9}$$

[10] developed a DP formulation to solve the problem. It is essentially the following DP. Let $H(d, n)$ be the optimal cost for querying cells $p_1, \ldots, p_n$ using exactly d rounds. Denote $W_{n,j} = n \sum_{l=j+1}^{n} p_l$ the contribution to (9) of one round that queries $p_{j+1}, \ldots, p_n$. Then

$$
H(d, n) = \begin{cases} \sum_{l=1}^{n} l p_l & n = d \\ \infty & d = 0, \ n \geq 1 \\ \min_{d-1 \leq j \leq n-1} (H(d-1, j) + W_{n,j}), & 1 \leq d < n \end{cases}
$$

[10] applied the naive approach to solve the DP in $O(DN^2)$ time. Actually, this DP satisfies the online Monge property. Since

$$
W_{n,j} - W_{n-1,j} = n p_n + \sum_{l=j+1}^{n-1} p_l,
$$

we can set $c_n = n p_n + \sum_{l=1}^{n-1} p_l$, $\delta_j = -\sum_{l=1}^{j} p_l$ and $\beta_n = 1$, satisfying (8). So, the DP can be solved in $O(DN)$ time, using either the SMAWK algorithm or the technique in this paper. However, in this problem, there is no physical interpretation to the meaning of the online situation. But, due to the simplicity of our algorithm, it runs faster than the SMAWK algorithm in practice, as suggested by the experiments in [11], and is therefore more suitable for real time applications.

References

1. Burkard, R.E., Klinz, B., Rudolf, R.: Perspectives of Monge properties in optimization. Discrete Applied Mathematics **70**(2) (1996) 95–161
2. Aggarwal, A., Klawe, M.M., Moran, S., Shor, P.W., Wilber, R.E.: Geometric applications of a matrix-searching algorithm. Algorithmica **2** (1987) 195–208
3. Wilber, R.: The concave least-weight subsequence problem revisited. Journal of Algorithms **9**(3) (1988) 418–425
4. Eppstein, D., Galil, Z., Giancarlo, R.: Speeding up dynamic programming. In: Proceedings of the 29th Annual Symposium on Foundations of Computer Science. (1988) 488–496
5. Galil, Z., Giancarlo, R.: Speeding up dynamic programming with applications to molecular biology. Theoretical Computer Science **64**(1) (1989) 107–118
6. Galil, Z., Park, K.: A linear-time algorithm for concave one-dimensional dynamic programming. Information Processing Letters **33**(6) (1990) 309–311
7. Larmore, L.L., Schieber, B.: On-line dynamic programming with applications to the prediction of RNA secondary structure. Journal of Algorithms **12**(3) (1991) 490–515
8. Woeginger, G.J.: Monge strikes again: Optimal placement of web proxies in the Internet. Operations Research Letters **27**(3) (2000) 93–96
9. Fleischer, R., Golin, M.J., Zhang, Y.: Online maintenance of k-medians and k-covers on a line. Algorithmica **45**(4) (2006) 549–567

10. Krishnamachari, B., Gau, R.H., Wicker, S.B., Haas, Z.J.: Optimal sequential paging in cellular wireless networks. Wireless Networks **10**(2) (2004) 121–131
11. Bar-Noy, A., Feng, Y., Golin, M.J.: Efficiently paging mobile users under delay constraints. Unpublished manuscript (2006)

A   Dropping the Extra Condition

This appendix will show how to drop the condition that

- the values of δ_j can be computed in $O(1)$ time, provided that the values of $h(i)$ for $1 \leq i < j$ are known.

In real applications, this doesn't seem to be an issue. For example, in both of the applications in Section 3, the value of δ_j can easily be computed in $O(1)$ time when needed, and in neither of the applications does δ_j depend on the previously-computed values of $h(i)$ for $1 \leq i < j$. It is a theoretical issue, though, so in this appendix, we will show how to dispense with the condition.

Recall (2) from Theorem 1. It is true that we cannot compute δ_n from other values available at step n, since the constraints containing δ_n will only appear from step $n + 1$. However, it suffices to compute δ_n at step $n + 1$, since we can modify the algorithm a little bit. The only place that uses δ_n in step n of the algorithm is in the addition of new line $L_n^n(x)$ to the lower envelope. After that, the algorithm computes $h(n)$ by evaluating the value of the lower envelope at $x = 0$, and then precedes to step $n + 1$. So, we can postpone the addition of line $L_n^n(x)$ to the beginning of step $n + 1$, after we compute δ_n. To compute $h(n)$ at step n, we can evaluate the value of the lower envelope *without* $L_n^n(x)$ at $x = 0$, compare it with $L_n^n(0) = a(n, n)$, and take the smaller of the two. Hence, what is left is to show

Lemma 5. *A feasible value of δ_n can be computed in $O(1)$ time at step $n + 1$.*

Proof. We will show an algorithm that computes c_n and β_n at step n, and computes δ_n at step $n+1$. There are actually many feasible solutions of c_n, β_n and δ_j for (2). Consider a particular solution c_n, β_n and δ_j. If we set $c_n' = c_n + x\beta_n$, $\beta_n' = \beta_n$ and $\delta_j' = \delta_j - x$ for some arbitrary value x, then the new solution c_n', β_n' and δ_j' still satisfies (2). This gives us the degree of freedom to choose δ_1. We choose $\delta_1 = 0$ and immediately get

$$c_n = a(n, 1) - a(n - 1, 1), \qquad \forall \, 1 < n \leq N.$$

So, we can compute c_n at step n.

What is left is to compute β_n and δ_j. The constraints (2) become $\forall \, 1 < j < n \leq N$,

$$\delta_j \beta_n = a(n, j) - a(n - 1, j) - c_n. \tag{10}$$

β_2 does not show up in the constraints (10). In fact, the value of β_2 will not affect the algorithm. So, we can choose an arbitrary value for it, e.g. $\beta_2 = 0$. All other values, β_n ($3 \leq n \leq N$) and δ_j ($2 \leq j \leq N$), appear in the constraints (10), but we still have one degree of freedom. Consider a particular solution β_n and δ_j to the constraints (10). If we set $\beta'_n = \beta_n/x$, and $\delta'_j = \delta_j \cdot x$ for some $x > 0$, then we obtain another feasible solution. So, we can choose δ_2 to be an arbitrary negative value, e.g. $\delta_2 = -1$. The rest is easy. In step n, we can compute β_n by

$$\beta_n = [a(n,2) - a(n-1,2) - c_n]/\delta_2,$$

and in step $n+1$, we compute δ_n by

$$\delta_n = [a(n+1,n) - a(n,n) - c_{n+1}]/\beta_{n+1}.$$

Hence, the lemma follows.

Covering Many or Few Points with Unit Disks[*]

Mark de Berg[1], Sergio Cabello[2], and Sariel Har-Peled[3]

[1] Department of Computer Science, TU Eindhoven, the Netherlands
[2] Department of Mathematics, FMF, University of Ljubljana, and Department of Mathematics, IMFM, Slovenia
[3] Department of Computer Science, University of Illinois, USA

Abstract. Let P be a set of n weighted points. We study approximation algorithms for the following two continuous facility-location problems.

In the first problem we want to place m unit disks, for a given constant $m \geqslant 1$, such that the total weight of the points from P inside the union of the disks is maximized. We present a deterministic algorithm that can compute, for any $\varepsilon > 0$, a $(1 - \varepsilon)$-approximation to the optimal solution in $O(n \log n + \varepsilon^{-4m} \log^{2m}(1/\varepsilon))$ time.

In the second problem we want to place a single disk with center in a given constant-complexity region X such that the total weight of the points from P inside the disk is minimized. Here we present an algorithm that can compute, for any $\varepsilon > 0$, with high probability a $(1 + \varepsilon)$-approximation to the optimal solution in $O(n(\log^3 n + \varepsilon^{-4} \log^2 n))$ expected time.

1 Introduction

Let P be a set of n points in the plane, where each point $p \in P$ has a given weight $w_p > 0$. For any $P' \subseteq P$, let $w(P') = \sum_{p \in P'} w_p$ denote the sum of the weights over P'. We consider the following two geometric optimization problems:

- $\mathcal{M}(P, m)$. Here we are given a weighted point set P and a parameter m, where m is an integer constant with $m \geqslant 1$. The goal is to place m unit disks that maximize the sum of the weights of the covered points. With a slight abuse of notation, we also use $\mathcal{M}(P, m)$ to denote the value of an optimal solution, that is,

$$\mathcal{M}(P, m) = \max \{w(P \cap U) \mid U \text{ is the union of } m \text{ unit disks}\}.$$

- $\min(P, X)$. Here we are given a weighted point set P and a region X of constant complexity in the plane. The goal is to place a single unit disk with center in X that minimizes the sum of the weights of the covered points. Note that the problem is not interesting if $X = \mathbb{R}^2$. We use $\min(P, X)$ as the value of an optimal solution, that is,

$$\min(P, X) = \min \{w(P \cap D) \mid D \text{ is a unit disk whose center is in } X\}.$$

[*] MdB was supported by the Netherlands' Organisation for Scientific Research (NWO) under project no. 639.023.301. SC was partially supported by the European Community Sixth Framework Programme under a Marie Curie Intra-European Fellowship, and by the Slovenian Research Agency, project J1-7218.

T. Erlebach and C. Kaklamanis (Eds.): WAOA 2006, LNCS 4368, pp. 55–68, 2006.

The problems under consideration naturally arise in the context of locational analysis, namely when considering placement of facilities that have a fixed area of influence, such as antennas or sensors. $\mathcal{M}(P, m)$ models the problem of placing m of such new facilities that maximize the number of covered clients, while $\min(P, X)$ models the placement of a single obnoxious facility. $\min(P, X)$ also models the placement of a facility in an environment of obnoxious points.

Related work and other variants. Facility location has been studied extensively in many different variants and it goes far beyond the scope of our paper to review all the work in this area. We confine ourselves to discussing the work that is directly related to our variant of the problem. For a general overview of facility-location problems in the plane, we refer to the survey by Plastria [16].

The problem $\mathcal{M}(P, m)$ for $m = 1$ was introduced by Drezner [10]. Later Chazelle and Lee [7] gave an $O(n^2)$-time algorithm for this case. An approximation algorithm has also been given: Agarwal *et al.* [1] provided a Monte-Carlo $(1 - \varepsilon)$-approximation algorithm for $\mathcal{M}(P, 1)$ when P is an unweighted point set. If we replace each point $p \in P$ by a unit disk centered at p, then $\mathcal{M}(P, 1)$ is reduced to finding a point of maximum depth in the arrangement of disks. This implies that the recent results of Aronov and Har-Peled [2] give a Monte-Carlo $(1 - \varepsilon)$-approximation algorithm that runs in $O(n\varepsilon^{-2} \log n)$ time. Both the running time and the approximation factor hold with high probability. Although the algorithm is described for the unweighted case, it can be extended to the weighted case, giving an $O(n\varepsilon^{-2} \log n \log(n/\varepsilon))$ time algorithm.

Somewhat surprisingly, the problem $\mathcal{M}(P, m)$ seems to have not been studied so far for $m > 1$. For $m = 2$, however, Cabello *et al.* [4] have shown how to solve a variant of the problem where the two disks are required to be disjoint. (This condition changes the problem significantly, because now issues related to packing problems arise.) Their algorithm runs in $O(n^{8/3} \log^2 n)$ time.

The problem $\min(P, X)$ was first studied by Drezner and Wesolowsky [11], who gave an $O(n^2)$-time algorithm. Note that if as before we replace each point by a unit disk, the problem $\min(P, X)$ is reduced to finding a point with minimum depth in an arrangement of disks restricted to X. This means that for unweighted points sets, we can use the results of Aronov and Har-Peled [2] to get a $(1 + \varepsilon)$-approximation algorithm for $\min(P, X)$ in $O(n\varepsilon^{-2} \log n)$ expected time. For technical reasons, however, this algorithm cannot be trivially modified to handle weighted points.

The extension of $\min(P, X)$ to the problem of placing of m unit disks, without extra requirements, would have a solution consisting of m copies of the same disk. Hence, we restrict our attention to the case $m = 1$. (Following the paper by Cabello *et al.* [4] mentioned above one could study this problem under the condition that the disks be disjoint, but in the current paper we are interested in possibly overlapping disks.)

There are several papers studying these problems for other shapes than unit disks. The problem $\min(P, X)$ for unit squares—this problem was first considered by Drezner and Wesolowsky [11]—turns out to be significantly easier than for disks and one can get subquadratic exact algorithms: Katz, Kedem, and Segal

[13] gave an optimal $O(n \log n)$ algorithm that computes the exact optimum. For disks this does not seem to be possible: Aronov and Har-Peled [2] showed that for disks $\min(P, X)$ and also $\mathcal{M}(P, 1)$ are 3SUM-HARD [12], that is, these problems belong to a class of problems for which no subquadratic algorithms are known. (For some problems from this class, an $\Omega(n^2)$ lower bound has been proved in a restricted model of computation.) The problem $\mathcal{M}(P, 1)$ has also been studied for other shapes [1]. We will limit our discussion to disks from now on. Our algorithms can be trivially modified to handle squares, instead of disks, or other fixed shapes of constant description; only the logarithmic factors are affected.

Our results. As discussed above, $\mathcal{M}(P, m)$ is 3SUM-HARD for $m = 1$ and also $\min(P, X)$ is 3SUM-HARD. Since we are interested in algorithms with near-linear running time we therefore focus on approximation algorithms. For $\mathcal{M}(P, m)$ we aim to achieve $(1 - \varepsilon)$-approximation algorithms; given a parameter $\varepsilon > 0$, such algorithms compute a set of m disks such that the total weight of all points in their union is at least $(1 - \varepsilon)\mathcal{M}(P, m)$. Similarly, for $\min(P, X)$ we aim for computing a disk such that the total weight of the covered points is at most $(1 + \varepsilon)\min(P, X)$. When stating our bounds we consider $m \geqslant 1$ to be a constant and we assume a model of computation where the floor function takes constant time.

For $\mathcal{M}(P, m)$ with $m \geqslant 1$ we give a deterministic $(1 - \varepsilon)$-approximation algorithm that runs in $O(n \log n + n\varepsilon^{-4m} \log^{2m}(1/\varepsilon))$ time. As a byproduct of our approach, we also consider an exact algorithm to compute $\mathcal{M}(P, m)$; it runs in $O(n^{2m-1} \log n)$ time. For $m = 1$, we improve [1,2]. For $m > 1$, we obtain the first near-linear time algorithms.

For $\min(P, X)$ we give a randomized algorithm that runs in $O(n(\log^3 n + \varepsilon^{-4} \log^2 n))$ expected time and gives a $(1 + \varepsilon)$-approximation with high probability. This is the first near-linear time approximation algorithm for this problem that can handle weighted points.

2   Notation and Preliminaries

It will be convenient to define a *unit disk* as a closed disk of diameter 1. Let $s := \sqrt{2}/2$, so that a square of side s has diagonal of unit length and can be covered by a unit disk, and let $\Delta = 3ms$. (Recall that m is the number of disks we want to place.) We assume without loss of generality that no coordinate of the points in P is a multiple of s. For a positive integer I we use the notation $[I]$ to denote the set $\{0, 1, 2, \ldots, I\}$. For a pair $(a, b) \in [3m]^2$, we use $G_{(a,b)}$ to denote the grid of spacing Δ such that (as, bs) is one of the grid vertices, and we define $G := G_{(0,0)}$. We consider the cells of a grid to be open. Finally, we let $L_{(a,b)}$ denote the set of grid lines that define $G_{(a,b)}$. Thus $L_{(a,b)}$ is given by

$$\{(x, y) \in \mathbb{R}^2 \mid y = bs + k \cdot \Delta \text{ and } k \in \mathbb{Z}\} \cup \{(x, y) \in \mathbb{R}^2 \mid x = as + k \cdot \Delta \text{ and } k \in \mathbb{Z}\}$$

The following lemma follows from an easy counting argument.

Lemma 1. *Let $U := D_1 \cup \cdots \cup D_m$ be the union of m unit disks. There is some $(a,b) \in [3m]^2$ such that $L_{(a,b)}$ does not intersect U so that each disk D_i is fully contained in a cell of $G_{(a,b)}$.* □

Throughout the paper we use the expression *with high probability*, or *whp* for short, to indicate that, for any given constant $c > 0$, the failure probability can be bounded by n^{-c}. (In our algorithms, the value c affects the constant factor in the O-notation expressing the running time.)

An *integer-weighted* point set Q is a weighted point set with integer weights. We can see Q as a multiset where each point is repeated as many times as its weight. We use P for arbitrary weighted point sets and Q for integer-weighted point sets. A *p-sample* R of Q, for some $0 \leqslant p \leqslant 1$ is obtained by adding each point of the multiset Q to R with probability p, independently. If R is a p-sample of Q and $p \cdot w(Q) \geqslant c \log n$, for an appropriate constant c, then it follows from Chernoff bounds that R has $\Theta(p \cdot w(Q))$ points whp.

3   Approximation Algorithms for $\mathcal{M}(P, m)$

Our algorithm uses $(1/r)$-approximations [5,6]. In our application they can be defined as follows. Let $\mathcal{U}$ be the collection of sets $U \subset \mathbb{R}^2$ that are the union of m unit disks, and let P be a weighted point set. A weighted point set A is a $(1/r)$-*approximation* for P if for each $U \in \mathcal{U}$ we have: $|w(U \cap A) - w(U \cap P)| \leqslant w(P)/r$. The following result is due to Matoušek [15].

Lemma 2. *Let P be a weighted point set with n points and $1 \leqslant r \leqslant n$. We can construct in $O(n(r^2 \log r)^{2m})$ time a $(1/r)$-approximation A for P consisting of $O(r^2 \log r)$ points.* □

At first sight it may seem that this solves our problem: compute a $(1/r)$-approximation for $r = 1/\varepsilon$, and solve the problem for the resulting set of $O(\varepsilon^{-2} \log(1/\varepsilon))$ points. Unfortunately, this is not true: the error in the approximation is $w(P)/r$, not $w(U \cap P)/r$. Hence, when $w(P)$ is significantly larger than $w(U \cap P)$ we do not get a good approximation. Indeed, to obtain a good approximation we need to choose $r = w(P)/(\varepsilon \cdot \mathcal{M}(P, m))$. But now r may become quite large—in fact $\Theta(n)$ in the worst case—and it seems we do not gain anything. Nevertheless, this is the route we take. The crucial fact is that, even though the size of the approximation may be $\Theta(n)$, we can still gain something: we can ensure that any cell of $G = G(0,0)$ contains only a few points. This will allow us to compute the optimal solution within a cell quickly. By combining this with a dynamic-programming approach and using several shifted grids, we can then obtain our result. We start with a lemma guaranteeing the existence of an approximation with a few points per grid cell.

Lemma 3. *Let $0 < \varepsilon < 1$ be a parameter and let P be a set with n weighted points. Let $r := w(P)/(\varepsilon \cdot \mathcal{M}(P, m))$; note that the value of r is not known. We can find in $O(n \log n + n\varepsilon^{-4m} \log^{2m}(1/\varepsilon))$ time a $(1/2r)$-approximation A for P consisting of at most n points and such that each cell of G contains $O(\varepsilon^{-2} \log(1/\varepsilon))$ points from A.*

Proof. Let $\mathcal{C}$ be the collection of cells from G that contain some point of P. For a cell $C \in \mathcal{C}$, define $P_C := P \cap C$. Set $r' := 72m^2/\varepsilon$. For each cell $C \in \mathcal{C}$, compute a $(1/r')$-approximation A_C for P_C. We next show that the set $A := \bigcup_{C \in \mathcal{C}} A_C$ is a $(1/2r)$-approximation for P with the desired properties.

For any cell C we have $w(P_C) \leqslant 9m \cdot \mathcal{M}(P, m)$ because C can be decomposed into $9m$ rectangles of size $s \times ms$, and for each of these rectangles R we have $w(R \cap P) \leqslant \mathcal{M}(P, m)$. Since A_C is a $(1/r')$-approximation for P_C, we therefore have for any $U \in \mathcal{U}$,

$$|w(U \cap A_C) - w(U \cap P_C)| \leqslant \frac{w(P_C)}{r'} \leqslant \frac{9m \cdot \mathcal{M}(P, m)}{72m^2/\varepsilon} = \frac{\varepsilon}{8m} \cdot \mathcal{M}(P, m).$$

A unit disk of $U \in \mathcal{U}$ can intersect at most 4 cells of G, and therefore any $U \in \mathcal{U}$ can intersect at most $4m$ cells of G. If $\mathcal{C}_U$ denotes the cells of G intersected by U, we have $|\mathcal{C}_U| \leqslant 4m$, so

$$|w(U \cap A) - w(U \cap P)| = \left| \sum_{C \in \mathcal{C}_U} (w(U \cap A_C) - w(U \cap P_C)) \right|$$

$$\leqslant \sum_{C \in \mathcal{C}_U} |w(U \cap A_C) - w(U \cap P_C)|$$

$$\leqslant \sum_{C \in \mathcal{C}_U} \frac{\varepsilon}{8m} \cdot \mathcal{M}(P, m) \quad \leqslant (\varepsilon/2) \cdot \mathcal{M}(P, m).$$

We conclude that A is indeed a $(1/2r)$-approximation for P. For constructing the set A, we can classify the points P by cells of G in $O(n \log n)$ time, and then for each non-empty cell C apply Lemma 2 to get a $(1/r')$-approximation A_C for P_C. Since m is a fixed constant, we have $r' = O(1/\varepsilon)$, and according to Lemma 2, A_C will contain $O((r')^2 \log(r')) = O(\varepsilon^{-2} \log(1/\varepsilon))$ points. Also, computing A_C takes $O(|P_C| \cdot (r'^2 \log r')^{2m}) = O(|P_C| \cdot (\varepsilon^{-2} \log(1/\varepsilon))^{2m})$ time, and adding the time over all cells $C \in \mathcal{C}$, we obtain the claimed running time. $\square$

It is not hard to show that choosing the value of r as in Lemma 3 indeed leads to a $(1 - \varepsilon)$-approximation.

Lemma 4. *Let $0 < \varepsilon < 1$ be a parameter and let P be a set with n weighted points. Let A be a $(1/2r)$-approximation for P, where $r = w(P)/(\varepsilon \cdot \mathcal{M}(P, m))$. If U_A^* is an optimal solution for $\mathcal{M}(A, m)$, then $w(P \cap U_A^*) \geqslant (1 - \varepsilon) \cdot \mathcal{M}(P, m)$.* $\square$

It remains to find an optimal solution U_A^* for A. For a point set B, an integer m, and a cell C, define $\mathcal{M}(B, m, C)$ to be the maximum sum of the weights of B that m disks inside the cell C can cover. Let us assume that we have an algorithm $Exact(B, m, C)$—later we will provide such an algorithm—that finds the exact value $\mathcal{M}(B, m, C)$ in $T(k, m)$ time. For technical reasons, we also assume that $T(k, m)$ has the following two properties: $T(k, j) \leqslant T(k, m)$ for $j \leqslant m$ and $T(k, m)$ is superlinear but polynomially bounded for any fixed m. The next lemma shows that we can then compute the optimal solution for A quickly, using a dynamic-programming approach.

Lemma 5. *Let A be a point set with at most n points such that each cell of G contains at most k points. We can find $\mathcal{M}(A, m)$ in $O(n \log n + (n/k) \cdot T(k, m))$ time.*

Proof. For each $(a, b) \in [3m]^2$, let $\mathcal{M}_{(a,b)}(A, m)$ be the optimal weight we can cover with m unit disks that are disjoint from $L_{(a,b)}$. We have $\mathcal{M}(A, m) = \max_{(a,b) \in [3m]^2} \mathcal{M}_{(a,b)}(A, m)$ by Lemma 1. We will show how to compute each $\mathcal{M}_{(a,b)}(A, m)$ in $O(n \log n + (n/k) \cdot T(k, m))$ time, which proves our statement because $m^2 = O(1)$. First we give the algorithm, and then discuss its time bound.

Consider a fixed $(a, b) \in [3m]^2$. Let $\mathcal{C} = \{C_1, \ldots, C_t\}$ be the cells of $G_{(a,b)}$ that contain some point from P; we have $|\mathcal{C}| = t \leqslant n$. For any cell $C_i \in \mathcal{C}$, define $A_i = A \cap C_i$.

For each cell $C_i \in \mathcal{C}$ and each $j \in \{1, \ldots, m\}$, compute $\mathcal{M}(A_i, j, C_i)$ by calling the procedure $Exact(A_i, j, C_i)$. From the values $\mathcal{M}(A_i, j, C_i)$ we can compute $\mathcal{M}_{(a,b)}(A, m)$ using dynamic programming across the cells of $\mathcal{C}$, as follows. Define $B_i = A_1 \cup \cdots \cup A_i$. We want to compute $\mathcal{M}_{(a,b)}(B_i, j)$ for all i, j. To this end we note that an optimal solution $\mathcal{M}_{(a,b)}(B_i, j)$ will have ℓ disks inside A_i, for some $0 \leqslant \ell \leqslant j$, and the remaining $j - \ell$ disks spread among the cells $C_1, \ldots, C_{i-1}$. This leads to the following recursive formula:

$$\mathcal{M}_{(a,b)}(B_i, j) = \begin{cases} \mathcal{M}(A_1, j, C_1) & \text{if } i = 1 \\ \max_{0 \leqslant \ell \leqslant j}\{\mathcal{M}(A_i, \ell, C_i) + \mathcal{M}_{(a,b)}(B_{i-1}, j - \ell)\} & \text{otherwise} \end{cases}$$

Since $\mathcal{M}_{(a,b)}(B_t, m) = \mathcal{M}_{(a,b)}(A, m)$, we end up computing the desired value $\mathcal{M}_{(a,b)}(A, m)$. This finishes the description of the algorithm.

The time used to compute $\mathcal{M}_{(a,b)}(A, m)$ can be bounded as follows. Firstly, observe that constructing A_i for all $C_i \in \mathcal{C}$ takes $O(n \log n)$ time. For computing the values $\mathcal{M}(A_i, j, C_i)$ for all i, j we need time

$$\sum_{C_i \in \mathcal{C}} \sum_{j=1}^{m} T(|A_i|, j) \leqslant \sum_{C_i \in \mathcal{C}} m \cdot T(|A_i|, m) = O\left(\sum_{C_i \in \mathcal{C}} T(|A_i|, m)\right),$$

where the first inequality follows because for $j \leqslant m$ we have $T(k, j) \leqslant T(k, m)$, and the second one follows since m is a constant. We have $|A_i| \leqslant 4k$ for any $C_i \in \mathcal{C}$ because C_i intersects at most 4 cells of G. Moreover, because $T(k, m)$ is superlinear in k for fixed m, the sum is maximized when the points concentrate in as few sets A_i as possible. Therefore, the needed time can be bounded by

$$O\left(\sum_{C_i \in \mathcal{C}} T(|A_i|, m)\right) \leqslant O\left(\sum_{i=1}^{\lceil n/4k \rceil} T(4k, m)\right) = O((n/k) \cdot T(k, m)),$$

where we have used that $T(4k, m) = O(T(k, m))$ because T is polynomially bounded. Once we have the values $\mathcal{M}(A_i, j, C_i)$ for all i, j, the dynamic programming requires computing $O(tm) = O(n)$ values $\mathcal{M}_{(a,b)}(B_i, j)$, and each element requires $O(m) = O(1)$ time. Therefore, the dynamic programming takes $O(n)$ time. We conclude that finding $\mathcal{M}_{(a,b)}(A, m)$ takes $O(n \log n + (n/k) \cdot T(k, m))$ time for any $(a, b) \in [3m]^2$. $\qquad \square$

Putting everything together, we obtain the following result.

Lemma 6. *For any weighted point set P with n points, we can find in time $O(n \log n + n\varepsilon^{-4m} \log^{2m}(1/\varepsilon) + (n/k) \cdot T(k, m))$ a set of m disks that cover a weight of at least $(1 - \varepsilon)\,\mathcal{M}(P, m)$, where $k = O(\varepsilon^{-2} \log(1/\varepsilon))$.*

Proof. Given P and a parameter ε, consider the (unknown) value $r = \frac{w(P)}{\varepsilon \cdot \mathcal{M}(P,m)}$. We use Lemma 3 to compute a point set A with at most n points and such that A is a $(1/2r)$-approximation for P and any cell of G contains $O(\varepsilon^{-2} \log(1/\varepsilon))$ points.

We then use Lemma 5 to find an optimal solution U_A^* for $\mathcal{M}(A, m)$. It takes $O(n \log n + (n/k) \cdot T(k, m))$ time, where $k = O(\varepsilon^{-2} \log(1/\varepsilon))$. From Lemma 4, we know that $w(U_A^* \cap P) \geqslant (1 - \varepsilon)\,\mathcal{M}(P, m)$, and the result follows.   □

Theorem 2 below states there is an algorithm for the exact problem that uses $T(k, m) = O(k^{2m-1} \log k)$ time for $m > 1$. For $m = 1$, we have $T(k, 1) = O(k^2)$ because of Chazelle and Lee [7]. We can then use the previous lemma to obtain our final result.

Theorem 1. *Let $m \geqslant 1$ be a fixed positive integer constant. Given a parameter $0 < \varepsilon < 1$ and a weighted point set P with n points, we can find a set of m disks that cover a weight of at least $(1 - \varepsilon)\,\mathcal{M}(P, m)$ in time $O(n \log n + n\varepsilon^{-4m} \log^{2m}(1/\varepsilon))$ time.*   □

Exact algorithms for $\mathcal{M}(P, m, C)$. We want to find the set of m disks contained in a cell C of a grid that maximize the sum of the weights of the covered points. Let X be the set of possible centers for a unit disk contained in C—the domain X is simply a square with the same center as C and of side length $\Delta - 1$ instead of Δ.

For a point $p \in P$, let D_p be the unit disk centered at p. The weight of D_p is w_p, the weight of p. Let $\mathcal{D}_P := \{D_p : p \in P\}$ be the set of all disks defined by P. For a point $q \in \mathbb{R}^2$ and a set $\mathcal{D}$ of weighted disks, we define depth$(q, \mathcal{D})$ to be the sum of the weights of the disks from $\mathcal{D}$ that contain q. Let $\mathcal{A}$ denote the arrangement induced by the disks from $\mathcal{D}_P$. For any point q inside a fixed cell c of $\mathcal{A}$, the function depth$(q, \mathcal{D}_P)$ is constant; we denote its value by depth$(c, \mathcal{D}_P)$. Because each disk D_p has the same size, the arrangement $\mathcal{A}$ can be constructed in $O(n^2)$ time [7]. Moreover, a simple traversal of $\mathcal{A}$ allows us to compute depth$_P(c)$ for all cells $c \in \mathcal{A}$ in $O(n^2)$ time.

Let $V_{\mathcal{A}}$ be the set of vertices of $\mathcal{A}$, let V_X be the intersection points of the boundary of X with the boundary of some disk D_p, $p \in P$, and let V_{left} be set of leftmost points from each disk D_p, $p \in P$. Finally, let $V = (V_{\mathcal{A}} \cup V_X \cup V_{\text{left}}) \cap X$. See Figure 1, left. If $V = \emptyset$, then X is contained in some cell of $\mathcal{A}$ and the problem can trivially be solved. Otherwise we have

$$\mathcal{M}(P, m, C) = \max\{w(P \cap U) \mid U \text{ union of } m \text{ unit disks with centers at } V\},$$

that is, we only need to consider disks whose centers are in V. Based on this observation, we can solve $\mathcal{M}(P, m, C)$ for $m > 1$. We first consider the case $m = 2$.

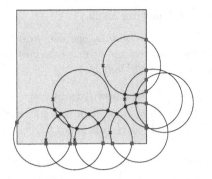 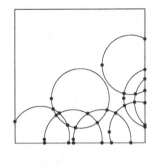

Fig. 1. Left: Example showing the points V. The dots indicate $V_\mathcal{A} \cap X$, the squares indicate V_X, and the crosses indicate $V_{\text{left}} \cap X$. Right: planar graph G_V with V as vertices and connected using portions of $\mathcal{A}$ or the boundary of X as edges.

Lemma 7. *We can compute* $\mathcal{M}(P, 2, C)$ *in* $O(n^3 \log n)$ *time.*

Proof. Our approach is similar to the one used by Katz and Sharir [14]. Let $\mathcal{A}^*$ the arrangement induced by the set $\mathcal{D}_P$ of disks and the sets X and V. Let G be the plane graph obtained by considering the restriction of $\mathcal{A}^*$ to X: the vertices of G are the vertices of $\mathcal{A}^*$ contained in X and the edges of G are the edges of $\mathcal{A}^*$ fully contained in X—see Figure 1, right. For simplicity, let's assume that each vertex in G has degree 4, meaning that no three points of P are cocircular. This condition can be lifted at the cost of making the discussion more tedious, but without extra ideas. Consider a spanning tree of G and double each edge to obtain an Euler path π. The path π has $O(n^2)$ edges and it visits each vertex of V at least once and at most four times.

The idea of the algorithm is as follows. We want to find two vertices $q, v \in V$, such that $P \cap (D_q \cup D_v)$ has maximum weight. If we fix q and let $\mathcal{D}_P(q) \subset \mathcal{D}_P$ denote the disks in $\mathcal{D}_P$ *not* containing q, then the best pair q, v (for this choice of q) covers a weight of $\text{depth}(q, \mathcal{D}_P) + \max_{v \in V} \text{depth}(v, \mathcal{D}_P(q))$. So our approach is to walk along the tour π to visit all possible vertices $q \in V$, and maintain the set $\mathcal{D} := \mathcal{D}_P(q)$—we call this the set of *active disks*—such that we can efficiently perform the following operations: (i) report a vertex $v \in V$ maximizing $\text{depth}(v, \mathcal{D})$, and (ii) insert or delete a disk into $\mathcal{D}$. Then we can proceed as follows. Consider two vertices q', q'' that are connected by an edge of π. The sets $\mathcal{D}_P(q')$ and $\mathcal{D}_P(q'')$ of active disks can differ by at most two disks. So while we traverse π, stepping from a vertex q' to an adjacent one q'' along an edge of π, we can update $\mathcal{D}$ with at most two insertions/deletions, and then report a vertex $v \in V$ maximizing $\text{depth}(v, \mathcal{D})$. Next we show how to maintain $\mathcal{D}$ such that both operations—reporting and updating—can be performed in $O(n \log n)$ time. Since π has $O(n^2)$ vertices, the total time will then be $O(n^3 \log n)$, as claimed.

The main problem in maintaining the set of active disks $\mathcal{D}$ is that the insertion or deletion of a disk can change $\text{depth}(v, \mathcal{D})$ for $\Theta(n^2)$ vertices $v \in V$. Hence, to obtain $O(n \log n)$ update time, we cannot maintain all the depths explicitly. Instead we do this implicitly, as follows.

Let $\mathcal{T}$ be a balanced binary tree on the path π, where the leftmost leaf stores the first vertex of π, the next leaf the second vertex of π, and so on. Thus the tree $\mathcal{T}$ has $O(n^2)$ nodes. For an internal node ν we denote by $\mathcal{T}_\nu$ the subtree of $\mathcal{T}$ rooted at ν. Furthermore, we define $\pi(\nu)$ to be the subpath of π from the leftmost vertex in $\mathcal{T}_\nu$ to the rightmost vertex in $\mathcal{T}_\nu$. Note that if μ_1 and μ_2 are the children of ν, then $\pi(\nu)$ is the concatenation of $\pi(\mu_1)$ and $\pi(\mu_2)$. Also note that $\pi(\mathrm{root}(\mathcal{T})) = \pi$. Finally, note that any subpath from π can be expressed as the concatenation of the subpaths $\pi(\nu_1), \pi(\nu_2), \ldots$ of $O(\log n)$ nodes—this is similar to the way a segment tree [9] works.

Now consider some disk $D_p \in \mathcal{D}_P$. Since D_p has $O(n)$ vertices from V on its boundary, the part of π inside D_p consists of $O(n)$ subpaths. Hence, there is a collection $N(D_p)$ of $O(n \log n)$ nodes in $\mathcal{T}$—we call this set the *canonical representation* of D_p—such that $\pi \cap D_p$ is the disjoint union of the set of paths $\{\pi(\nu) : \nu \in N(D_p)\}$. We store at each node ν the following two values:

- Cover(ν): the total weight of all disks $D_p \in \mathcal{D}$ (that is, all active disks) such that $\nu \in N(D_p)$.
- MaxDepth(ν): the value $\max\{\mathrm{depth}(v, \mathcal{D}(\nu)) : v \in \pi(\nu)\}$, where $\mathcal{D}(\nu) \subset \mathcal{D}$ is the set of all active disks whose canonical representation contains a node μ in $\mathcal{T}_\nu$.

Notice that $\mathrm{MaxDepth}(\mathrm{root}(\mathcal{T})) = \max_{v \in V} \mathrm{depth}(v, \mathcal{D})$, so $\mathrm{MaxDepth}(\mathrm{root}(\mathcal{T}))$ is exactly the value we want to report. Hence, it remains to describe how to maintain the values Cover(ν) and MaxDepth(ν) when $\mathcal{D}$ is updated. Consider the insertion of a disk D_p into $\mathcal{D}$; deletions are handled similarly. First we find in $O(n \log n)$ time the set $N(D_p)$ of nodes in $\mathcal{T}$ that forms the canonical representation of D_p. The values Cover(ν) and MaxDepth(ν) are only influenced for nodes ν that are in $N(D_p)$, or that are an ancestor of such a node. More precisely, for $\nu \in N(D_p)$, we need to add the weight of D_p to Cover(ν) and to MaxDepth(ν). To update the values at the ancestors we use that, if μ_1 and μ_2 are the children of a node ν, then we have

$$\mathrm{MaxDepth}(\nu) = \mathrm{Cover}(\nu) + \max(\mathrm{MaxDepth}(\mu_1), \mathrm{MaxDepth}(\mu_2)).$$

This means we can update the values in a bottom-up fashion in $O(1)$ time per ancestor, so in $O(n \log n)$ time in total. This finishes the description of the data structure—see Katz et al [13] or Bose et al [3] for similar ideas, or how to reduce the space requirements. □

Theorem 2. *For fixed $m > 1$, we can compute $\mathcal{M}(P, m, C)$ in $O(n^{2m-1} \log n)$ time.*

Proof. For $m > 2$, fix any $m - 2$ vertices $v_1, \ldots, v_{m-2} \in V$, compute the point set $P' = P \setminus (D_{v_1} \cup \cdots \cup D_{v_{m-2}})$, and compute $\mathcal{M}(P', 2, C)$ in $O(n^3 \log n)$ time using the previous lemma. We obtain a placement of disks covering a weight of $w(P \setminus P') + \mathcal{M}(P', 2, C)$. This solution is optimal under the assumption that the first $m - 2$ disks are placed at $v_1, \ldots, v_{m-2}$. Iterating this procedure over the $O(|V|^{m-2}) = O(n^{2m-4})$ possible tuples of vertices $v_1, \ldots, v_{m-2}$, it is clear that we obtain the optimal solution for placing m disks inside C. The time we spend can be bounded as $O(n^{2m-4}n^3 \log n) = O(n^{2m-1} \log n)$. □

4   Approximation Algorithms for $\min(P, X)$

We now turn our attention to the problem $\min(P, X)$ where we wish to place a single disk D in X so as to minimize the sum of the weights of the points in $P \cap D$. The approach consists of two stages. First, we make a binary search to find a value T that is a constant factor approximation for $\min(P, X)$. For this to work, we give a decision procedure that drives the search for the value T. Second, we compute a $(1+\varepsilon)$-approximation of $\min(P, X)$ using a random sample of appropriate density. In both cases, we will use the following lemma; a similar result was obtained by Aronov and Har-Peled [2]. Recall that D_a denotes the unit disk centered at a point a.

Lemma 8. *Let Q be an unweighted point set with at most n points, let X be a domain of constant complexity, let A be a set of at most n points, and let κ be a non-negative integer. We can decide in $O(n\kappa + n \log n)$ expected time if $\min\left(Q, X \setminus \left(\bigcup_{a \in A} D_a\right)\right) \leqslant \kappa$ or $\min\left(Q, X \setminus \left(\bigcup_{a \in A} D_a\right)\right) > \kappa$. In the former case we can also find a unit disk D that is optimal for $\min\left(Q, X \setminus \left(\bigcup_{a \in A} D_a\right)\right)$. The running time is randomized, but the result is always correct.*

Proof. Let $\mathcal{A}$ be the arrangement induced by the $O(n)$ disks D_a, $a \in A$, and D_q, $q \in Q$, and let $\mathcal{A}_\kappa$ be the portion of $\mathcal{A}$ that has depth at most κ. The portion $\mathcal{A}_\kappa$ has complexity $O(n\kappa)$ [17] and it can be constructed using a randomized incremental construction, in $O(n\kappa + n \log n)$ expected time [8]. Then, we just discard all the cells of $\mathcal{A}_\kappa$ that are covered by any disk D_a with $a \in A$, and for the remaining cells we check if any has depth over κ and intersects X. Since X has constant complexity, in each cell we spend time proportional to its complexity, and the result follows. □

The following combinatorial lemma is very similar to [2, Lemma 3.1].

Lemma 9. *Let Q be an integer-weighted point set with n points, let X be any domain, and let $\Delta_Q = \min(Q, X)$. Given a value k, set $p = \min\{1, ck^{-1} \log n\}$, where $c > 0$ is an appropriate constant. If R is a p-sample of Q and $\Delta_R = \min(R, X)$, then whp it holds*

(i) *if $\Delta_Q \geqslant k/2$, then $\Delta_R \geqslant kp/4$;*
(ii) *if $\Delta_Q \leqslant 2k$, then $\Delta_R \leqslant 3kp$;*
(iii) *if $\Delta_Q \notin [k/8, 6k]$, then $\Delta_R \notin [kp/4, 3kp]$.*

□

The idea for the decision version is to distinguish between heavy and light points. The heavy points have to be avoided, while the light ones can be approximated by a set of n integer-weighted points. Then we can use the previous lemma to decide.

Lemma 10. *Let X be a domain with constant complexity. Given a weighted point set P with n points and a value T, we can decide in $O(n \log^2 n)$ expected time whether (i) $\min(P, X) < T$, or (ii) $\min(P, X) > 2T$, or (iii) $\min(P, X) \in (T/10, 10T)$, where the decision is correct whp.*

Proof. First we describe the algorithm then discuss its running time and finally show its correctness.

Algorithm. We compute the sets $A = \{p \in P \mid w_p > 2T\}$ and $\tilde{P} = P \setminus A$, as well as the domain $Y = X \setminus \bigcup_{a \in A} D_a$. If $Y = \emptyset$, then we can report $\min(P, X) > 2T$, since any disk with center in X covers some point with weight at least $2T$. If $Y \neq \emptyset$, we construct the integer-weighted point set Q obtained by picking each point from $\tilde{P}$ with weight $\lfloor 2nw_p/T \rfloor$. Define $k = 2n$, and $p = \min\{1, ck^{-1} \log n\}$, where c is the same constant as in the previous lemma. We compute a p-sample R of Q, and decide as follows: If $\min(R, Y) < kp/4$ we decide $\min(P, X) < T$; if $\min(R, Y) > 3kp$ we decide $\min(P, X) > 2T$; otherwise we decide $\min(P, X) \in (T/10, 10T)$.

Running time. We can compute $A, \tilde{P}, Q, R$ in linear time, and check if $Y = \emptyset$ in $O(n \log n)$ time by constructing $\bigcup_{a \in A} D_a$ explicitly using a randomized incremental construction. Each point in $\tilde{P}$ has weight at most $2T$, and therefore a point in Q has an integer weight bounded by $\lfloor 2n \cdot 2T/T \rfloor = O(n)$. We conclude that Q is an integer-weighted point set with n points and weight $w(Q) = O(n^2)$. Since $p = O(n^{-1} \log n)$ and $kp = O(\log n)$, it follows that R has $\Theta(w(Q) \cdot p) = \Theta(n \log n)$ points whp.

Because $Y = X \setminus \bigcup_{a \in A} D_a$, we can use Lemma 8 to find if $\Delta_R = \min(R, Y) > 3kp$ or otherwise compute Δ_R exactly, in $O(|R| \log |R| + |R|kp) = O(n \log^2 n)$ time. The running time follows.

Correctness. We next show that, whp, the algorithm gives a correct answer. For any unit disk D centered in Y we have

$$w(D \cap P) - T/2 \leqslant \sum_{p \in D \cap P} \left(w_p - \frac{T}{2n} \right) \leqslant \sum_{p \in D \cap P} \left\lfloor \frac{2nw_p}{T} \right\rfloor \cdot \frac{T}{2n} = w(D \cap Q) \cdot \frac{T}{2n} \quad (1)$$

and

$$w(D \cap Q) \cdot \frac{T}{2n} = \sum_{p \in D \cap P} \left\lfloor \frac{2nw_p}{T} \right\rfloor \cdot \frac{T}{2n} \leqslant \sum_{p \in D \cap P} \frac{2nw_p}{T} \cdot \frac{T}{2n} = w(D \cap P). \quad (2)$$

We conclude that $w(D \cap P) - T/2 \leqslant (T/2n) \cdot w(D \cap Q) \leqslant w(D \cap P)$. Using the notation $\Delta_R = \min(R, Y)$, $\Delta_Q = \min(Q, Y)$, and $\Delta_P = \min(P, Y)$, we have

$$\Delta_P - T/2 \leqslant \frac{T}{2n} \cdot \Delta_Q \leqslant \Delta_P. \quad (3)$$

The value Δ_R provides us information as follows:

- If $\Delta_R < kp/4$, then $\Delta_Q < k/2 = n$ whp because of Lemma 9(i). We then have

$$\Delta_P \leqslant \frac{T}{2n} \Delta_Q + \frac{T}{2} < \frac{T}{2n} \cdot n + \frac{T}{2} = T.$$

- If $\Delta_R > 3kp$, then $\Delta_Q > 2k = 4n$ whp because of Lemma 9(ii). We then have

$$\Delta_P \geqslant \frac{T}{2n} \cdot \Delta_Q > \frac{T}{2n} 4n = 2T.$$

- If $\Delta_R \in [kp/4, 3kp]$, then $\Delta_Q \in [k/8, 6k] = [n/4, 12n]$ whp by Lemma 9(iii). We then have

$$\Delta_P \leqslant \frac{T}{2n} \cdot \Delta_Q + T/2 < 10T \quad \text{and} \quad \Delta_P \geqslant \frac{T}{2n} \cdot \Delta_Q \geqslant \frac{T}{2n} > \frac{T}{10}.$$

It follows that the algorithm gives the correct answer whp. $\square$

Lemma 11. *Let X be a domain with constant complexity. Given a weighted point set P with n points, we can find in $O(n \log^3 n)$ expected time a value T that, whp, satisfies $T/10 < \min(P, X) < 10T$.*

Proof. The idea is to make a binary search. For this, we will use the previous lemma for certain values T. Note that, if at any stage, the previous lemma returns that $\min(P, X) \in (T/10, 10T)$, then we have found our desired value T, and we can finish the search. In total, we will make $O(\log n)$ calls to the procedure of Lemma 10, and therefore we obtain the claimed expected running time. Also, the result is correct whp because we make $O(\log n)$ calls to procedures that are correct whp.

Define the interval $I_p = [w_p, (n+1) \cdot w_p)$ for any point $p \in P$, and let $I = \bigcup_{p \in P} I_p$. It is clear that $\min(P, X) \in I$, since the weight of the heaviest point covered by an optimal solution can appear at most n times in the solution. Consider the values $B = \{w_p, (n+1) \cdot w_p \mid p \in P\}$, and assume that $B = \{b_1, \ldots, b_{2n}\}$ is sorted increasingly. Note that for the (unique) index i such that $\min(P, X) \in [b_i, b_{i+1})$, it must hold that $b_{i+1} \leqslant (n+1)b_i$

We first perform a binary search to find two consecutive elements b_i, b_{i+1} such that $\min(P, X) \in [b_i, b_{i+1})$. Start with $\ell = 1$ and $r = 2n$. While $r \neq \ell + 1$, set $m = \lfloor (\ell + r)/2 \rfloor$ and use the previous lemma with $T = b_m$:

- if $\min(P, X) < T$, then set $r = m$.
- if $\min(P, X) > 2T$, then set $\ell = m$.
- if $T/10 < \min(P, X) < 10T$, then we just return T as the desired value.

Note that during the search we maintain the invariant $\min(P, X) \in [b_\ell, b_r)$. Since we end up with two consecutive indices $\ell = i, r = i + 1$, it must hold that $\min(P, X) \in [b_i, b_{i+1})$.

Next, we perform another binary search in the interval $[b_i, b_{i+1})$ as follows. Start with $\ell = b_i$ and $r = b_{i+1}$. While $r/\ell > 10$, set $m = (\ell + r)/2$ and call the procedure of Lemma 10 with $T = m$:

- if $\min(P, X) < T$, then set $r = m$.
- if $\min(P, X) > 2T$, then set $\ell = m$.
- if $T/10 < \min(P, X) < 10T$, then we just return T as the desired value.

Since $b_{i+1} \leqslant (n+1)b_i$, it takes $O(\log n)$ iterations to ensure that $r/\ell \leqslant 10$. During the search we maintain the invariant that $\min(P, X) \in [\ell, r]$, and therefore we can return the last value ℓ as satisfying $\min(P, X) \in (\ell/10, 10\ell)$.   □

When we have a constant factor approximation to the value $\min(P, X)$, we can then round the weights accordingly and take a random sample of appropriate size to obtain a $(1 + \varepsilon)$-approximation. The precise statement is as follows.

Lemma 12. *Let $0 < \varepsilon < 1$ be a parameter, let P be a weighted point set with n points, and let T be a value such that $T/10 < \min(P, X) < 10T$. We can find in $O((n/\varepsilon^4) \log^2 n)$ expected time a unit disk D that, whp, satisfies $w(D \cap P) \leqslant (1 + \varepsilon) \min(P, X)$.*

Proof. First we describe the algorithm, then show its correctness, and finally discuss its running time. The ideas are similar to the ones used in Lemma 10. However, now we also need to take into account the parameter ε.

Algorithm. We compute the sets $A = \{p \in P \mid w_p > 10T\}$ and $\tilde{P} = P \setminus A$, as well as the domain $Y = X \setminus \bigcup_{a \in A} D_a$. Since $\min(P, X) < 10T$ by hypothesis, we know that $\min(P, X) = \min(\tilde{P}, Y) = \min(P, Y)$, because any disk with center in $\bigcup_{a \in A} D_a$ covers some point of A. We construct an integer-weighted point set Q by picking each point from $\tilde{P}$ with weight $\lfloor 20nw_p/\varepsilon T \rfloor$. Define $k = \lfloor 20n/\varepsilon \rfloor$, and let $p = \min\{1, ck^{-1}\varepsilon^{-2} \log n\}$, where c is an appropriate constant. Finally, compute a p-sample R of Q, find a best disk D_R for $\min(R, X)$, and report the disk D_R as solution.

Correctness. The correctness is seen using the same approach as in Lemma 9 and [2, Lemma 3.1].

Running time. For the running time, observe that $A, \tilde{P}, Q, R, Y$ can be computed in $O(n \log n)$ time, like in Lemma 10. Each point in $\tilde{P}$ has weight at most $10T$, and therefore each point of Q has an integer weight bounded by $\lfloor 20n \cdot 10T/\varepsilon T \rfloor = O(n/\varepsilon)$. We conclude that Q is an integer-weighted point set with n points and total weight $w(Q) = O(n^2/\varepsilon)$. Note that $p = O(n^{-1}\varepsilon^{-1} \log n)$ and $kp = O(\varepsilon^{-2} \log n)$. Therefore, the set R has $O(w(Q) \cdot p) = O((n/\varepsilon^2) \log n)$ points whp.

Using Chernoff bound, one can see that whp $\Delta_R = O(p \Delta_Q)$. Substituting p and using that $\Delta_Q \leqslant \frac{20n}{\varepsilon T} \Delta_P$, we obtain

$$\Delta_R = O\left(n^{-1}\varepsilon^{-1} \log n \frac{20n}{\varepsilon T} \Delta_P\right) = O\left(\frac{\Delta_P \log n}{\varepsilon^2 T}\right) = O\left(\varepsilon^{-2} \log n\right).$$

Since $Y = X \setminus \bigcup_{a \in A} D_a$, we can use Lemma 8 to find a best disk for $\min(R, Y)$ in

$$O(|R| \log |R| + |R|\Delta_R) = O(n\varepsilon^{-2} \log n \log(n/\varepsilon) + n\varepsilon^{-4} \log^2 n) = O(n\varepsilon^{-4} \log^2 n)$$

expected time, as the lemma claims.   □

By applying Lemma 11 to obtain a constant factor approximation, and then Lemma 12 to obtain a $(1 + \varepsilon)$-approximation, we obtain our final result.

Theorem 3. *Given a domain X of constant complexity, a parameter $0 < \varepsilon < 1$, and a weighted point set P with n points, we can find in $O(n(\log^3 n + \varepsilon^{-4} \log^2 n))$ expected time a unit disk that, with high probability, covers a weight of at most $(1 + \varepsilon) \min(P, X)$.* □

References

1. P. K. Agarwal, T. Hagerup, R. Ray, M. Sharir, M. Smid, and E. Welzl. Translating a planar object to maximize point containment. In *ESA 2002*, LNCS 2461, 2002.

2. B. Aronov and S. Har-Peled. On approximating the depth and related problems. In *SODA 2005*, pages 886–894, 2005.

3. P. Bose, M. van Kreveld, A. Maheshwari, P. Morin, and J. Morrison. Translating a regular grid over a point set. *Comput. Geom. Theory Appl.*, 25:21–34, 2003.

4. S. Cabello, J. M. Díaz Báñez, C. Seara, J.A. Sellarès, J. Urrutia, and I. Ventura. Covering point sets with two disjoint disks or squares. Manuscript available at http://www.fmf.uni-lj.si/~cabello/publications/. Preliminary version appeared at EWCG'05.

5. B. Chazelle. *The Discrepancy Method: Randomness and Complexity*. Cambridge University Press, New York, 2001.

6. B. Chazelle. The discrepancy method in computational geometry. In *Handbook of Discrete and Computational Geometry*, pages 983–996. CRC Press, 2004.

7. B. Chazelle and D. T. Lee. On a circle placement problem. *Computing*, 36:1–16, 1986.

8. K. L. Clarkson and P. W. Shor. Applications of random sampling in computational geometry, II. *Discrete Comput. Geom.*, 4:387–421, 1989.

9. M. de Berg, M. van Kreveld, M. Overmars, and O. Schwarzkopf. *Computational Geometry: Algorithms and Applications*. Springer-Verlag, Berlin, Germany, 2nd edition, 2000.

10. Z. Drezner. On a modified one-center model. *Management Science*, 27:848–851, 1991.

11. Z. Drezner and G. O. Wesolowsky. Finding the circle or rectangle containing the minimum weight of points. *Location Science*, 2:83–90, 1994.

12. A. Gajentaan and M. H. Overmars. On a class of $O(n^2)$ problems in computational geometry. *Comput. Geom. Theory Appl.*, 5:165–185, 1995.

13. M. J. Katz, K. Kedem, and M. Segal. Improved algorithms for placing undesirable facilities. *Computers and Operations Research*, 29:1859–1872, 2002.

14. M. J. Katz and M. Sharir. An expander-based approach to geometric optimization. *SIAM J. Computing*, 26:1384–1408, 1997.

15. J. Matoušek. Approximations and optimal geometric divide-an-conquer. *J. Comput. Syst. Sci.*, 50:203–208, 1995.

16. F. Plastria. Continuous covering location problems. In H. Hamacher and Z. Drezner, editors, *Location Analysis: Theory and Applications*, chapter 2, pages 39–83. Springer, 2001.

17. M. Sharir. On k-sets in arrangements of curves and surfaces. *Discrete Comput. Geom.*, 6:593–613, 1991.

On the Minimum Corridor Connection Problem and Other Generalized Geometric Problems⋆

Hans Bodlaender[1], Corinne Feremans[2], Alexander Grigoriev[2],
Eelko Penninkx[1], René Sitters[3], and Thomas Wolle[4]

[1] Institute of Information and Computing Sciences, Utrecht University,
P.O.Box 80.089, 3508 TB Utrecht, The Netherlands
{hansb,penninkx}@cs.uu.nl
[2] Department of Quantitative Economics, Maastricht University,
P.O.Box 616, 6200 MD Maastricht, The Netherlands
{c.feremans,a.grigoriev}@ke.unimaas.nl
[3] Department of Algorithms and Complexity,
Max-Planck-Institute for Computer Science,
Stuhlsatzenhausweg 85, 66123 Saarbrücken, Germany
sitters@mpi-inf.mpg.de
[4] National ICT Australia Ltd⋆⋆,
Locked Bag 9013, Alexandria NSW 1435, Australia
thomas.wolle@nicta.com.au

Abstract. In this paper we discuss the complexity and approximability of the minimum corridor connection problem where, given a rectilinear decomposition of a rectilinear polygon into "rooms", one has to find the minimum length tree along the edges of the decomposition such that every room is incident to a vertex of the tree. We show that the problem is strongly NP-hard and give an subexponential time exact algorithm. For the special case of k-outerplanar graphs the running time becomes $O(n^3)$. We develop a polynomial time approximation scheme for the case when all rooms are fat and have nearly the same size. When rooms are fat but are of varying size we give a polynomial time constant factor approximation algorithm.

Keywords: minimum corridor connection, generalized geometric problems, complexity, exact algorithms, approximations.

1 Introduction

MCC and other generalized geometric problems. We consider the following geometric problem. Given a rectilinear decomposition of a rectilinear polygon (a subdivision into n "rooms"), find the minimum length tree ("*corridor*") along

---

⋆ This work was supported by the Netherlands Organisation for Scientific Research NWO (project *Treewidth and Combinatorial Optimisation*).

⋆⋆ National ICT Australia is funded through the Australian Government's Backing Australia's Ability initiative, in part through the Australian Research Council.

the edges of the decomposition ("*walls*") such that every room is incident to a vertex of the tree (has access to the corridor); for an illustration see Figure 1 which is borrowed from [6]. Let us refer to this problem as the *minimum corridor connection* (MCC) problem.

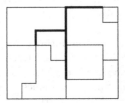

Fig. 1. Minimum length tree along the corridors to connect the rooms

This problem belongs to the class of so-called *generalized geometric problems* where given a collection of objects in the plane, one has to find a minimum length network satisfying certain properties that hits each object at least once. In particular, MCC can be viewed as a special case of the generalized geometric Steiner tree problem where given a set of disjoint groups of points in the plane, the problem is to find a shortest (in some metric space) interconnection tree which includes at least one point from each group.

The most studied generalized geometric problem is the following generalization of the classic Euclidean Traveling Salesman Problem (ETSP). Assume that a salesman has to visit n customers. Each customer has a set of specified locations in the plane (referred to as a *region* or *neighborhood*) where the customer is willing to meet the salesman. The objective is to find a shortest (closed) salesman tour visiting each customer. If each region is a single point, the problem becomes the classic ETSP. This described generalization of ETSP is known as the Generalized ETSP [12], or Group-ETSP, or ETSP with neighborhoods [5,8], or the Geometric Covering Salesman Problem [1]. For short, we shall refer to this problem as GTSP. In a similar way one can define generalizations for Minimum Steiner Tree (GSTP), Minimum Spanning Tree (GMST) and many other geometric problems.

Applications. Applications for the minimum corridor connection problem and other generalized geometric problems are naturally encountered in telecommunications, and VLSI design. For instance, a metropolitan area is divided by streets and avenues into rectilinear blocks and the blocks must be interconnected by an optical fiber network containing a gateway from each block. For easy maintenance the optical cables must be placed in the collector system which goes strictly under the streets and avenues. The problem is to find the minimum length network connecting all blocks. In Section 4.5 we discuss how our techniques can be applied to even more generalized variants of this problem. For the related problems and for the extended list of applications see Feremans [10], Feremans, Labbé and Laporte [11], Mitchell [23], Reich and Widmayer [25].

3D-applications of the generalized geometric problems, particularly MCC, appear also in constructions where, e.g., wiring has to be installed along the walls, floors and ceilings of a multistory building such that each room has electricity, phone lines, etc.

Related Work. To the best of our knowledge, before this article nothing was known on complexity and approximability of the minimum corridor connection problem; see list of open problems from the 12th Canadian Conference on Computational Geometry CCCG 2000 [6].

For GTSP it is known that the problem cannot be efficiently approximated within $(2 - \varepsilon)$ unless P=NP, see [26]. Constant factor approximations for GTSP were developed for the special cases where neighborhoods are disjoint convex fat objects [5,9], for disjoint unit discs [1], and for intersecting unit discs [8]. For the general GTSP, Mata and Mitchell [22] gave an $O(\log n)$-approximation. The closest related work to this article is the paper by Dumitrescu and Mitchell [8], where the authors have investigated the case of GTSP with regions given by pairwise disjoint unit disks, and they developed a polynomial time approximation scheme (PTAS) for this problem.

For the general GSTP, Helvig, Robins, and Zelikovsky [17] developed a polynomial time n^ε-approximation algorithm where $\varepsilon > 0$ is any fixed constant. For GSTP, GMST and several other generalized geometric problems exact search methods and heuristics have been developed, see e.g., Zachariasen and Rohe [28], and Feremans, Labbé and Laporte [11].

Our results and paper organization. The remainder of this extended abstract is organized as follows. In Section 2 we show that the problem is strongly NP-hard, answering an open question from CCCG 2000 on the complexity of the minimum corridor connection, see [6].

In Section 3 we present a subexponential time exact algorithm for MCC and a cubic time algorithm for the special case when the room connectivity graph is k-outerplanar. (We follow [20] for the definition of "subexponential".)

Then, in Section 4 we construct a PTAS for MCC with fat rooms having nearly the same size, that partially solves another open question from CCCG 2000 on the approximability of MCC, see [6]. More precisely, we consider the problem where a square of side length q can be inscribed in each room and the perimeter of each room is bounded from above by cq where c is a constant. In fact, we present a framework for construction the PTASs for a variety of generalized geometric problems restricted to (almost) disjoint fat object of nearly the same size. We refer to this restriction as *geographic clustering* since one can associate disjoint fat objects with countries on a map where all countries have comparable (up to a constant factor) border lengths.

The framework for PTASs presented in this paper is based on Arora's algorithm for ETSP [2]. In particular, this framework allows to construct PTASs for GTSP, GSTP, and GMST restricted to geographic clustering. The main advantage of our techniques compared to the recent approximation scheme by Dumitrescu and Mitchell [8] for GTSP on disjoint unit discs is that it leads to a more efficient approximation scheme running in time $n(\log n)^{O(1/\varepsilon)}$ compared to $n^{O(1/\varepsilon)}$ in [8]. Moreover, our techniques are applicable to many other norms (e.g., the one which is used in MCC) and to any fixed dimensional spaces, which resolves one of the open questions in [8].

Finally, in Section 5 we show how the algorithm for GTSP from Elbassioni et al. [9] can be used to derive a polynomial time constant approximation algorithm for MCC with fat rooms of varying sizes that complements our partial answer on the open question from CCCG 2000 on the approximability of MCC, see [6].

2   Complexity of MCC

In this section, we show that the decision version of MCC is strongly NP-complete. To show this result, we use a transformation from the CONNECTED VERTEX COVER problem for planar graphs with maximum degree four. In this later problem, given a planar graph $G = (V, E)$ such that each vertex in V has degree at most 4, and a positive integer $R \leq |V|$, the question is whether there exists a connected vertex cover of size at most R for G, i.e., does there exist a subset $W \subseteq V$ with $|W| \leq R$ such that the subgraph induced by W is connected and for each edge $\{u, v\} \in E$, $u \in W$ or $v \in W$? It is well known that CONNECTED VERTEX COVER for planar graphs with maximum degree four is NP-complete, see [14,15]. Now we state the main result of this section. Because of space constraints, we will omit proofs in this extended abstract.

Theorem 1. *The minimum corridor connection problem is NP-complete, even when coordinates of corner points are given in unary.*

3   Exact Algorithms with Branchwidth

In this section, we discuss how the problem can be solved exactly exploiting the notion of branchwidth and k-outerplanarity.

A *branch decomposition* of a graph $G = (V, E)$ is a pair (T, σ), with T an unrooted ternary tree and σ a bijection between the leaves of T and the edge set E. For each edge e in T, consider the two subtrees T_1 and T_2 obtained by removing e from T. Let $G_{e,1}$ ($G_{e,2}$) be the subgraph of G, formed by the edges associated with leaves in T_1 (T_2). The *middle set* of an edge e in T is the set of vertices that are in both $G_{e,1}$ and $G_{e,2}$. The *width* of a branch decomposition is the maximum size over all middle sets, and the *branchwidth* of a graph is the minimum width over all branch decompositions.

A *noose* is a closed simple curve on the plane that intersects the plane graph G only at vertices. To a noose, we can associate two regions of the plane (the "inside" and the "outside"), and likewise two subgraphs: the part of G drawn inside the noose, and the part of G drawn outside the noose. These subgraphs intersect precisely in the vertices on the noose.

A branch decomposition (T, σ) is a *sphere cut decomposition* or *sc-decomposition*, if for every edge e in T, there is a noose of G such that the two subgraphs associated with it are exactly $G_{e,1}$ and $G_{e,2}$, and the noose touches each face of G at most once. Necessarily, the set of the vertices on the noose is the middle set of e.

A sphere cut decomposition of a plane graph of minimum width can be found in $O(n^3)$ time with the ratcatcher algorithm of Seymour and Thomas [27], see [7]. See also [16,18,19] for a necessary improvements to the original algorithm and implementation issues.

Dynamic programming with a branch decomposition. Instead of the MCC problem, we consider a small generalization, which we call FACE COVER TREE: given a plane graph $G = (V, E)$, with edge weights $w : E \to \mathbf{N}$, find a subtree T of G of minimum total weight such that each interior face has at least one vertex on T.

We now give an algorithm that solves the FACE COVER TREE problem using a sphere cut decomposition of G.

Theorem 2. *Suppose a plane graph is given together with a sphere cut decomposition of width at most k. Then the FACE COVER TREE problem can be solved in $O((3 + \sqrt{5})^k k \cdot n)$ time.*

To obtain this result, we use techniques from Dorn et al. [7]. The basic idea is that we build a table for each edge in the branch decomposition. Assuming a root for T, we associate to each edge $e \in E(T)$, the subgraph formed by the edges of G associated with the leaves in T that are below e in the tree. This is one of the subgraphs $G_{e,1}$ or $G_{e,2}$; w.l.o.g., we will assume that this is always $G_{e,1}$. A forest T' that is a subgraph of $G_{e,1}$ can be extended to a solution of the FACE COVER TREE if each face of $G_{e,1}$ that does not intersect the noose is touched by T' and each subtree of T' contains at least one vertex in the middle set of e. We can characterize such forests of the second type by the set of vertices in the middle set that belong to the forest, an equivalence relation on these vertices which are connected by the forest, the information which faces that intersect the noose are touched by the forest, and (of course), the total length of all edges in the forest. Having this information is also sufficient to see how the forest can be extended.

Thus, in our dynamic programming algorithm, we tabulate for each edge e in the branch decomposition tree, for each triples (S, R, X), where S is a subset of the middle set of e, R is an equivalence relation on S, and X is a subset of the faces intersecting the noose of e, if there is at least one forest T' in $G_{e,1}$ such that S is the set of vertices in the middle set that belong to T', R is the relation on S that there is a path in T', and X is the set of faces intersecting the noose of e that are touched by e, the minimum total weight of such a forest.

Using counting techniques from [7], we can show that for a middle set of size ℓ, such a table contains at most $(3 + \sqrt{5})^\ell$ entries. (For instance, let R form a non-crossing partition on S. We only need to distinguish whether faces are touched whose two incident middle set vertices do not belong to S.)

It is trivial to compute the table for an edge in T incident to a leaf. For other edges e, we combine the two tables for the two edges incident to the lower endpoint of e. Basically, we try to combine each table entry of the left table with each table entry of the right table; in $O(k)$ time, we can verify whether these give a new table entry, and of what signature. Thus, the table for an edge can be computed in $O((14 + 6\sqrt{5})^k \cdot k)$ time.

From the table of the edge to the root, we can then determine the answer to the problem. We computed $O(n)$ tables, and hence used $O((14 + 6\sqrt{5})^k \cdot k \cdot n)$ time. Note that $14 + 6\sqrt{5} = 2^{4.7770}$.

Consequences. Given a plane graph $G = (V, E)$, we can divide the vertices of G into layers. All vertices incident to the exterior face are in layer L_1. For $i \geq 1$, all vertices incident to the exterior face after we removed all vertices in layers $L_1, \ldots, L_i$ are in layer L_{i+1}. A planar graph G is k-outerplanar, if it has a planar embedding with at most k non-empty layers. It is well known that a k-outerplanar graph has branchwidth at most $2k$; this can be proved in the same way as the proof in [4] that k-outerplanar graphs have treewidth at most $3k - 1$.

It is interesting to note that in some applications, graphs with small outer-planarity will arise in a natural way. For instance, for many buildings, the wall structure of one floor will have bounded outerplanarity, as usually, each room is adjacent to a corridor, and each corridor is adjacent to a room with a window, and thus, unless there is an open air part not at the exterior, this gives small outerplanarity.

It is well long known that planar graphs have branchwidth (and treewidth) $O(\sqrt{n})$. (This statement can be seen to be equivalent to the Lipton-Tarjan planar separator theorem [4,21].) The best known bound to our knowledge is the following.

Theorem 3 (Fomin and Thilikos [13]). *A planar graph with n vertices has branchwidth at most $\sqrt{4.5 \cdot n}$.*

Thus we have the following consequences, where we expect that the actual running times of these algorithms will be better in practice.

Corollary 1. *The* FACE COVER TREE, *and hence also the MCC problem can be solved in $O(n^3 + 2^{9.5539k})$ time on k-outerplanar graphs, and in $O^*(2^{10.1335\sqrt{n}})$ time on planar graphs.*

4   A PTAS for MCC with Geographic Clustering

To construct a polynomial time approximation scheme for MCC, we modify Arora's algorithm for ETSP [2,3]. We assume that the corner points of each of the n rooms have integer coordinates, that each room encloses a $q \times q$ square and has perimeter at most cq, for some constant $c \geq 4$.

4.1   Perturbation and Curved Dissection

Arora's algorithm for ETSP starts with perturbation of the instance that, without great increase of the optimum, ensures that in the resulting new instance all nodes lie on the unit grid, and the maximum internode distance is at most

$poly(n)$. In MCC, perturbation is not necessary. All corner points are already on the integer grid. Further, since all rooms are connected and the perimeter of a room is at most cq the smallest *bounding box* (the smallest axis parallel square containing all rooms) has side length at most cqn. Let the size of the bounding box be $L \in [cqn, 2cqn]$ such that L/cq is a power of 2. A simple packing argument shows that the value of the optimal solution is $OPT = \Omega(qn)$.

First we define the *straight dissection* of the bounding box. We stop the partitioning when the side length of the square is cq. Since $L \leq 2cqn$ the depth of the dissection tree is $O(\log n)$. Let the *level* of a square be its depth from the root in the straight dissection tree and the *level i dissection lines* are the straight lines participating in the division of the level $i - 1$ square into level i sub-squares.

A dissection line can cut a room into two or more parts. This causes troubles for the dynamic programming since we have to determine for each room in which square of the dissection it gets connected. To solve this problem we introduce a *curved dissection.*

Consider a horizontal level dissection line. We replace the line by a dissection curve by walking from left to right and whenever we hit the boundary of a room we follow the boundary (in arbitrary direction) until the dissection line is hit again. The obtained curve may go through some boundary segments twice. We shortcut the curve and obtain a simple path partitioning the set of rooms in an upper and lower set. Vertical dissection curves are defined in a similar way. Moreover, we can easily do this such that each horizontal curve crosses each vertical curve exactly once, i.e., the intersection is one point or a simple path. (See Figures 2 and 3.) Notice that no two horizontal (vertical) dissection curves intersect since, at any point on the curve, the deviation from the dissection line is strictly less than $cq/2$.

The transformation of lines to curves maps each node of the straight dissection tree onto a polygon which we denote by *node polygons* of the *curved dissection* tree of the bounding box.

In Figure 2, dissection lines are depicted by dotted lines and dissection curves are depicted by fat piece-wise linear curves. Notice that the middle room is crossed by vertical and horizontal dissection lines.

4.2   Portals and Portal Respecting Trees

Let a level i dissection curve have $2^i m$ special points equally spaced on that curve. By equally spaced we mean that the piece-wise linear fragments of the

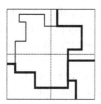

Fig. 2. Curved dissection

curve between two consecutive points have the same length. We refer to these points as *portals* and to *m* as *portal parameter* (to be defined later).

Remember that the intersection of a horizontal and vertical curve is in general a path. The definition above leads to two sets of portals on such paths. We keep only the portals of the highest level curve and pick one set arbitrarily if levels are equal. Further, we define one portal on both endpoints of each path of intersection which we call *corner portals*.

To make the dynamic programming work we have to assume that if some segment of the tree coincides with a dissection curve, it can only connect rooms on one side of the curve. To serve rooms at the other side it has to cross the curve. (See Figure 3.) We call a feasible tree *portal respecting* if these crossings only appear at portals. We refer to the boundary segment of the node polygon belonging to the dissection curve as the *side* of the node polygon. Notice that sides may overlap. A portal respecting tree is *k-light* if it crosses each side of each node polygon at most *k* times.

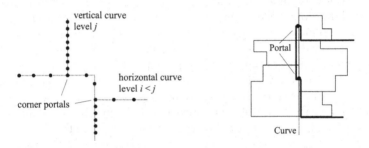

Fig. 3. Portals and a feasible portal respecting tree

4.3   The Algorithm

First we construct the bounding box with the dissection curves. Since each room is adjacent to at most two curves the construction can be done in $O(n)$ time. Next we choose $a, b \in \{1, 2, \ldots, L/(cq)\}$ at random and make the a-th horizontal and b-th vertical dissection curve the level zero curves. The curved dissection tree is now build in a wrap-around manner as in Arora [3]. By removing from the 4-ary tree all branches consisting of empty node polygons, we obtain a tree having at most $O(n)$ leaves and $O(n \log n)$ node polygons. Then we define the portals as in Section 4.2. Starting at the leaves of the dissection tree in a bottom-up way, we update the dynamic programming table. For each node polygon, for each k-elementary subset of the portals on the boundary of the polygon, and for each partition $B_1, \ldots, B_p$ of these k portals, we store the length of the optimal forest consisting of p trees which together touch all rooms and the i-th tree connects all portals in B_i.

For the node polygons in the leaves of the dissection tree we simply enumerate all such forests, since these polygons contain at most c^2 rooms. For the root polygon we guess the information for the portals on the two level one dissection curves separating the root polygon. We make sure that the four forests together form one tree. The number of different problems in one node polygon is $O(m^{O(k)}f(k))$ for some function f. Taking $m = O(\frac{\log n}{\varepsilon})$ and $k = O(\frac{1}{\varepsilon})$ the size of the look up table is $O(n \log^\gamma n)$, for some constant γ.

4.4 Performance Guarantee

The performance guarantee follows from the following theorem.

Theorem 4 (Structure Theorem). *Let $OPT_{a,b,k,m}$ be the length of the minimum k-light portal respecting tree when the portal parameter is m.*

$$E[OPT_{a,b,k,m} - OPT] \leq \left(O\left(\frac{\log n}{m}\right) + O\left(\frac{1}{k-4}\right)\right) OPT,$$

where $E[\cdot]$ is over the random choice of (a,b)-shift.

The proof is omitted and is slightly more complicated than in Arora [3]. Taking $m = O(\frac{\log n}{\varepsilon})$ and $k = O(\frac{1}{\varepsilon})$ we derive the following result.

Theorem 5. *The randomized algorithm described above returns a feasible tree of length at most $(1 + \varepsilon)OPT$ in time $n(\log n)^{O(1/\varepsilon)}$.*

To derandomize the algorithm, we can simply go through all possible choices for a and b. More sophisticated derandomization techniques are described in Rao and Smith [24]. In fact, a straightforward adaption of a more careful analysis presented in [24] can also significantly improve the running times presented in this extended abstract. For two dimensional space this would even imply an $O(n \log n)$ time and $O(n)$ space PTAS for MCC and other geometric problems with geographic clustering.

4.5 Extensions of the PTAS

As in Arora [2,3] we did not use much of the specifics of MCC. The basic idea to tackle the generalized geometric problems with geographic clustering is to introduce the curved dissection, new stoppage criteria and then to use the fact that under geographic clustering the lengths of the dissection curves only differ by a constant factor from the lengths of the dissection lines, yielding the same (up to a constant factor) charges to the objective function as in non-generalized versions of the geometric problems. In this way, with slight modifications in the analysis of the algorithm, we can derive PTASs for GTSP, GSTP, GMST and many other generalized geometric problems. Moreover, the approach is naturally applicable to many other norms, e.g., we can straightforwardly adopt the approximation scheme to any L_p norm. Also notice, that the requirement that the partition of the polygon must be rectilinear is not crucial. It is sufficient to assume that the walls of each room are given by a sequence of line segments forming a simple closed walk in the plane (here, the only critical assumption is that all rooms must be fat and have comparable sizes, i.e., for each room its perimeter must be bounded by cq where q is the minimum size over all rooms of the maximum inscribed square or ball and c is a fixed constant).

Dumitrescu and Mitchell in [8] pointed out that in their approximation scheme for GTSP only some of the arguments for disjoint discs can be lifted to higher dimensions and, naturally, one of the open questions they listed was: "What approximation bounds can be obtained in higher dimensions?" It is well known, see e.g., [2,3,24], that Arora's algorithm for ETSP is applicable also in higher fixed dimensional spaces. Using literally the same argumentation as in [2] and our construction for MCC with geographic clustering, one can derive the following theorem.

Theorem 6. *If the corner points of the rooms are in $\mathcal{R}^d$, the MCC with geographic clustering admits a randomized PTAS running in $n(\log n)^{(O(\sqrt{d}/\varepsilon))^{d-1}}$ time. Derandomization of the algorithm in this case will cost an additional factor of $O(n^d)$ leading to overall running time of $n^{d+1}(\log n)^{(O(\sqrt{d}/\varepsilon))^{d-1}}$.*

The same holds for GTSP, GSTP and GMST. This resolves the open question from Dumitrescu and Mitchell [8].

5 An Approximation Algorithm for MCC with Rooms of Varying Sizes

Elbassioni et al. [9] give a simple constant factor approximation algorithm for GTSP, where the factor depends on the fatness of the regions. Here we modify their algorithm and proof to obtain a constant factor approximation algorithm for MCC.

For any room R_i, $i \in \{1, \ldots, n\}$, we define its *size* ρ_i as the side length of the smallest enclosing square of the room. We restrict to rooms for which the perimeter is bounded by the size of the room, lets say at most $4\rho_i$. A room R is said to be *α-fat* if for any square Q whose boundary intersects R and whose center lies in R, the area of the intersection of R and Q is at least $\alpha/4$ times the area of Q. Note that the fatness of a square is 1 and in general $\alpha \in [0, 1]$.

Algorithm GREEDY:

(1) Pick the corner points $p_i \in R_i$, $i \in \{1, \ldots, n\}$, that minimize $\sum_{i=2}^{n} d(p_1, p_i)$, where $d(x, y)$ is the shortest distance between x and y along the walls.

(2) Let G be a graph with a vertex v_i for every room R_i and $d(v_i, v_j) = d(p_i, p_j)$. Find a minimum spanning tree T in G.

(3) Construct a solution to MCC as follows. For every edge (v_i, v_j) in T, let the minimum length (p_i, p_j)-path belong to the corridor. If the resulting corridor is not a tree, break the cycles (removing edges) arbitrarily.

Lemma 1. *Algorithm GREEDY gives an $(n-1)$-approximate solution for MCC.*

Proof. Consider an optimal solution and let OPT be its length. Identify for each room R_i a point p_i' in the room that is connected to the optimal tree. The optimal tree contains a path from p_1' to p_i' for all $i \in \{2, \ldots, n\}$. Therefore, $(n-1)OPT \geq \sum_{i=2}^{n} d(p_1', p_i') \geq \sum_{i=2}^{n} d(p_1, p_i)$, which is at most the length of the tree constructed by the algorithm.   □

Lemma 2. *The length of the shortest corridor that connects k rooms is at least $\rho_{\min}(k\alpha/2 - 2)$, where $\rho_{\min}$ is the size of the smallest of these rooms.*

Proof. Let P be a connecting corridor and let $d(P)$ denote its length (along the walls). Let the center of a square with side length $2\rho_{\min}$ move along the corridor P. The total area A covered by the moving square is at most $(2\rho_{\min})^2 + 2\rho_{\min} \cdot d(P)$. Assume a room is connected with P at point p. Putting the center of the square in point p we see that its boundary intersects the room. By definition of α at least a fraction $\alpha/4$ of the room is contained in the square. Therefore, $k(2\rho_{\min})^2\alpha/4$ is a lower bound on the area A. We have $k(2\rho_{\min})^2\alpha/4 \leq A \leq (2\rho_{\min})^2 + 2\rho_{\min} \cdot d(P)$, yielding $d(P) \geq \rho_{\min}(k\alpha/2 - 2)$, which completes the proof. $\qquad\square$

Algorithm CONNECT:

(1) Order the rooms by their sizes $\rho_1 \leq \rho_2 \leq \ldots \leq \rho_n$. Pick any p_1 on the boundary of R_1. For $i = 2$ up to n pick the point p_i in R_i that minimizes $\min\{d(p_i, p_1), d(p_i, p_2), \ldots, d(p_i, p_{i-1})\}$, i.e., pick the point that is closest to the already chosen points.

(2) Let G be a graph with a vertex v_i for every room R_i and $d(v_i, v_j) = d(p_i, p_j)$. Find a minimum spanning tree T in G.

(3) Construct a solution to MCC as follows. For every edge (v_i, v_j) in T, let the minimum length (p_i, p_j)-path belongs to the corridor. If the resulting corridor is not a tree, break the cycles (removing edges) arbitrarily. Output the minimum of the obtained tree and the tree constructed by algorithm GREEDY.

Theorem 7. *Algorithm CONNECT gives a $(16/\alpha - 1)$-approximate solution for the minimum corridor connection problem in which the fatness of every room is at least α.*

Proof. If $n - 1 \leq 16/\alpha - 1$ then GREEDY guarantees the approximation ratio for smaller values of n. So assume $n \geq 16/\alpha$. Denote the set of points chosen by CONNECT as $P' = \{p_1, \ldots, p_n\}$. Let p_i^* be the point from $\{p_1, \ldots, p_{i-1}\}$ that is at minimum distance from p_i. Denote the distance $d(p_i, p_i^*)$ by x_i.

Consider some *closed* walk Ω connecting all rooms and assume its length is minimum. The length of this walk is clearly an upper bound on OPT. For each room R_i, $i \in \{1, \ldots, n\}$, we define one connection point r_i on Ω in which it hits the room. Consider one of the two possible directions of Ω and assume that the tour connects the rooms in the order $1, 2, \ldots, n$. Let $k \in \{1, \ldots, n\}$. We define T_i as the part of this directed walk that connects exactly k rooms at their connection points and starts from point r_i. Let t_i be the length of the (not necessarily simple) path T_i. We have $OPT \leq d(\Omega) = \sum_{i=1}^{n} t_i/(k - 1)$.

Consider some $i \in \{1, \ldots, n\}$ and let $R_{h(i)}$ be the smallest room among those from the k rooms on the path T_i. Since R_i is on this path T_i and we ordered the rooms by their size we may assume $1 \leq h(i) \leq i$. We partition the rooms into two sets. Let F be the set of rooms for which $h(i) = i$ and let H contain the remaining rooms. Let T' be an MST on the point set P' restricted to the rooms in F. Then $d(T') \leq OPT + 2\sum_{i \in F} \rho_i$. The connected graph that we construct consists of the edges of T' and for all rooms i in H we add the path (p_i, p_i^*) which has length x_i. Note that the resulting graph is indeed connected and has total length at most

$$OPT + \sum_{i \in F} 2\rho_i + \sum_{i \in H} x_i.$$

We define $\gamma = k\alpha/2 - 2$. From Lemma 2 we know

$$t_i \geq \gamma \rho_i, \text{ for all } i \in F. \tag{1}$$

If $i \in H$, then we argue as follows. Since the algorithm picked point p_i we know that the distance from any point in R_i to the point $p_{h(i)}$ (which is chosen before p_i) is at least x_i. Hence, the distance from any point in R_i to any point in $R_{h(i)}$ is at least $x_i - 2\rho_{h(i)}$, implying $t_i \geq x_i - 2\rho_{h(i)}$. Additionally, we know from Lemma 2 that $t_i \geq \gamma \rho_{h(i)}$. Combining the two bounds we get

$$t_i \geq \max\{\gamma \rho_{h(i)}, x_i - 2\rho_{h(i)}\} \geq \frac{\gamma}{\gamma + 2} x_i, \text{ for all } i \in H. \tag{2}$$

Combining (1) and (2) we see that the MST given by the algorithm has length at most

$$
\begin{aligned}
OPT + \sum_{i \in F} 2/\gamma t_i + \sum_{i \in H} (1 + 2/\gamma) t_i &\leq OPT + \sum_{i=1}^{n} (1 + 2/\gamma) t_i \\
&\leq OPT + (1 + 2/\gamma)(k-1)OPT \\
&= OPT + (1 + 2/(k\alpha/2 - 2))(k-1)OPT \\
&= OPT + \frac{k(k-1)}{k - 4/\alpha} OPT
\end{aligned}
$$

It is easy to show that $k(k-1)/(k-4/\alpha)$ equals $16/\alpha - 2$ for $k = 8/\alpha - 1$ and also for $k = 8/\alpha$. Furthermore, it is strictly smaller for any value in between. Hence, there is an integer $k \in [8/\alpha - 1, 8/\alpha]$ such that $k(k-1)/(k-4/\alpha) \leq 16/\alpha - 2$. Note that by the assumption in the first line of the proof we satisfy $k \in \{1, \ldots, n\}$. We conclude that the length of the tree given by the algorithm is at most $(16/\alpha - 1)OPT$. $\qquad\square$

Acknowledgments

We thank Joe Mitchell, Sándor Fekete, and Mark de Berg for useful discussions on the GTSP with geographic clustering, and also thanks to Sergio Cabello.

References

1. E. M. Arkin and R. Hassin. Approximation algorithms for the geometric covering salesman problem. *Discrete Applied Mathematics*, 55:197–218, 1994.
2. S. Arora. Nearly linear time approximation schemes for Euclidean TSP and other geometric problems. *Journal of the ACM*, 45:1–30, 1998.

3. S. Arora. Approximation schemes for NP-hard geometric optimization problems: A survey. *Mathematical Programming*, 97:43–69, 2003.
4. H. L. Bodlaender. A partial k-arboretum of graphs with bounded treewidth. *Theoretical Computer Science*, 209:1–45, 1998.
5. M. de Berg, J. Gudmundsson, M. Katz, C. Levcopoulos, M. Overmars, and A. van der Stappen. TSP with neighborhoods of varying size. *Journal of Algorithms*, 57:22–36, 2005.
6. E. D. Demaine and J. O'Rourke. Open problems from CCCG 2000. http://theory.lcs.mit.edu/~edemaine/papers/CCCG2000pen/, 2000.
7. F. Dorn, E. Penninkx, H. L. Bodlaender, and F. V. Fomin. Efficient exact algorithms on planar graphs: Exploiting sphere cut branch decompositions. In *Algorithms - ESA 2005, 13th Annual European Symposium*, pages 95–106. LNCS 3669, Springer, October 2005.
8. A. Dumitrescu and J. S. B. Mitchell. Approximation algorithms for TSP with neighborhoods in the plane. *Journal of Algorithms*, 48:135–159, 2003.
9. K. Elbassioni, A. V. Fishkin, N. H. Mustafa, and R. Sitters. Approximation algorithms for Euclidean group TSP. In *Automata, Languages and Programming: 32nd International Colloquium, ICALP*, pages 1115–1126. LNCS 3580, Springer, July 2005.
10. C. Feremans. *Generalized Spanning Trees and Extensions*. PhD thesis, Université Libre de Bruxelles, Brussels, 2001.
11. C. Feremans, M. Labbé, and G. Laporte. Generalized network design problems. *European Journal of Operational Research*, 148:1–13, 2003.
12. M. Fischetti, J. J. Salazar, and P. Toth. A branch-and-cut algorithm for the symmetric generalized traveling salesman problem. *Operations Research*, 45:378–394, 1997.
13. F. V. Fomin and D. M. Thilikos. New upper bounds on the decomposability of planar graphs. *Journal of Graph Theory*, 51:53–81, 2006.
14. M. R. Garey and D. S. Johnson. The rectilinear Steiner tree problem is NP-complete. *SIAM Journal of Applied Mathematics*, 32:826–834, 1977.
15. M. R. Garey and D. S. Johnson. *Computers and intractability: A guide to the theory of NP-completeness*. W. H. Freeman, San Francisco, 1979.
16. Q. Gu and H. Tamaki. Optimal branch-decomposition of planar graphs in $O(n^3)$ time. In *Automata, Languages and Programming, 32nd International Colloquium, ICALP 2005*, pages 373–384. LNCS 3580, Springer, July 2005.
17. C. S. Helvig, G. Robins, and A. Zelikovsky. An improved approximation scheme for the Group Steiner Problem. *Networks*, 37:8–20, 2001.
18. I. V. Hicks. Planar branch decompositions I: The ratcatcher. *INFORMS Journal on Computing*, 17:402–412, 2005.
19. I. V. Hicks. Planar branch decompositions II: The cycle method. *INFORMS Journal on Computing*, 17:413–421, 2005.
20. R. Impagliazzo, R. Paturi, and F. Zane. Which problems have strongly exponential complexity? *J. Comput. Syst. Sci.*, 63(4):512–530, 2001.
21. R. J. Lipton and R. E. Tarjan. A separator theorem for planar graphs. *SIAM J. Appl. Math*, 36:177–189, 1979.
22. C. S. Mata and J. S. B. Mitchell. Approximation algorithms for geometric tour and network design problems. In *ACM SoCG 1995: Vancouver, BC, Canada*, pages 360–369. ACM, 1995.
23. J. S. B. Mitchell. *Handbook of Computational Geometry*, chapter Geometric shortest paths and network optimization, pages 633–701. Elsevier, North-Holland, Amsterdam, 2000.

24. S. Rao and W. D. Smith. Approximating geometric graphs via "spanners" and "banyans". In *ACM STOC 1998: Dallas, Texas, USA*, pages 540–550. ACM, 1998.
25. G. Reich and P. Widmayer. Beyond Steiner's problem: a VLSI oriented generalization. In *Graph-Theoretic Concepts in Computer Science, 15th International Workshop, WG '89*, pages 196–210. LNCS 411, Springer, 1990.
26. S. Safra and O. Schwartz. On the complexity of approximating TSP with neighborhoods and related problems. In *Algorithms - ESA 2003, 11th Annual European Symposium*, pages 446–458. LNCS 2832, Springer, September 2003.
27. P. D. Seymour and R. Thomas. Call routing and the ratcatcher. *Combinatorica*, 14:217–241, 1994.
28. M. Zachariasen and A. Rohe. Rectilinear group Steiner trees and applications in VLSI design. *Mathematical Programming*, 94:407–433, 2003.

Online k-Server Routing Problems

Vincenzo Bonifaci[1,2,*] and Leen Stougie[1,3,**]

[1] Department of Mathematics and Computer Science
Eindhoven University of Technology
Den Dolech 2 – PO Box 513, 5600 MB Eindhoven, The Netherlands
v.bonifaci@tue.nl, l.stougie@tue.nl
[2] Department of Computer and Systems Science
University of Rome "La Sapienza"
Via Salaria, 113 – 00198 Roma, Italy
bonifaci@dis.uniroma1.it
[3] CWI, Kruislaan 413, 1098 SJ Amsterdam, The Netherlands
stougie@cwi.nl

Abstract. In an online k-server routing problem, a crew of k servers has to visit points in a metric space as they arrive in real time. Possible objective functions include minimizing the makespan (k-Traveling Salesman Problem) and minimizing the average completion time (k-Traveling Repairman Problem). We give competitive algorithms, resource augmentation results and lower bounds for k-server routing problems on several classes of metric spaces. Surprisingly, in some cases the competitive ratio is dramatically better than that of the corresponding single server problem. Namely, we give a $1 + O((\log k)/k)$-competitive algorithm for the k-Traveling Salesman Problem and the k-Traveling Repairman Problem when the underlying metric space is the real line. We also prove that similar results cannot hold for the Euclidean plane.

1 Introduction

In a k-server routing problem, k servers (vehicles) move in a metric space in order to visit a set of points (cities). Given a schedule, that is, a sequence of movements of the servers, the time at which a city is visited for the first time by one of the servers is called the *completion time* of the city. The objective is to find a schedule that minimizes some function of the completion times.

We study k-server routing problems in their *online* version, where decisions have to be taken without having any information about future requests. New requests may arrive while processing previous ones. This online model is often called the *real time* model, in contrast to the *one-by-one* model, which is the more common model in texts about online optimization [5], but inadequate for server routing problems. The same real time model is also the natural model and indeed

---

* Partly supported by the Dutch Ministry of Education, Culture and Science through a Huygens scholarship.
** Partly supported by MRT Network ADONET of the European Community (MRTN-CT-2003-504438) and the Dutch BSIK/BRICKS project.

T. Erlebach and C. Kaklamanis (Eds.): WAOA 2006, LNCS 4368, pp. 83–94, 2006.
© Springer-Verlag Berlin Heidelberg 2006

is used for machine scheduling problems [22]. In fact, many of the algorithms for online routing problems are adaptations of online machine scheduling algorithms.

Competitive analysis [5] has become the standard way to study online optimization problems: an online algorithm A is said to be *c-competitive* if, for any instance σ, the cost of A on σ is at most c times the offline optimum cost on σ. This worst-case measure can be seen as the outcome of a game between the online algorithm and an offline *adversary*, that is trying to build input instances for which the cost ratio is as large as possible.

There is an abundant amount of literature on offline server routing problems, both in past and recent times [7, 10, 12, 14, 18]. Online single server routing problems have a recent but growing literature. The first paper by Ausiello et al. [3] introduced the model for the online traveling salesman problem. Later works investigated competitiveness of the more general dial-a-ride problems [1, 11] and studied different objective functions or different adversarial models [2, 4, 13, 16, 17, 20]. A summary of single server results is contained in the thesis [19].

Prior to this publication, there was essentially no work on online multi-server routing problems, except for some isolated algorithms [1, 4]. We give competitive algorithms and negative results for online multi-server routing problems, with the objective of minimizing either *makespan* or *average completion time*. In the case of makespan we consider the variant known as *nomadic*, in which the servers are not required to return at the origin after serving all requests; the above cited previous results refer to the other variant, known as the *homing* traveling salesman problem. Apart from being the first paper dedicated to multi-server online routing problems, the results are somewhat unexpected. We give the first results of online problems for which multiple server versions admit lower competitive ratios than their single server counterparts. This is typically not the case for problems in the one-by-one model; for example, it is known that in the famous *k-server* problem [21] the competitive ratio necessarily grows linearly with k.

It may also be useful to draw a comparison with machine scheduling, which is closer to routing problems in many ways. In scheduling a lot of research has been conducted to online multiple machine problems [22]. In the one-by-one model competitive ratios increase with increasing number of machines. In real time online scheduling nobody has been able to show smaller competitive ratios for multiple machine problems than for the single machine versions, though here lower bounds do not exclude that such results exist (and indeed people suspect they do) [8, 9].

The rest of our paper is structured as follows. After introducing our model in Section 2, we give in Section 3 competitive algorithms and lower bounds for both the k-Traveling Salesman and the k-Traveling Repairman in general spaces. For these algorithms, the upper bounds on the competitive ratio match those of the best known algorithms for the single server versions. In Section 4, we show that in the case of the real line we have an almost optimal algorithm for large k. The same result cannot hold in the Euclidean plane, as we show in Section 5. We give our conclusions in Section 6.

2   Preliminaries

We assume a real time online model, in which requests arrive over time in a metric space $\mathbb{M}$. Every *request* is a pair $(r, x) \in \mathbb{R}_+ \times \mathbb{M}$ where r is the *release date* of the request and x the location of the request. All the information about a request with release date r, including its existence, is revealed only at time r. Thus, an online algorithm does not know when all requests have been released.

An algorithm controls k vehicles or *servers*. Initially, at time 0, all these servers are located in a distinguished point $o \in \mathbb{M}$, the origin. The algorithm can then move the servers around the space at speed at most 1. (We do not consider the case in which servers have different maximum speeds; in compliance with machine scheduling vocabulary we could say that the servers are identical and work in parallel.) To process, or *serve*, a request, a server has to visit the associated location, but not earlier than the release date of the request.

We consider so-called *path metric* spaces, in which the distance d between two points is equal to the length of the shortest path between them. We also require the spaces to be continuous, in the sense that $\forall x, y \in \mathbb{M} \ \forall a \in [0, 1]$ there is $z \in \mathbb{M}$ such that $d(x, z) = ad(x, y)$ and $d(z, y) = (1 - a)d(x, y)$. A discrete space, like a weighted graph, can be extended to a continuous path metric space in the natural way; the continuous space thus obtained is said to be *induced* by the original space. We recall that a function $d : \mathbb{M}^2 \rightarrow \mathbb{R}_+$ is a *metric* if satisfies: definiteness $(\forall x, y \in \mathbb{M}, \ d(x, y) = 0 \Leftrightarrow x = y)$; symmetry $(\forall x, y \in \mathbb{M}, \ d(x, y) = d(y, x))$; triangle inequality $(\forall x, y, z \in \mathbb{M}, \ d(x, z) + d(z, y) \geq d(x, y))$. When referring to a *general space*, we mean any element of our class of continuous, path metric spaces. We will also be interested in special cases, namely the real line $\mathbb{R}$ and the real halfline $\mathbb{R}_+$, both with the origin o at 0, and the plane $\mathbb{R}^2$, with o at $(0, 0)$.

Defining the *completion time* of a request as the time at which the request has been served, the *k-traveling salesman problem* (k-TSP) has objective minimizing the *maximum completion time*, the *makespan*, and the *k-traveling repairman problem* (k-TRP) has objective minimizing the *average completion time*.

We will use σ to denote a sequence of requests. Given σ, a feasible *schedule* for σ is a sequence of moves of the servers such that all requests in σ are served. $\text{OL}(\sigma)$ is the cost online algorithm OL incurs on σ, and $\text{OPT}(\sigma)$ the optimal offline cost on σ. OL is said to be c-*competitive* if $\forall \sigma \ \text{OL}(\sigma) \leq c \cdot \text{OPT}(\sigma)$.

We use $s_1, \ldots, s_k$ to denote the k servers, and write $s_j(t)$ for the position of server s_j at time t, and $d_j(t)$ for $d(s_j(t), o)$. Finally, given a path P in $\mathbb{M}$, we denote its length by $|P|$.

All the lower bounds we prove hold for randomized algorithms against an oblivious adversary [5]. In order to prove these results, we frequently resort to the following form of Yao's principle [6, 23].

Theorem 2.1 (Yao's principle). *Let $\{\text{OL}_y : y \in \mathcal{Y}\}$ denote the set of deterministic online algorithms for an online minimization problem. If X is a distribution over input sequences $\{\sigma_x : x \in \mathcal{X}\}$ such that*

$$\inf_{y \in \mathcal{Y}} \mathbb{E}_X[\text{OL}_y(\sigma_x)] \geq c \, \mathbb{E}_X[\text{OPT}(\sigma_x)]$$

Algorithm 1. Group Return Home (GRH)

Divide the servers into $g = \lfloor k/k^* \rfloor$ disjoint sets (*groups*) of k^* servers each. Any remaining server is not used by the algorithm.

Initially, all servers wait at o. Every time a new request arrives, all servers not at o return to the origin at full speed. Once all of the servers in one of the groups, say group G (ties broken arbitrarily), are at o, compute a set of k^* paths $\{P_1, \ldots, P_{k^*}\}$ starting at o, covering all unserved requests and minimizing $\max_i |P_i|$. Then, for $i = 1, \ldots, k^*$, the i-th server in G follows path P_i at the highest possible speed while remaining at a distance at most αt from o at any time t, for some constant $\alpha \in (0, 1]$. Servers in other groups continue to head towards o (or wait there) until a new request is released.

for some real number $c \geq 1$, then c is a lower bound on the competitive ratio of any randomized algorithm against an oblivious adversary.

3   Algorithms for General Metric Spaces

In this section, we give competitive algorithms and lower bounds for the k-TSP and the k-TRP in general spaces. Our results will be formulated in a more general *resource augmentation* framework [15]. We define the (k, k^*)-TSP and (k, k^*)-TRP exactly as the k-TSP and the k-TRP, except that we measure the performance of an online algorithm with k servers relative to an optimal offline algorithm with $k^* \leq k$ servers. Throughout the section, we let $g = \lfloor k/k^* \rfloor$.

Sections 3.1 and 3.2 give an algorithm for the (k, k^*)-TSP and the (k, k^*)-TRP respectively. A lower bound for both problems is proved in Section 3.3.

3.1   The k-Traveling Salesman Problem

Theorem 3.1. *There is a deterministic online algorithm for the (k, k^*)-TSP with competitive ratio*

$$1 + \sqrt{1 + 1/2^{\lfloor k/k^* \rfloor - 1}}.$$

The algorithm achieving this bound is called Group Return Home (Algorithm 1). Define the *distance of a group to the origin* at time t as the maximum distance of a server in the group to o at time t.

Lemma 3.1. *At any time t, in the schedule generated by GRH, let $G_1(t), \ldots, G_g(t)$ be the g groups in order of nondecreasing distance to o. Then the distance of $G_i(t)$ to o is at most $2^{i-g}\alpha t$.*

Proof. We prove the lemma by induction on the number of requests. That is, we show that if the lemma holds at the release date t of some request, it will hold until the release date $t + \delta$ of the next request. Obviously, the lemma is true up to the time the first request is given, since all servers remain at o.

Suppose a request is given at time t. By induction, we know that there are groups $G_1(t), \ldots, G_g(t)$ such that each server of group $G_i(t)$ is at distance at most

$2^{i-g}\alpha t$ from o. For the rest of the proof we fix the order of the groups as the order they have at time t and write G_i instead of $G_i(t)$. Let $D_i(\tau) = \max_{s \in G_i} d(s(\tau), o)$.

Between time t and $t' = t + D_1(t)$, the lemma holds since all servers are getting closer to o. We show that the lemma holds at $t' + \delta$ for all $\delta > 0$. Notice that $D_1(t' + \delta) \le \delta$ since every server moves at most at unit speed.

If $\delta \in (0, 2^{1-g}\alpha t]$, we know that $D_1(t' + \delta) \le 2^{1-g}\alpha t$, so the lemma holds with the groups in the same order as before.

Now, let $\delta \in (2^{i-1-g}\alpha t, 2^{i-g}\alpha t]$ for $2 \le i \le g$. Then at time $t' + \delta$, group G_j is already at o for each $1 < j < i$. For group G_i, $D_i(t' + \delta) \le 2^{i-g}\alpha t - 2^{i-1-g}\alpha t = 2^{i-1-g}\alpha t$. For group G_1, $D_1(t' + \delta) \le 2^{i-g}\alpha t$. For groups G_{i+1} through G_g, $D_{i+1}(t' + \delta) \le 2^{i+1-g}\alpha t, \ldots, D_g(t' + \delta) \le 2^0\alpha t$. So the lemma holds for these values of δ.

The last case is $\delta > \alpha t$. In this case all groups except G_1 are at o, and because of the speed constraint $D_1(t' + \delta) \le \alpha(t' + \delta)$. Thus the lemma holds.   □

Proof (of Theorem 3.1). Let t be the release date of the last request and let G_1 be the group minimizing the distance to the origin at time t. Using Lemma 3.1 we know that $D_1(t) \le 2^{1-g}\alpha t$. Group G_1 will return to the origin and then follow the offline set of paths $\{P_1, \ldots, P_{k^*}\}$. Notice that $\text{OPT}(\sigma) \ge t$, since no schedule can end before the release date of a request, and $\text{OPT}(\sigma) \ge \max_i |P_i|$ because of the optimality of the P_i.

Let s be the server in G_1 that achieves the makespan. If s does not limit its speed after time t, we have $\text{OL}(\sigma) \le t + D_1(t) + \max_i |P_i| \le (2 + 2^{1-g}\alpha)\text{OPT}(\sigma)$.

Otherwise, let t' be the last time at which s is moving at limited speed. It is not difficult to see that s must serve some request at that time. Let x_0 be the location of this request. Then $t' = (1/\alpha)d(x_0, o)$ and s continues following the remaining part of its path, call it P', at full speed. Hence, $\text{OL}(\sigma) = t' + |P'|$. Since $\text{OPT}(\sigma) \ge \max_i |P_i| \ge d(o, x_0) + |P'|$ this yields $\text{OL}(\sigma) \le (1/\alpha)\text{OPT}(\sigma)$.

Thus, the competitive ratio is at most $\max\{2 + 2^{1-g}\alpha, 1/\alpha\}$ and choosing α in order to minimize it gives $\alpha = \sqrt{2^{g-1}(2^{g-1} + 1)} - 2^{g-1}$ and the desired competitive ratio.   □

Corollary 3.1. *There is a deterministic $(1 + \sqrt{2})$-competitive online algorithm for the k-TSP.*

3.2 The k-Traveling Repairman Problem

Theorem 3.2. *There is a deterministic online algorithm for the (k, k^*)-TRP with competitive ratio $2 \cdot 3^{1/\lfloor k/k^* \rfloor}$.*

We call the algorithm achieving the bound Group Interval (Algorithm 2), as it can be seen as a multi-server generalization of algorithm Interval [16]. The algorithm is well defined since the time between two departures of the same group is enough for the group to complete its first schedule and return to the origin: $B_{i+g} - B_i = 2B_i$.

To sketch the proof of Theorem 3.2, we start with two auxiliary lemmas.

Algorithm 2. Group Interval (GI)

Divide the servers into $g = \lfloor k/k^* \rfloor$ disjoint sets (groups) of k^* servers each. Any remaining server is not used by the algorithm.

Let L be the earliest time that any request can be completed (wlog $L > 0$). For $i = 0, 1, \ldots$, define $B_i = \alpha^i L$ where $\alpha = 3^{1/g}$.

At time B_i, compute a set of paths $S_i = \{P_1^i, \ldots, P_{k^*}^i\}$ for the set of yet unserved requests released up to time B_i with the following properties:

(i) every P_j^i starts at the origin o;

(ii) $\max_j |P_j^i| \leq B_i$;

(iii) S_i maximizes the number of requests served among all schedules satisfying the first two conditions.

Starting at time B_i, the j-th server in the $(i \bmod g)$-th group follows path P_j^i, then returns to o at full speed.

Lemma 3.2 ([16]). *Let $a_i, b_i \in \mathbb{R}$ for $i = 1, \ldots, p$, for which*

(i) $\sum_{i=1}^p a_i = \sum_{i=1}^p b_i$, and

(ii) $\sum_{i=1}^{p'} a_i \geq \sum_{i=1}^{p'} b_i$ for all $1 \leq p' \leq p$.

Then the $\sum_{i=1}^p \tau_i a_i \leq \sum_{i=1}^p \tau_i b_i$ for any nondecreasing sequence of real numbers $0 \leq \tau_1 \leq \tau_2 \leq \ldots \leq \tau_p$.

Lemma 3.3. *Let R_i be the set of requests served by the set of paths S_i computed by Group Interval at time B_i, $i = 1, 2, \ldots$ and let R_i^* be the set of requests in the optimal offline solution that are completed in the time interval $(B_{i-1}, B_i]$. Then*

$$\sum_{i=1}^q |R_i| \geq \sum_{i=1}^q |R_i^*| \quad \text{for all } q = 1, 2, \ldots.$$

Proof. We omit the proof, as it is basically the same as that of Lemma 4 in [16]. $\square$

Proof (of Theorem 3.2). Let $\sigma = \sigma_1 \ldots \sigma_m$ be any sequence of requests. By construction of Group Interval, each request in R_i is served at most at time $2B_i$. Now, let p be such that the optimal offline schedule completes in the interval $(B_{p-1}, B_p]$. Summing over all phases $1, \ldots, p$ yields

$$\text{OL}(\sigma) \leq 2 \sum_{i=1}^p B_i |R_i| = 2 \cdot 3^{1/g} \sum_{i=1}^p B_{i-1} |R_i|. \tag{1}$$

From Lemma 3.3 we know that $\sum_{i=1}^q |R_i| \geq \sum_{i=1}^q |R_i^*|$ for $q = 1, 2, \ldots$ We also know that $\sum_{i=1}^p |R_i| = \sum_{i=1}^p |R_i^*|$. Applying Lemma 3.2 to the sequences $a_i := |R_i|$, $b_i := |R_i^*|$, $\tau_i := B_{i-1}$, $i = 1, \ldots, p$ yields in (1)

$$\text{OL}(\sigma) \leq 2 \cdot 3^{1/g} \sum_{i=1}^p B_{i-1} |R_i| \leq 2 \cdot 3^{1/g} \sum_{i=1}^p B_{i-1} |R_i^*|. \tag{2}$$

Let C_j^* be the optimal off-line completion time of request σ_j. For each σ_j denote by $(B_{\phi_j}, B_{\phi_{j+1}}]$ the interval that contains C_j^*. This inserted in (2) yields

$$\text{OL}(\sigma) \leq 2 \cdot 3^{1/g} \sum_{j=1}^{m} B_{\phi_j} \leq 2 \cdot 3^{1/g} \sum_{j=1}^{m} C_j^* = 2 \cdot 3^{1/g} \cdot \text{OPT}(\sigma).$$

□

Corollary 3.2. *There is a deterministic 6-competitive online algorithm for the k-TRP.*

We can improve the bounds slightly such as to match the $(1 + \sqrt{2})^2$-competitive algorithm [16] for the TRP but at the expense of increased technical details.

3.3 Lower Bounds

Theorem 3.3. *Any randomized c-competitive online algorithm for the (k, k^*)-TSP or the (k, k^*)-TRP has $c \geq 2$.*

Proof. Consider the metric space induced by a star graph with m unit-length rays, the origin being the center of the star. No request is given until time 1. At time 1, the adversary gives a request on an edge chosen uniformly at random, at distance 1 from the origin. The expected makespan for the adversary is 1. For the online algorithm, we say that a server *guards* a ray if at time 1 the server is located on the ray, but not at the center of the star. Then the makespan is at least 2 if no server guards the ray where the request is released, and at least 1 otherwise. But k servers can guard at most k rays, so

$$\mathbb{E}[\text{OL}(\sigma)] \geq 2 \cdot \left(1 - \frac{k}{m}\right) + 1 \cdot \frac{k}{m} \geq 2 - \frac{k}{m}$$

and the result follows by Yao's principle, since m can be arbitrarily large.   □

Notice that this lower bound is independent of the values k and k^*. A consequence of this is that the upper bounds of Sections 3.1 and 3.2 are essentially best possible when $k \gg k^*$, as in that case they both approach 2.

4 Algorithms for the Real Line

4.1 An Asymptotically Optimal Algorithm

Theorem 4.1. *There is a deterministic online algorithm with competitive ratio $1 + O((\log k)/k)$ for both the k-TSP and the k-TRP on the real line.*

As a preliminary, we prove a similar result on the *halfline*. Let g_k be the unique root greater than 1 of the equation $z^k(z-1) = 3z - 1$.

Lemma 4.1. *GPS (Algorithm 3) is g_k-competitive for k-TSP and k-TRP on the halfline.*

Algorithm 3. Geometric Progression Speeds (GPS)

As a preprocessing step, the algorithm delays every request (r, x) for which $x \geq r$ to time x; that is, the release date of each request (r, x) is reset at $r' := \max\{r, x\}$ (the *modified release date*).

Then, let g_k be the unique root greater than 1 of the equation $g_k^k = \frac{3g_k - 1}{g_k - 1}$ and define $\alpha_j = g_k^{j-k-1}$ for $j \in \{2, 3, \ldots, k\}$. For every $j > 1$, server s_j departs at time 0 from o at speed α_j and never turns back. The first server s_1 waits in o until the first request (r_0, x_0) is released with $0 < x_0 < s_2(r_0')$. For $i \geq 0$, define $t_i = g_k^i r_0'$. During any interval $[t_{i-1}, t_i]$, s_1 moves at full speed first from o to $\frac{g_k - 1}{2} t_{i-1}$ and then back to o.

Proof. First, notice that the modified release date of a request is a lower bound on its completion time. Thus it is enough to prove that, for every request (r, x), the time at which it is served is at most $g_k r'$.

For $1 < j < k$, we say that a request (r, x) is in *zone* j if $\alpha_j \leq x/r' < \alpha_{j+1}$. We also say that a request is in zone 1 if $x/r' < \alpha_2$, and that it is in zone k if $x/r' \geq \alpha_k$. By construction, every request is in some zone and a request in zone j will be eventually served by server s_j.

For a request (r, x) in a zone j with $1 < j < k$, since the request is served by server s_j at time x/α_j and since $x \leq \alpha_{j+1} r$, the ratio between completion time and modified release date is at most $\alpha_{j+1}/\alpha_j = g_k$. Similarly, for a request in zone k, since $x \leq r'$, the ratio between completion time and modified release date is at most $1/\alpha_k = g_k$.

It remains to give a bound for requests in zone 1. Take any such request, i.e., a request (r, x) such that $x < \alpha_2 r'$ and suppose it is served at time $\tau \in [t_{i-1}, t_i]$ for some i. If $r' \geq t_{i-1}$, then, since $\tau \leq t_i$, the ratio between τ and r' is at most g_k by definition of t_i, $i \geq 0$.

If $r' < t_{i-1}$, then, since $\tau > t_{i-1}$, only two possible cases remain. First, the situation that $x > \frac{g_k - 1}{2} t_{i-2}$. Since $\tau = t_{i-1} + x$ and $r' \geq x/\alpha_2$, we have

$$\frac{\tau}{r'} \leq \frac{x + t_{i-1}}{x/\alpha_2} \leq \alpha_2 \left(1 + \frac{2g_k t_{i-2}}{(g_k - 1)t_{i-2}}\right) = \alpha_2 \frac{3g_k - 1}{g_k - 1} = \alpha_2 g_k^k = g_k.$$

In the second situation, $x \leq \frac{g_k - 1}{2} t_{i-2}$. Then r' must be such that s_1 was already on its way back to 0 during $[t_{i-2}, t_{i-1}]$, in particular $r' \geq g_k t_{i-2} - x$. Thus,

$$\tau/r' \leq \frac{g_k t_{i-2} + x}{g_k t_{i-2} - x} \leq \frac{3g_k - 1}{g_k + 1} \leq g_k. \qquad \square$$

The algorithm for the real line simply splits the k servers evenly between the two halflines, and uses GPS on each halfline.

Lemma 4.2. *For any $k \geq 2$, SGPS (Algorithm 4) is $g_{\lfloor k/2 \rfloor}$-competitive for the k-TSP and the k-TRP on the line.*

Proof. The only lower bounds on the offline cost that we used in the proof of Lemma 4.1 were the distance of every request from o and the release date of

Algorithm 4. Split Geometric Progression Speeds (SGPS)

Arbitrarily assign $\lceil k/2 \rceil$ servers to $\mathbb{R}_+$ and $\lfloor k/2 \rfloor$ servers to $\mathbb{R}_-$. On each of the two halflines, apply Algorithm 3 independently (i.e., ignoring the requests and the servers in the other halfline).

every request. They are valid independent of the number of offline servers. In particular, they hold if the number of offline servers is twice the number of online servers. Thus, we can analyze the competitiveness of the online servers on each of the two halflines separately and take the worst of the two competitive ratios. □

Lemma 4.3. *For any $k \geq 1$, $g_k \leq 1 + \frac{2 \log k + 3}{k}$.*

Proof. We defined g_k as the unique root greater than 1 of $z^k = 1 + \frac{2z}{z-1}$. Since $\lim_{z \to \infty} z^k > \lim_{z \to \infty} 1 + \frac{2z}{z-1}$, it suffices to prove that $z_0 := 1 + \frac{2 \log k + 3}{k}$ satisfies $z_0^k \geq 1 + \frac{2z_0}{z_0-1}$. The binomial theorem and the standard fact that $\binom{k}{j} \geq \frac{k^j}{j^j}$ yield

$$z_0^k - 1 = \sum_{j=1}^{k} \binom{k}{j} \frac{(2 \log k + 3)^j}{k^j} \geq \sum_{j=1}^{k} \frac{(2 \log k + 3)^j}{j^j} \geq \sum_{j=1}^{\lfloor \log k \rfloor + 1} \left(\frac{2 \log k + 3}{j} \right)^j$$

$$\geq \sum_{j=1}^{\lfloor \log k \rfloor + 1} 2^j \geq 2^{\log k + 1} - 2 = 2k - 2.$$

Now it can be verified that for all $k > 2$, $2k - 2 > \frac{2k}{2 \log k + 3} + 2 = \frac{2z_0}{z_0 - 1}$. Finally, the bound also holds for $k \in \{1, 2\}$ as seen by explicitly finding g_1 and g_2. □

Theorem 4.1 now follows from Lemma 4.2 and Lemma 4.3.

4.2 Lower Bounds

Theorem 4.2. *Any randomized c-competitive online algorithm for the k-TSP or the k-TRP on the line has $c \geq 1 + 1/2k = 1 + \Omega(1/k)$.*

Proof. The adversary gives a single request at time 1, in a point drawn uniformly at random from the interval $[-1, 1]$. The expected optimal cost is obviously 1. Thus, by Yao's principle it suffices to show that $\mathbb{E}[\text{OL}(\sigma)] \geq 1 + 1/2k$.

In order to bound $\mathbb{E}[\text{OL}(\sigma)]$, let $f(x) = \min_{j \in \{1,\dots,k\}} d(x, s_j(1))$. Notice that $1 + f(x)$ is a lower bound on the cost paid by the online algorithm, assuming that the request was given at x. In terms of expected values,

$$\mathbb{E}[\text{OL}(\sigma)] \geq \mathbb{E}[1 + f(x)] = 1 + \frac{1}{2} \int_{-1}^{1} f(x) dx.$$

Thus, we want to find the minimum value of the area below f in $[-1, 1]$. That area is minimized when the servers are evenly spread inside the interval and at distance $1/k$ from the extremes, in which case its value is $1/k$. □

5 Lower Bounds on the Plane

Comparing the results in Section 3 with those in Section 4, we see that while in general spaces the competitive ratio of both the k-TSP and the k-TRP always remains lower bounded by 2, on the real line we can achieve $1 + o(1)$ asymptotically. A natural question is whether on a low-dimensional space like the Euclidean plane we can also achieve $1 + o(1)$ competitiveness. In this section we answer this question negatively.

Theorem 5.1. *Any randomized c-competitive online algorithm for the k-TSP on the plane has $c \geq 4/3$.*

Proof. As a crucial ingredient of the proof we introduce a new kind of request, which is located in a single point x of the space but has an arbitrarily long processing time p (this processing time can be divided among the servers processing the request). We show how this can be emulated in the Euclidean plane with arbitrarily good precision by giving a high enough number of requests packed inside an arbitrarily small square around x.

Fix some arbitrary $\epsilon > 0$. Consider a square with sidelength $s = \sqrt{\epsilon p}$ centered around x. The square can be partitioned in s^2/ϵ^2 smaller squares of sidelength ϵ. In the center of each of these smaller squares we give a request. Notice that the distance between any pair of such requests is at least ϵ. Thus, the sum of the times required for any k servers to serve all requests is at least $(\frac{s^2}{\epsilon^2} - k)\epsilon$, no matter where the servers start (the $-k\epsilon$ term reflects the possible saving each server could have by starting arbitrarily close to the *first* request he serves).

For ϵ tending to zero, the requests converge to the point x and the total processing time needed converges to p. If the starting points of the servers are most favourable, an algorithm could finish serving all requests in time p/k.

We show how to use such a "long" request to achieve our lower bound. At time 1, the adversary gives a long request of processing time $p = 2k$ in a point drawn uniformly at random from $\{(1,0), (-1,0)\}$. The expected optimal cost is $1 + p/k = 3$. By Yao's principle, it remains to prove that $\mathbb{E}[\text{OL}(\sigma)] \geq 4$.

Since there is a single long request, we can assume wlog that all the online servers will move to the request and contribute to serving it. Since $p = 2k$, the server that will contribute most to the service will have to spend time at least $2k/k = 2$ in x, and this is enough for any other server to arrive and give a contribution (since at time 1 no server can be farther than 2 from x).

Suppose wlog that the servers are numbered in order of nondecreasing distance to x and let $d_i = d(x, s_i(1))$. $\text{OL}(\sigma) \geq 1 + t_0$, with t_0 the time needed for the servers to completely serve the request, i.e., the time when its remaining processing time is zero. Thus, t_0 satisfies $\sum_{i=1}^{k-1} i(d_{i+1} - d_i) + k(t_0 - d_k) = p$, since during interval $[d_i, d_{i+1})$ exactly i servers are processing the request. Hence,

$$kt_0 = p + kd_k - \sum_{i=1}^{k-1} i(d_{i+1} - d_i) = p + \sum_{i=1}^{k} d_i.$$

Now consider the positions of the online servers at time 1 inside the ball of radius 1 around the origin. Regarding points as vectors in $\mathbb{R}^2$, d_i can be written as $||s_i(1) - x||$ (here $|| \cdot ||$ denotes the L_2 norm). Then

$$\sum_{i=1}^{k} d_i = \sum_i ||s_i(1) - x|| \geq || \sum_i (s_i(1) - x)||$$

$$= k || \frac{1}{k} \sum_i s_i(1) - x|| = k ||b - x|| = k \cdot d(b, x),$$

where $b = \frac{1}{k} \sum_i s_i(1)$ is the centroid of the $s_i(1)$. Hence,

$$\mathbb{E}[\text{OL}(\sigma)] \geq 1 + \mathbb{E}[t_0] \geq 1 + p/k + \mathbb{E}[d(b, x)] =$$
$$= 3 + (1/2) \, d(b, (1, 0)) + (1/2) \, d(b, (-1, 0))$$
$$\geq 3 + (1/2) \, d((1, 0), (-1, 0)) = 4. \qquad \square$$

A similar technique gives an analogous lower bound for the k-TRP on the plane.

Theorem 5.2. *Any randomized c-competitive online algorithm for the k-TRP on the plane has $c \geq 5/4$.*

6 Conclusions and Open Problems

After analyzing the differences between multiple and single server variants, we can conclude that sometimes having multiple servers is more beneficial to the online algorithm than to the offline adversary. In some cases, including the traveling repairman problem on the line, the online algorithms can approach the offline cost when there are enough servers. In more general spaces, these extremely favorable situation cannot occur. Still in some intermediate cases, like the Euclidean plane, it is conceivable that the competitive ratios become lower than those of the corresponding single server problems. We leave the analysis of the competitive ratio in these situations as an open problem.

References

[1] N. Ascheuer, S. O. Krumke, and J. Rambau. Online dial-a-ride problems: Minimizing the completion time. In H. Reichel and S. Tison, editors, *Proc. 17th Symp. on Theoretical Aspects of Computer Science*, volume 1770 of *Lecture Notes in Computer Science*, pages 639–650. Springer-Verlag, 2000.

[2] G. Ausiello, V. Bonifaci, and L. Laura. The on-line asymmetric traveling salesman problem. In F. Dehne, A. López-Ortiz, and J. Sack, editors, *Proc. 9th Workshop on Algorithms and Data Structures*, volume 3608 of *Lecture Notes in Computer Science*, pages 306–317. Springer-Verlag, 2005.

[3] G. Ausiello, E. Feuerstein, S. Leonardi, L. Stougie, and M. Talamo. Algorithms for the on-line travelling salesman. *Algorithmica*, 29(4):560–581, 2001.

[4] M. Blom, S. O. Krumke, W. E. de Paepe, and L. Stougie. The online TSP against fair adversaries. *INFORMS Journal on Computing*, 13(2):138–148, 2001.

[5] A. Borodin and R. El-Yaniv. *Online Computation and Competitive Analysis.* Cambridge University Press, 1998.

[6] A. Borodin and R. El-Yaniv. On randomization in online computation. *Information and Computation,* 150:244–267, 1999.

[7] K. Chaudhuri, B. Godfrey, S. Rao, and K. Talwar. Paths, trees, and minimum latency tours. In *Proc. 44th Symp. on Foundations of Computer Science,* pages 36–45, 2003.

[8] B. Chen and A. P. A. Vestjens. Scheduling on identical machines: How good is LPT in an on-line setting? *Operations Research Letters,* 21:165–169, 1998.

[9] J. R. Correa and M. R. Wagner. LP-based online scheduling: From single to parallel machines. In *Integer programming and combinatorial optimization,* volume 3509 of *Lecture Notes in Computer Science,* pages 196–209. Springer-Verlag, 2005.

[10] J. Fakcharoenphol, C. Harrelson, and S. Rao. The k-traveling repairman problem. In *Proc. 14th Symp. on Discrete Algorithms,* pages 655–664, 2003.

[11] E. Feuerstein and L. Stougie. On-line single-server dial-a-ride problems. *Theoretical Computer Science,* 268(1):91–105, 2001.

[12] G. N. Frederickson, M. S. Hecht, and C. E. Kim. Approximation algorithms for some routing problems. *SIAM Journal on Computing,* 7(2):178–193, 1978.

[13] D. Hauptmeier, S. O. Krumke, and J. Rambau. The online dial-a-ride problem under reasonable load. In G. Bongiovanni, G. Gambosi, and R. Petreschi, editors, *Proc. 4th Italian Conference on Algorithms and Complexity,* volume 1767 of *Lecture Notes in Computer Science,* pages 125–136. Springer-Verlag, 2000.

[14] R. Jothi and B. Raghavachari. Minimum latency tours and the k-traveling repairmen problem. In M. Farach-Colton, editor, *Proc. 6th Symp. Latin American Theoretical Informatics,* volume 2976 of *Lecture Notes in Computer Science,* pages 423–433. Springer-Verlag, 2004.

[15] B. Kalyanasundaram and K. Pruhs. Speed is as powerful as clairvoyance. *Journal of the ACM,* 47:214–221, 2000.

[16] S. O. Krumke, W. E. de Paepe, D. Poensgen, and L. Stougie. News from the online traveling repairman. *Theoretical Computer Science,* 295(1-3):279–294, 2003.

[17] S. O. Krumke, L. Laura, M. Lipmann, A. Marchetti-Spaccamela, W. E. de Paepe, D. Poensgen, and L. Stougie. Non-abusiveness helps: an $O(1)$-competitive algorithm for minimizing the maximum flow time in the online traveling salesman problem. In K. Jansen, S. Leonardi, and V. V. Vazirani, editors, *Proc. 5th Int. Workshop on Approximation Algorithms for Combinatorial Optimization,* volume 2462 of *Lecture Notes in Computer Science,* pages 200–214. Springer-Verlag, 2002.

[18] E. L. Lawler, J. K. Lenstra, A. Rinnooy Kan, and D. B. Shmoys. *The Traveling Salesman Problem: A Guided Tour of Combinatorial Optimization.* Wiley, Chichester, England, 1985.

[19] M. Lipmann. *On-Line Routing.* PhD thesis, Technical Univ. Eindhoven, 2003.

[20] M. Lipmann, X. Lu, W. E. de Paepe, R. A. Sitters, and L. Stougie. On-line dial-a-ride problems under a restricted information model. *Algorithmica,* 40:319–329, 2004.

[21] M. Manasse, L. A. McGeoch, and D. Sleator. Competitive algorithms for server problems. *Journal of Algorithms,* 11:208–230, 1990.

[22] J. Sgall. On-line scheduling. In A. Fiat and G. J. Woeginger, editors, *Online Algorithms: The State of the Art,* pages 196–231. Springer, 1998.

[23] L. Stougie and A. P. A. Vestjens. Randomized on-line scheduling: How low can't you go? *Operations Research Letters,* 30:89–96, 2002.

Theoretical Evidence for the Superiority of LRU-2 over LRU for the Paging Problem*

Joan Boyar, Martin R. Ehmsen, and Kim S. Larsen

Department of Mathematics and Computer Science
University of Southern Denmark, Odense, Denmark
{joan,ehmsen,kslarsen}@imada.sdu.dk

Abstract. The paging algorithm LRU-2 was proposed for use in database disk buffering and shown experimentally to perform better than LRU [O'Neil, O'Neil, and Weikum, 1993]. We compare LRU-2 and LRU theoretically, using both the standard competitive analysis and the newer relative worst order analysis. The competitive ratio for LRU-2 is shown to be $2k$ for cache size k, which is worse than LRU's competitive ratio of k. However, using relative worst order analysis, we show that LRU-2 and LRU are asymptotically comparable in LRU-2's favor, giving a theoretical justification for the experimental results.

1 Introduction

On many layers in a computer system, one is faced with maintaining a subset of memory units from a relatively slow memory in a significantly smaller fast memory. For ease of terminology, we refer to the fast memory as the *cache* and to the memory units as *pages*. The cache will have size k, meaning that it can hold at most k pages at one time. Pages are requested by the user (possibly indirectly by an operating or a database system) and the requests must be treated one at a time without knowledge of future requests. This makes the problem an *on-line* problem [2]. If a requested page is already in cache, this is referred to as a *hit*. Otherwise, we have a *page fault*. The treatment of a request must entail that the requested page reside in cache. Thus, the only freedom is the choice of a page to evict from cache in order to make room for the requested page in the case of a page fault. An algorithm for this problem is referred to as a *paging algorithm*. Other names for this in the literature are "eviction strategy" or "replacement policy". Various cost models for this problem have been studied. We focus on the classic model of minimizing the number of page faults. The problem is of great importance in database systems where it is often referred to as the *database disk buffering problem*. See [2] for an overview of the paging problem, cost models, and paging algorithms in general.

Probably the most well-known paging algorithm is LRU (Least-Recently-Used), which on a page fault evicts the least recently used page from cache.

---

* This work was supported in part by the Danish Natural Science Research Council (SNF).

T. Erlebach and C. Kaklamanis (Eds.): WAOA 2006, LNCS 4368, pp. 95–107, 2006.
© Springer-Verlag Berlin Heidelberg 2006

The experience from real-life request sequences is that overall LRU performs better than all other paging algorithms which have been proposed up until the introduction of LRU-2 [12]. On a page fault, LRU-2 evicts the page with the least recent second to last request (if there are pages in cache which have been requested only once, the least recently used of these is evicted). Compelling empirical evidence is given in [12] in support of the superiority of LRU-2 over LRU in database systems. We return to this issue below. Since the introduction of LRU-2, there have been other proposals for better paging algorithms; for example [8].

In the on-line community, there are, to our knowledge, no published results on LRU-2. We assume that this is because it has not been possible to explain the experimental results theoretically. In this paper, we provide a theoretical justification of LRU-2's superiority over LRU. More specifically, we show using relative worst order analysis [4] that LRU-2 and LRU are asymptotically comparable in LRU-2's favor. In establishing this result, we prove a general result giving on upper bound on how well any algorithm can perform relative to LRU.

It is well-known that analysis of the paging problem is particularly problematic for the most standard quality measure for on-line algorithms, the competitive ratio [9,13]. This has lead researchers to investigate alternative methods. See a long list of these in [1]. However, these methods are mostly only applicable to the paging problem.

In contrast, it has been demonstrated that the relative worst order ratio is generally applicable. In most cases, the relative worst order ratio makes the same distinction between algorithms as the competitive ratio does. However, the following is a list of results, where the relative worst order ratio has distinguished algorithms in a situation where the competitive ratio cannot distinguish or even in some cases favors the "wrong" algorithm. This is not an exclusive list; we merely highlight one result from each of these on-line problems:

- For classical bin packing, Worst-Fit is better than Next-Fit [4].
- For dual bin packing, First-Fit is better than Worst-Fit [4].
- For paging, LRU is better than FWF (Flush-When-Full) [5].
- For scheduling, minimizing makespan on two related machines, a post-greedy algorithm is better than scheduling all jobs on the fast machine [7].
- For bin coloring [11], a natural greedy-type algorithm is better than just using one open bin at a time [10].
- For proportional price seat reservation, First-Fit is better than Worst-Fit [6].

We refer the reader to the referenced papers for details and more results. Here, we merely want to point out that the relative worst order ratio is an appropriate tool to apply to on-line problems in general and the paging problem in particular.

LRU-2, along with previous results and testing of the algorithm, is described in Section 2. Its competitive ratio is proven to be $2k$ in Section 3, showing that LRU-2 has a suboptimal competitive ratio, in comparison to LRU's competitive ratio of k. However, in Section 4, relative worst order analysis is applied showing that LRU-2 is asymptotically comparable to LRU in LRU-2's favor, providing the theoretical justification for LRU-2's superiority. A result which may be of independent interest bounds $c_{LRU,A}$, the factor by which any algorithm A can be better than LRU using relative worst order analysis: $c_{LRU,A} \leq \frac{k+1}{2}$.

2   LRU-2 and Experimental Results

In [12], a new family of paging algorithms, LRU-K, is defined. Here, K is a constant which defines the algorithm. On a page fault, LRU-K evicts the page with the least recent K'th last request (for $K = 1$, LRU-K is LRU). If there are pages in cache which have been requested fewer than K times, then some subsidiary policy must be employed for those pages. In [12], LRU is suggested as a possible subsidiary policy. However, it would also be natural to recursively use LRU-$(K - 1)$. For the case of $K = 2$, this is the same.

The authors' motivation for considering LRU-2 (or LRU-K in general for various K) is that LRU does not discriminate between pages with very frequent versus very infrequent references. Both types will be held in cache for a long time once they are brought in. This can be at the expense of pages with very frequent references.

The algorithm LFU (Least-Frequently-Used) which evicts the page which is least frequently used is the ultimate algorithm in the direction of focusing on frequency, but this algorithm appears to adjust too slowly to changing patterns in the request sequence [13]. The family of algorithms, LRU = LRU-1, LRU-2, LRU-3, ... with recursive subsidiary policies can be viewed as approaching the behavior of LFU.

A conscientious testing in [12] of particularly LRU-2 and LRU-3 up against LRU and LFU lead the authors to conclude that LRU-2 is the algorithm of choice.

The algorithms are tested in a real database system environment using random references from a Zipfian distribution, using real-life data from a CODASYL database system, and finally using data generated to simulate request sequences which would arise from selected applications where LRU-2 is expected to improve performance.

LRU-2 and LRU-3 perform very similarly and in all cases significantly better than the other algorithms. Many test results are reported which can be viewed in different ways. If one should summarize the results in one sentence, we would say that LRU and LFU need 50–100% extra cache space in order to approach the performance of LRU-2.

3   Competitive Ratio Characterizations

Let $A(I)$ denote the number of page faults A has on request sequence I. The standard measure for the quality of on-line algorithms is the *competitive ratio*: CR(A) of A is CR(A) = inf $\{c \mid \exists b \colon \forall I \colon A(I) \leq c \cdot \text{OPT}(I) + b\}$, where OPT denotes an optimal off-line algorithm [9,13].

LRU is known to be both a conservative algorithm [14] and a marking algorithm [3]. Both types of algorithms have competitive ratio k. The following request sequence, $\langle (p_1, p_1), (p_2, p_2), ..., (p_{k+1}, p_{k+1}), (p_1, p_2, p_1, p_2) \rangle$, shows that LRU-2 belongs to neither of these classes, since it faults on all of the last four requests. Thus, it is not obvious that its competitive ratio is k. In fact the lemma below shows that it is larger than k.

Lemma 1. *The competitive ratio of* LRU-2 *is at least* $2k$ *for* k *even.*

Proof. Assume that there are $k+1$ distinct pages, $p_1, p_2, ..., p_{k+1}$, in slow memory, and that k is even.

Let

$$P_1 = \langle (p_2, p_2), (p_3, p_3), \ldots, (p_{k+1}, p_{k+1}) \rangle$$
$$P_2 = \langle (p_2, p_2), (p_3, p_3), \ldots, (p_k, p_k), (p_1, p_1) \rangle$$

and define the request sequence I_l by

$$\langle P_1, (p_1, p_2, p_1, p_2), (p_3, p_4, p_3, p_4), \ldots, (p_{k-1}, p_k, p_{k-1}, p_k),$$
$$P_2, (p_{k+1}, p_2, p_{k+1}, p_2), (p_3, p_4, p_3, p_4), \ldots, (p_{k-1}, p_k, p_{k-1}, p_k) \rangle^l.$$

After LRU-2 processes P_1, the page p_1 will not be in cache. Considering any block $(p_i, p_{i+1}, p_i, p_{i+1})$, for $1 \leq i \leq k+1$ in I_l (where $(k+1)+1$ will be considered 2), it follows inductively that the page p_i is not in LRU-2's cache on the first request in that block. The faults on p_i cause p_{i+1} to be evicted and vice versa until the second fault on p_{i+1}, which causes p_{i+2} (or p_3, if $i = k+1$) to be evicted. Thus, LRU-2 faults k times during the first occurrence of P_1, never on P_1 or P_2 after that, and on all $4kl$ of the remaining requests. OPT, on the other hand, faults k times during the first occurrence of P_1. It also faults on requests to p_1 immediately following P_1 and evicts p_{k+1} each time. Similarly, on the requests to p_{k+1} immediately following P_2, it evicts p_1. Thus it faults $k + 2l$ times in all. Since l can be arbitrarily large, this gives a ratio of $2k$ asymptotically.   □

Lemma 2. LRU-2 *is* $2k$-*competitive.*

Proof. First notice that it is enough to prove that in each k-phase (a maximal subsequence of consecutive requests containing exactly k distinct pages) of any sequence I, LRU-2 faults at most two times on each of the k different pages requested in that phase. Suppose, for the sake of contradiction, that LRU-2 faults more than two times on some page in a phase P. Let p be the first page in P with more than two faults. At some point between the second and third faults on p, p must have been evicted by a request to some page q. The page q is one of the k pages in P. Thus, at this point there must be some page r in cache which is not in P. The second to last request to r must be before the start of P and thus before the second to last request to p. Hence, p could not have been evicted at this point. This gives a contradiction, so there are at most $2k$ faults in any k-phase.   □

The following theorem follows immediately from the previous two results:

Theorem 1. $CR(LRU\text{-}2) = 2k$.

4 Relative Worst Order Characterizations

Now we show the theoretical justification for the empirical result that LRU-2 performs better than LRU. In order to do this, we use a different measure for the quality of on-line algorithms, the relative worst order ratio [4,5], which has previously [4,5,7,10,6] proven capable of differentiating between algorithms in other cases where the competitive ratio failed to give the "correct" result. Instead of comparing on-line algorithms to an optimal off-line algorithm (and then comparing their competitive ratios), two on-line algorithms are compared directly. However, instead of comparing their performance on the exact same sequence, they are compared on their respective worst permutations of the same sequence:

Definition 1. *Let $\sigma(I)$ denote a permutation of the sequence I, let $\mathbb{A}$ and $\mathbb{B}$ be algorithms for the paging problem, and let $\mathbb{A}_W(I) = \max_\sigma\{\mathbb{A}(\sigma(I))\}$. Let S_1 and S_2 be statements about algorithms $\mathbb{A}$ and $\mathbb{B}$ defined in the following way.*

$$S_1(c) \triangleq \exists b \colon \forall I \colon \mathbb{A}_W(I) \leq c\,\mathbb{B}_W(I) + b$$
$$S_2(c) \triangleq \exists b \colon \forall I \colon \mathbb{A}_W(I) \geq c\,\mathbb{B}_W(I) - b$$

The relative worst order ratio $\mathrm{WR}_{\mathbb{A},\mathbb{B}}$ of algorithm $\mathbb{A}$ to algorithm $\mathbb{B}$ is defined if $S_1(1)$ or $S_2(1)$ holds.

If $S_1(1)$ holds, then $\mathrm{WR}_{\mathbb{A},\mathbb{B}} = \sup\{r \mid S_2(r)\}$, and

if $S_2(1)$ holds, then $\mathrm{WR}_{\mathbb{A},\mathbb{B}} = \inf\{r \mid S_1(r)\}$.

The statements $S_1(1)$ and $S_2(1)$ check that the one algorithm is always at least as good as the other on every sequence (on their respective worst permutations). When one of them holds, the relative worst order ratio is a bound on how much better the one algorithm can be. In some cases, however, the first algorithm can do significantly better better than the second, while the second can sometimes do marginally better than the first. In such cases, we use the following definitions (from [5], but restricted to the paging problem here) and show that the two algorithms are asymptotically comparable in favor of the first algorithm.

Definition 2. *Let $\mathbb{A}$ and $\mathbb{B}$ be algorithms for the paging problem, and let the statement $S_1(c)$ be defined as above. If there exists a positive constant c such that $S_1(c)$ is true, let $c_{\mathbb{A},\mathbb{B}} = \inf\{r \mid S_1(r)\}$. Otherwise, $c_{\mathbb{A},\mathbb{B}}$ is undefined.*

- *If $c_{\mathbb{A},\mathbb{B}}$ and $c_{\mathbb{B},\mathbb{A}}$ are both defined, $\mathbb{A}$ and $\mathbb{B}$ are $(c_{\mathbb{A},\mathbb{B}}, c_{\mathbb{B},\mathbb{A}})$-related.*
- *If $c_{\mathbb{A},\mathbb{B}}$ is defined and $c_{\mathbb{B},\mathbb{A}}$ is undefined, $\mathbb{A}$ and $\mathbb{B}$ are $(c_{\mathbb{A},\mathbb{B}}, \infty)$-related.*
- *If $c_{\mathbb{A},\mathbb{B}}$ is undefined and $c_{\mathbb{B},\mathbb{A}}$ is defined, $\mathbb{A}$ and $\mathbb{B}$ are $(\infty, c_{\mathbb{B},\mathbb{A}})$-related.*

$\mathbb{A}$ and $\mathbb{B}$ are asymptotically comparable, if

$$\left(\lim_{k\to\infty}\{c_{\mathbb{A},\mathbb{B}}\} \leq 1 \wedge \lim_{k\to\infty}\{c_{\mathbb{B},\mathbb{A}}\} \geq 1\right) \vee \left(\lim_{k\to\infty}\{c_{\mathbb{A},\mathbb{B}}\} \geq 1 \wedge \lim_{k\to\infty}\{c_{\mathbb{B},\mathbb{A}}\} \leq 1\right)$$

where k is the size of the cache.

If $\mathbb{A}$ and $\mathbb{B}$ are asymptotically comparable algorithms, then $\mathbb{A}$ and $\mathbb{B}$ are asymptotically comparable in $\mathbb{A}$'s favor if $\lim_{k\to\infty}\{c_{\mathbb{B},\mathbb{A}}\} > 1$.

The relation, being asymptotically comparable in the first algorithm's favor, is transitive, so it gives a well defined means of comparing on-line algorithms.

Lemma 3. *Asymptotically comparable in an on-line algorithm's favor is a transitive relation.*

Proof. Assume that three algorithms $\mathbb{A}$, $\mathbb{B}$, and $\mathbb{C}$ are related such that $\mathbb{A}$ is asymptotically comparable to $\mathbb{B}$ in $\mathbb{A}$'s favor and $\mathbb{B}$ is asymptotically comparable to $\mathbb{C}$ in $\mathbb{B}$'s favor.

We need to show that $\mathbb{A}$ is asymptotically comparable to $\mathbb{C}$ in $\mathbb{A}$'s favor, i.e.,

$$\left(\lim_{k \to \infty} c_{\mathbb{A},\mathbb{C}} \le 1 \right) \wedge \left(\lim_{k \to \infty} c_{\mathbb{C},\mathbb{A}} > 1 \right).$$

Since $\mathbb{A}$ is asymptotically comparable to $\mathbb{B}$ in $\mathbb{A}$'s favor and $\mathbb{B}$ is asymptotically comparable to $\mathbb{C}$ in $\mathbb{B}$'s favor, we know that

$$\left(\lim_{k \to \infty} c_{\mathbb{A},\mathbb{B}} \le 1 \right) \wedge \left(\lim_{k \to \infty} c_{\mathbb{B},\mathbb{A}} > 1 \right)$$

and

$$\left(\lim_{k \to \infty} c_{\mathbb{B},\mathbb{C}} \le 1 \right) \wedge \left(\lim_{k \to \infty} c_{\mathbb{C},\mathbb{B}} > 1 \right).$$

It follows that

$$
\begin{aligned}
1 \ge &\left(\lim_{k \to \infty} c_{\mathbb{A},\mathbb{B}} \right)\left(\lim_{k \to \infty} c_{\mathbb{B},\mathbb{C}} \right) \\
= &\lim_{k \to \infty} \left(\inf\{c_1 : \exists b_1 \forall I : \mathbb{A}_W(I) \le c_1 \mathbb{B}_W(I) + b_1\} \cdot \right. \\
&\qquad\qquad \inf\{c_2 : \exists b_2 \forall I : \mathbb{B}_W(I) \le c_2 \mathbb{C}_W(I) + b_2\}) \\
= &\lim_{k \to \infty} \inf\{c_1 c_2 : \exists b_1, b_2 \forall I : \mathbb{A}_W(I) \le c_1 \mathbb{B}_W(I) + b_1 \wedge \\
&\qquad\qquad\qquad \mathbb{B}_W(I) \le c_2 \mathbb{C}_W(I) + b_2\} \\
= &\lim_{k \to \infty} \inf\{c : \exists b \forall I : \mathbb{A}_W(I) \le c \mathbb{C}_W(I) + b\} \\
= &\lim_{k \to \infty} c_{\mathbb{A},\mathbb{C}}.
\end{aligned}
$$

In the above, we use the fact that the c_1's and c_2's can be assumed to be non-negative.

A similar argument shows that $1 < \lim_{k \to \infty} c_{\mathbb{C},\mathbb{A}}$.   □

We proceed to show that LRU-2 and LRU are asymptotically comparable in LRU-2's favor. First, we show that LRU-2 can perform significantly better than LRU on some sets of input.

Theorem 2. *There exists a family of sequences I_n of page requests and a constant b such that*

$$\text{LRU}_W(I_n) \ge \frac{k+1}{2}\text{LRU-2}_W(I_n) - b,$$

and $\lim_{n \to \infty} \text{LRU}_W(I_n) = \infty$.

Proof. Let I_n consist of n phases, where in each phase, the first $k-1$ requests are to the $k-1$ pages $p_1, p_2, \ldots, p_{k-1}$, always in that order, and the last two requests are to completely new pages. LRU will fault on every page, so it will fault $n(k+1)$ times.

Regardless of the order this sequence is given in, LRU-2 will never evict any page $p' \in \{p_1, p_2, \ldots, p_{k-1}\}$ after the second request to p'. This follows from the fact that there are at most $k-1$ pages in cache with two or more requests at any point in time. Hence, when LRU-2 faults there is at least one page in cache with only one request in its history. By definition, LRU-2 must use its subsidiary policy when there exists pages in cache with less than two requests, and it must choose among these pages. This means that LRU-2 faults at most $2(k-1) + 2n$ times.

Asymptotically, the ratio is $\frac{k+1}{2}$.   □

The above ratio of $\frac{k+1}{2}$ cannot be improved. In fact no paging algorithm $\mathbb{A}$ can be $(c_{\mathbb{A},LRU}, c_{LRU,\mathbb{A}})$-related to LRU with $c_{LRU,\mathbb{A}} > \frac{k+1}{2}$.

Theorem 3. *For any paging algorithm $\mathbb{A}$,*

$$c_{LRU,\mathbb{A}} \leq \frac{k+1}{2}.$$

Proof. Suppose there exists a sequence I, where LRU faults s times on its worst permutation, I_{LRU}, $\mathbb{A}$ faults s' times on its worst permutation, $I_{\mathbb{A}}$, and $s > \frac{k+1}{2}s'$. As proven in [5], there exists a worst permutation I_f of I_{LRU} with respect to LRU where all faults appear before all hits. Let I_1 be the prefix of I_f consisting of the s faults. Partition the sequence I_1 into subsequences of length $k+1$ (except possibly the last which may be shorter). We process these subsequences one at time, possibly reordering some of them, so that $\mathbb{A}$ faults at least twice on all, except possibly the last. (Note that since $\mathbb{A}$ will fault on the first $k+1$ requests, if $k \geq 3$, this last incomplete subsequence can be ignored. Otherwise, it contributes at most an additive constant of k to the inequality in statement $S_1(\frac{k+1}{2})$.) The first subsequence need not be reordered. Suppose the first i subsequences have been considered and consider the $i+1$st, $I' = \langle r_1, r_2, \ldots, r_{k+1} \rangle$, of consecutive requests in I_1, where $\mathbb{A}$ faults at most once. Since LRU faults on every request, they must be to $k+1$ different pages, $p_1, p_2, \ldots, p_{k+1}$. Let p be the page requested immediately before I'. Clearly, p must be in $\mathbb{A}$'s cache when it begins to process I' (it is a paging algorithm). If r_{k+1} is not a request to p, then I' contains $k+1$ pages different from p, but at most $k-1$ of them are in $\mathbb{A}$'s cache when it begins to process I' (p is in its cache). Hence, $\mathbb{A}$ must fault at least twice on the requests in I'. On the other hand, if r_{k+1} is a request to p, there are exactly k requests in I' which are different from p. At least one of them, say p_i, must cause a fault, since at most $k-1$ of them could have been in $\mathbb{A}$'s cache just before it began processing I'. If $\mathbb{A}$ faults on no other page than p_i in I', then all the pages $p, p_1, p_2, \ldots, p_{i-1}, p_{i+1}, \ldots, p_k$ must be in $\mathbb{A}$'s cache just before it starts to process I'. Now, move the request to p_i to the beginning of I' which causes $\mathbb{A}$ to fault and evict one of the pages $p, p_1, p_2, \ldots, p_{i-1}, p_{i+1}, \ldots, p_k$. Hence, it must fault at least one additional time while processing the rest of this reordering of I'.   □

Next, we show that LRU can never do significantly better than LRU-2. In order to show this, we need some definitions and lemmas characterizing LRU-2's behavior.

Lemma 4. *For any request sequence I, there exists a worst ordering of I with respect to LRU-2 with all faults appearing before all hits.*

Proof. We describe how any permutation I' of I can be transformed, step by step, to a permutation $I_{\text{LRU-2}}$ with all hits appearing at the end of the sequence, without decreasing the number of faults LRU-2 will incur on the sequence. Let I' consist of the requests $r_1, r_2, \ldots, r_n$, in that order.

If all hits in I' appears after all the faults, we are done. Otherwise consider the first hit r_i in I' with respect to LRU-2. We construct a new ordering by moving r_i later in I'.

Let p denote the page requested by r_i. First, we remove r_i from I' and call the resulting sequence I''.

If LRU-2 never evicts p in I'' or evicts p at the same requests in I'' as it does in I', then insert r_i after r_n in I''. This case is trivial since the behavior of LRU-2 on I' and I'' is the same.

Thus, we need only consider the case where p is evicted at some point after r_{i-1} in I'', and is not evicted at the same point in I'. Let r_j, $j > i$, denote the first request causing p to get evicted in I'' but not evicted in I'. Insert r_i just after r_j in I''. The resulting request sequence I'' is shown in Fig. 1 where $r_{p,1}$ and $r_{p,2}$ denote the next two requests to p (if they exist).

$$I' : \langle \ldots, r_{i-1}, r_i, r_{i+1}, \ldots, r_j, \ldots, r_{p,1}, \ldots, r_{p,2}, \ldots \rangle$$
$$I'' : \langle \ldots, r_{i-1}, r_{i+1}, \ldots, r_j, r_i, \ldots, r_{p,1}, \ldots, r_{p,2}, \ldots \rangle$$

Fig. 1. The request sequence I'' after moving r_i

First note that moving a request to p within the sequence only affects p's position in the queue that LRU-2 evicts from. The relative order of the other pages stays the same. Just before r_{i+1} the content of LRU-2's cache is the same for both sequences. Therefore, for I'', the behavior of LRU-2 is the same as for I' until p is evicted at r_j. Just after this eviction in I'', p is requested by r_i in I''. Thus, just before r_{j+1}, the cache contents are again the same for both sequences. This means that all pages that are in cache just before r_{j+1}, except p, are evicted no later for I'' than for I'. Hence, no faults are removed on requests to pages different from p, so we only need to count the faults removed on requests to p.

No faults on requests to p are removed on requests after $r_{p,2}$ since after that request the second to last request to p occurs at the same relative position in I' as in I'', so LRU-2 cannot evict it in one and not the other. Hence, the only potential faults that could have been removed are at the two requests $r_{p,1}$ and $r_{p,2}$.

The only case that needs special care is the case where $r_{p,1}$ and $r_{p,2}$ both are faults in I' but both are hits in I''. In all other cases at most one fault is removed which is counterbalanced by the fault created on r_i.

Consider the case where $r_{p,1}$ and $r_{p,2}$ both are faults in I' but neither is in I''. First remove $r_{p,1}$ from I'' and call the resulting sequence I'''. Since $r_{p,2}$ is a fault in I', p must get evicted in the subsequence $\langle r_{p,1}, \ldots, r_{p,2} \rangle$ based on its second to last request in that subsequence, which is the same request for both I' and I'''. Consequently, if p is evicted in that subsequence of I' it must also get evicted in that subsequence of I''' and it follows that $r_{p,2}$ is a fault in both sequences and no faults have been removed.

The situation we are facing with I''' is no different from the situation we faced with I''. We need to insert a (removed) request to p (for I'' it was r_i and for I''' it is $r_{p,1}$) without removing any faults, except that we now have increased the number of faults among the first j requests by at least one, and we have moved the problem of inserting a request to p later in the sequence. We now proceed inductively on I''' in the same manner as we did for I'' until we can insert the request to p without removing any faults or we reach the end of the sequence (in which case we place it there).

Thus, we obtain $I_{\text{LRU-2}}$ in a finite number of steps.   □

Thus, when considering a worst case sequence for LRU-2, one can assume that there is a prefix of the sequence containing all of the faults and no hits. In the remaining, we will only be considering such prefixes. We define LRU-2-phases, starting from any request in a request sequence I.

Definition 3. *Let $I = \langle r_1, r_2, \ldots, r_n \rangle$ be a request sequence for which LRU-2 faults on every request. The LRU-2-phase starting at request r_i is $P(r_i) = \langle r_i, r_{i+1}, \ldots, r_j \rangle$, where j is as large as possible under the restriction that no page should be requested three times in $P(r_i)$. A LRU-2-phase is complete if r_j is not the last request in the I, i.e., r_{j+1} is a page which occurs twice in $P(r_i)$.*

Lemma 5. *Each complete LRU-2-phase contains at least $2k + 1$ requests to at least $k + 1$ distinct pages. In addition, it contains two requests to each of at least k different pages.*

Proof. By definition, within a complete LRU-2-phase, P, there is a request to a page p immediately after that phase. This request causes a fault, and p was requested at least twice within the phase. In order for p to be evicted within the phase P, the second to last request to each of the $k-1$ other pages in cache must have occurred more recently than the first request to p in P. Thus, counting p, at least k distinct pages must have been requested twice in P. In addition, the page causing the second eviction of p within P cannot have been in cache at that point, so P consists of at least $2k + 1$ requests.   □

Lemma 6. *Let p be the page that starts a complete LRU-2-phase containing exactly $2k+1$ requests, then the following phase (if it exists) starts with a request to p.*

Proof. In general after a complete LRU-2-phase, LRU-2 has at least $k - 1$ of the pages requested twice in that phase in cache. If the phase ends with the second request to a page, then LRU-2 contains k of the pages requested twice. This fact follows from the observation that by the LRU-2 policy no page with two requests in a phase can get evicted if there is a page in cache with only one request in that phase.

This means that a phase containing only $2k + 1$ requests must end with a request to the only page with one request in that phase. Before that request the cache contains k pages which have all been requested twice in the current phase, hence p is the page with the earliest second request. This means p gets evicted on the last request in the phase and hence (by the construction of LRU-2-phases) it must be the page starting the next phase.   □

By induction the above shows that if there exist several consecutive phases, each containing $2k + 1$ requests, then they must all begin with a request to the same page. This then shows that if $p_1, p_2, \ldots, p_k$ are the k pages requested twice in a phase containing $2k + 1$ requests, then after the first request in the following phase (if it exists) all of the pages $p_1, p_2, \ldots, p_k$ are in LRU-2's cache.

Lemma 7. *For any sequence I of page requests,*

$$\text{LRU-2}_W(I) \le (1 + \frac{1}{2k + 2})\text{LRU}_W(I).$$

Proof. Consider any sequence I of requests. By Lemma 4, there exists a worst permutation, $I_{\text{LRU-2}}$, of I such that LRU-2 faults on each request of a prefix I_1 of $I_{\text{LRU-2}}$ and on no requests after I_1. Partition I_1 into LRU-2-phases. We will now inductively transform I_1 into a sequence I_1' such that

$$\text{LRU-2}(I_1) \le (1 + \frac{1}{2k + 2})\text{LRU}_W(I_1').$$

Start at the beginning of I_1, and consider the LRU-2-phase starting with the first request not already placed in a processed phase. By Lemma 6 each LRU-2-phase contains at least a total of $2k + 1$ requests to at least $k + 1$ distinct pages. Since each page requested in a LRU-2-phase is at most requested twice, a LRU-2-phase containing at least $2k + 2$ requests can be partitioned into two sets, each containing at least $k + 1$ pages, none of which are repeated. Each of these sets of requests can then be ordered so that LRU faults on every request.

Hence, suppose the current LRU-2-phase contains exactly $2k + 1$ requests. See Fig. 2 where | marks the beginning of a new phase in I_1 which contains exactly $2k + 1$ requests. Let $p_1, p_2, \ldots, p_k$ be the k pages requested twice and q_1 be the page requested once in that phase and let p_1 be the page which begin the following phase. The request r_i to p_1 which starts the following phase must evict q_1 and hence all the pages $p_1, p_2, \ldots, p_k$ are in LRU-2's cache just before the request to $r_{i+1} = q_2$. It follows that $q_2 \notin \{p_1, p_2, \ldots, p_k\}$.

$$\langle \ldots, |p_1, \ldots, q_1, |r_i = p_1, r_{i+1} = q_2, r_{i+2} \ldots \rangle$$

Fig. 2. A LRU-2-phase containing $2k + 1$ requests

By moving r_i to the end of the request sequence it follows from the above that the modified phase in question now contains at least $2(k+1)$ requests (the $2k+1$ requests and the request to q_2) and by the same argument as above it follows that it is possible to make LRU fault on every request. Hence in each such phase LRU faults on (possibly) one request less than LRU-2. The next LRU-2-phase to be processed starts with r_{i+2} or later.

Let l denote the total number of modified phases. For each modified phase i, there are $s_i \geq 2(k+1)$ requests, plus possibly one additional request which LRU-2 faulted on and has been moved to the end. Thus, LRU faults at least $\sum_{i=1}^{l} s_i$ times and LRU-2 faults at most $\sum_{i=1}^{l}(s_i + 1)$ times. It follows that

$$\text{LRU-2}_W(I) \leq \frac{\sum_{i=1}^{l}(s_i + 1)}{\sum_{i=1}^{l} s_i} \text{LRU}_W(I)$$

$$\leq \frac{l(2k + 3)}{l(2k + 2)} \text{LRU}_W(I)$$

$$= (1 + \frac{1}{2k + 2}) \text{LRU}_W(I)$$

□

Combining Theorem 2 and the lemma above gives the following:

Theorem 4. LRU-2 *and* LRU *are* $(1 + \frac{1}{2k+2}, \frac{k+1}{2})$*-related, i.e., they are asymptotically comparable in* LRU-2*'s favor.*

5   Concluding Remarks

In contrast to the results using competitive analysis, relative worst order analysis yields a theoretical justification for superiority of LRU-2 over LRU, confirming previous empirical evidence. It would be interesting to see if these results generalize to LRU-K for $K > 2$. Recently, we have shown that the competitive ratio for LRU-K is kK, and that the separation result showing that LRU-K can be better than LRU holds. The question is: Does the asymptotic comparability still hold.

Although it was shown here that LRU-2 and LRU are asymptotically comparable, it would be interesting to know if the stronger result, that LRU-2 and LRU are comparable using relative worst order analysis, holds. If they are, then the above results show that the relative worst order ratio of LRU to LRU-2 is $\frac{k+1}{2}$.

Note that any result showing that the relative worst order ratio is defined for two algorithms immediately gives a result showing that they are asymptotically comparable. Thus, the results from [5], showing that LRU is at least as good as any conservative algorithm and better than Flush-When-Full (FWF), combined with the results proven here, show that LRU-2 is asymptotically comparable to any conservative algorithm and FWF, in LRU-2's favor in each case.

An algorithm called RLRU was proposed in [5] and shown to be better than LRU using relative worst order analysis. We conjecture that LRU-2 is also asymptotically comparable to RLRU in LRU-2's favor. We have found a family of sequences showing that LRU-2 can be better than RLRU, but would also like to show that the algorithms are asymptotically comparable.

Acknowledgments

The authors would like to thank Peter Sanders for bringing LRU-2 to their attention. The second author would like to thank Troels S. Jensen for helpful discussions.

References

1. Susanne Albers. Online algorithms: A survey. In *Proceedings of the 18th International Symposium on Mathematical Programming*, pages 3–26, 2003.
2. Allan Borodin and Ran El-Yaniv. *Online Computation and Competitive Analysis*. Cambridge University Press, 1998.
3. Allan Borodin, Sandy Irani, Prabhakar Raghavan, and Baruch Schieber. Competitive Paging with Locality of Reference. *Journal of Computer and System Sciences*, 50(2):244–258, 1995.
4. Joan Boyar and Lene M. Favrholdt. The relative worst order ratio for on-line algorithms. In *Proceedings of the Fifth Italian Conference on Algorithms and Complexity*, volume 2653 of *Lecture Notes in Computer Science*, pages 58–69. Springer-Verlag, 2003. Extended version to appear in *ACM Transactions on Algorithms*.
5. Joan Boyar, Lene M. Favrholdt, and Kim S. Larsen. The Relative Worst Order Ratio Applied to Paging. In *Sixteenth Annual ACM-SIAM Symposium on Discrete Algorithms*, pages 718–727. ACM Press, 2005.
6. Joan Boyar and Paul Medvedev. The Relative Worst Order Ratio Applied to Seat Reservation. In *Proceedings of the Nineth Scandinavian Workshop on Algorithm Theory*, volume 3111 of *Lecture Notes in Computer Science*, pages 90–101. Springer-Verlag, 2004.
7. Leah Epstein, Lene M. Favrholdt, and Jens S. Kohrt. Separating Scheduling Algorithms with the Relative Worst Order Ratio. *Journal of Combinatorial Optimization*. To appear.
8. Amos Fiat and Ziv Rosen. Experimental Studies of Access Graph Based Heuristics: Beating the LRU Standard? In *Proceedings of the 8th Annual ACM-SIAM Symposium on Discrete Algorithms*, pages 63–72, 1997.
9. Anna R. Karlin, Mark S. Manasse, Larry Rudolph, and Daniel D. Sleator. Competitive Snoopy Caching. *Algorithmica*, 3:79–119, 1988.

10. Jens Svalgaard Kohrt. *Online Algorithms under New Assumptions.* PhD thesis, Department of Mathematics and Computer Science, University of Southern Denmark, Odense, Denmark, 2004.
11. Sven Oliver Krumke, Willem de Paepe, Jörg Rambau, and Leen Stougie. Online Bin Coloring. In *Proceedings of the Nineth Annual European Symposium on Algorithms*, volume 2161 of *Lecture Notes in Computer Science*, pages 74–85. Springer-Verlag, 2001.
12. Elizabeth J. O'Neil, Patrick E. O'Neil, and Gerhard Weikum. The LRU-K Page Replacement Algorithm for Database Disk Buffering. In *Proceedings of the ACM SIGMOD International Conference on Management of Data*, pages 297–306, 1993.
13. Daniel D. Sleator and Robert E. Tarjan. Amortized Efficiency of List Update and Paging Rules. *Communications of the ACM*, 28(2):202–208, 1985.
14. Neal Young. The k-Server Dual and Loose Competitiveness for Paging. *Algorithmica*, 11(6):525–541, 1994.

Improved Approximation Bounds for Edge Dominating Set in Dense Graphs

Jean Cardinal, Stefan Langerman*, and Eythan Levy

Computer Science Department
Université Libre de Bruxelles, CP212
B–1050 Brussels, Belgium
{jcardin,slanger,elevy}@ulb.ac.be

Abstract. We analyze the simple greedy algorithm that iteratively removes the endpoints of a maximum-degree edge in a graph, where the degree of an edge is the sum of the degrees of its endpoints. This algorithm provides a 2-approximation to the minimum edge dominating set and minimum maximal matching problems. We refine its analysis and give an expression of the approximation ratio that is strictly less than 2 in the cases where the input graph has n vertices and at least $\epsilon\binom{n}{2}$ edges, for $\epsilon > 1/2$. This ratio is shown to be asymptotically tight for $\epsilon > 1/2$.

1   Introduction

While there exist sophisticated methods yielding approximate solutions to many NP-hard combinatorial optimization problems, the methods that are the simplest to implement are often the most widely used. Among these methods, greedy strategies are extremely popular and certainly deserve thorough analyses.

We study the worst-case approximation factor of a simple greedy algorithm for the following two NP-hard problems.

Definition 1 (MINIMUM EDGE DOMINATING SET)
INPUT: *A graph $G = (V, E)$.*
SOLUTION: *A subset $M \subseteq E$ of edges such that each edge in E shares an endpoint with some edge in M.*
MEASURE: $|M|$.

Definition 2 (MINIMUM MAXIMAL MATCHING)
INPUT: *A graph $G = (V, E)$.*
SOLUTION: *A subset $M \subseteq E$ of disjoint edges such that each edge in E shares an endpoint with some edge in M.*
MEASURE: $|M|$.

It has been noted since long ago that MINIMUM EDGE DOMINATING SET (EDS) and MINIMUM MAXIMAL MATCHING (MMM) admit optimal solutions of the same size and that an optimal solution to EDS can be transformed in polynomial

---

* Chercheur qualifié du FNRS.

time into an optimal solution to MMM [13], the converse transformation being trivial.

The algorithm that we analyze in this paper uses the degree of the edges, with the degree of an edge being the sum of the degrees of its endpoints. It iteratively removes the highest-degree edge and updates the graph accordingly, as shown in Algorithm 1. The algorithm returns a maximal matching, which provides a

---

Algorithm 1. The greedy algorithm

$res \leftarrow \emptyset$
while $E(G) \neq \emptyset$ **do**
    $e \leftarrow \arg\max_{e \in E(G)} deg_G(e)$
    $res \leftarrow res \cup \{e\}$
    for each edge f adjacent to e **do**
        $E(G) \leftarrow E(G)\backslash\{f\}$
    end for
    $E(G) \leftarrow E(G)\backslash\{e\}$
end while
return res

---

solution to both our problems. The algorithm therefore guarantees exactly the same approximation ratios for the two problems.

It is well-known that any maximal matching M provides a 2-approximation for MMM, as each edge in the optimal solution can cover at most two edges of M. Our algorithm is thus clearly a 2-approximation algorithm and is expected to return small matchings as the greedy step always selects a high-degree edge. We however refine this analysis, and provide a tight approximation factor as a function of the density of the graph.

Our Contributions

We provide a new bound on the approximation ratio of the greedy heuristic for our problems in graphs with at least $\epsilon\binom{n}{2}$ edges (ϵ-dense graphs). This bound is asymptotic to $1/(1 - \sqrt{(1 - \epsilon)/2})$, which is smaller than 2 when ϵ is greater than $1/2$. We further provide a family of tight examples for our bound. No algorithm for ϵ-dense graphs with a better approximation ratio than the one shown in this paper seems to be known.

Related Works

The MMM and EDS problems go back a long way. Both problems are already referred to in the classical work of Garey and Johnson [6] on NP-completeness. Yannakakis and Gavril [13] then showed that EDS remains NP-hard when restricted to planar or bipartite graphs of maximum degree 3, and gave a polynomial-time algorithm for MMM in trees. Later, Horton et al. [8] and Srinivasan et al. [12] gave additional hard and polynomially solvable classes of graphs. More recently, Carr et al. [2] gave a $2\frac{1}{10}$-approximation algorithm for the weighted edge dominating set problem, a result which was later improved to 2 by Fujito et al. [5]. Finally, Chlebìk and Chlebìkovà [3] showed that it is NP-Hard to approximate EDS(and hence also MMM) within any factor better than 7/6.

Another recent trend of research on approximation algorithms deals with expressing approximation ratios as functions of some density parameters [4,7,9], related to the number of edges, or the minimum and maximum degrees. Not many such results have yet been obtained for our problems. It was nevertheless shown in [1] that MMM and EDS are approximable within ratios that are asymptotic to $\min\{2, 1/(1 - \sqrt{1 - \epsilon})\}$ for graphs having at least $\epsilon\binom{n}{2}$ edges, and to $\min\{2, 1/\epsilon\}$ for graphs having minimum degree at least ϵn.

2   Analysis of Algorithm 1

Definitions and Notations

Let $G = (V, E)$ be a (simple, loopless, undirected) graph, with $V = \{v_1, \ldots, v_n\}$. Let OPT be a fixed optimal solution to MMM in G and let T be the set of endpoints in OPT. Let $M = \{e_1, \ldots, e_\mu\}$ be a set of μ edges returned by an execution of the greedy algorithm on G. We assume that these edges are ordered according to the order in which they were chosen by the algorithm.

The definition of the algorithm ensures that M is a maximal matching. Since M is a matching, at least one endpoint of each edge e_i belongs to T. Let us call $\{v_1, \ldots, v_{2\mu}\}$ the endpoints of the edges of M, with $e_i = v_{2i-1}v_{2i}$ and $v_{2i-1} \in T$. Since the matching M is maximal, the set of vertices $\{v_{2\mu+1}, \ldots, v_n\}$ forms a stable set, i.e. a set of vertices sharing no edge. The set of vertices $V\backslash T$ also forms a stable set as the vertices in T are the endpoints of a maximal matching. Fig. 1 shows an example with $\mu = 6$ and $|OPT| = 5$. Our assumptions on the ordering of the vertices ensure that a vertex has a higher index when it is included later (or never) in the heuristic solution and that the vertex with lowest index in e_i belongs to T.

As can be seen in Fig. 1, there are two types of edges in M. Edges of the first type have only one endpoint in T. We let X be the set of these endpoints. Edges of the second type have both endpoints in T. Let a be the number of such edges. Let finally b be the number of vertices of T outside M. Fig. 1 also illustrates X, a and b. Note that in practice the two types of edges can be interleaved in M, whereas they are shown separated in the figure for the sake of clarity.

The approximation ratio is $\beta = \mu/|OPT|$. This quantity is fixed when M and OPT are given. In order to give an upper bound on β, we prove an upper bound on the number of edges in a graph when M and OPT are fixed. This bound is then inverted in order to obtain an upper bound on β as a function of the number of edges. Our results are expressed in terms of the *density* of our graphs, according to the following definitions. We define an ϵ-*dense graph* as a graph with at least $\epsilon\binom{n}{2}$ edges.

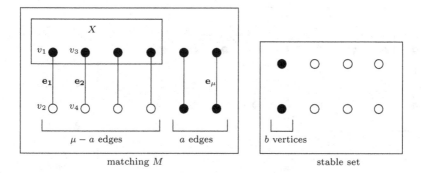

Fig. 1. An example with $\mu = 6$ and $|OPT| = 5$. Black vertices are the endpoints of the minimum maximal matching.

The following additional graph-theoretic notations will be useful. For any vertex set $W \subseteq V$ and vertex v, let $N_W(v)$ be the set of neighbors of v in set W and let $d_W(v) = |N_W(v)|$. Let an *anti-edge* xy be a pair of vertices x and y sharing no edge. Let $N_W^{<}(v_j)$ be the set of neighbors v_i of v_j with $i < j$ and $v_i \in W$, and let $d_W^{<}(v_j) = |N_W^{<}(v_j)|$. For any of these notations, the subscript W may be omitted when $W = V$. We also use the classical notation $G[X]$ for the subgraph of G induced by a vertex set X. Let $\bar{m}(G) = \binom{n}{2} - m(G)$ be the number of anti-edges in G. We omit the parameter G when it is clear from context. We define $G \times G'$, the *join* of graphs $G = (V, E)$ and $G' = (V', E')$ as a new graph that contains all the vertices and edges of G ang G' as well as all the possible edges joining both sets of vertices.

Upper Bound

Lemma 1 shows that a certain set of vertices has degree at most $|T|$. This result is then used by Lemma 2 in order to find an upper bound on the number of edges in the graph.

Lemma 1. *If $d_X^{<}(v_j) > 0$ for some vertex v_j, then $d(v_j) \leq |T|$.*

Proof. We call the vertices of T black vertices and the vertices outside of T white vertices. Let i be the smallest index such that $v_i \in X$ and $v_i v_j \in E$. Let $V_a^b = (v_a \ldots v_b)$. Fig. 2 illustrates these notations. We can express the degree of v_j as:

$$d(v_j) = d_{V_1^{i-1}}(v_j) + d_{V_i^n}(v_j).$$

Since v_j has no neighbor in $V_1^{i-1} \cap X$, we have $d_{V_1^{i-1}}(v_j) \leq |V_1^{i-1} \backslash X|$ and therefore

$$d(v_j) \leq |V_1^{i-1} \backslash X| + d_{V_i^n}(v_j). \tag{1}$$

It can easily be seen that $|V_1^{i-1} \backslash X| = |V_1^{i-1} \cap T|$ and therefore

$$d(v_j) \leq |V_1^{i-1} \cap T| + d_{V_i^n}(v_j).$$

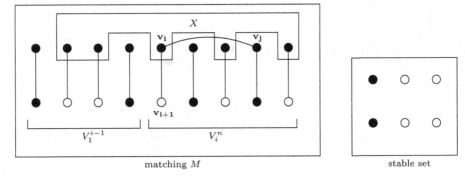

Fig. 2. Structure of the matching M. In this example, v_j was chosen inside the matching and outside X. Note that Lemma 1 also allows v_j to be in the stable set or in X.

The greedy algorithm ensures that edge $v_i v_{i+1}$ has maximum degree in $G[V_i^n]$, and therefore

$$d(v_j) \leq |V_1^{i-1} \cap T| + d_{V_i^n}(v_{i+1}). \tag{2}$$

It is worth noticing that this is the only place in the whole proof of Theorem 1 where this property is used. Finally, since v_{i+1} is a white vertex, it can only be adjacent to vertices in T, as the white vertices form a stable set. Therefore

$$d(v_j) \leq |V_1^{i-1} \cap T| + |V_i^n \cap T|$$
$$= |T|. \qquad \square$$

The following result provides a lower bound on the number of anti-edges in the graph, hence an upper bound on the number of edges. Its proof uses counting arguments that heavily rely on the bound given in Lemma 1. Recall that a is the number of edges of M having both endpoints in T, and that b is the number of vertices of T that are outside M.

Lemma 2

$$\bar{m} \geq 2 \binom{n/2 - a - b}{2}$$

Proof. Let $\bar{d}_W(v)$, the *anti-degree* of v, be the number of anti-edges between v and vertices of W. Thus

$$\bar{d}_W(v) = \begin{cases} |W| - d_W(v) & \text{if } v \notin W \\ |W| - 1 - d_W(v) & \text{otherwise.} \end{cases}$$

We first define a family of vertex sets $\{X_i\}$ and show a lower bound on $\bar{m}$ as a function of the sizes of these sets. We call the vertices in (resp. outside) T *black* (resp. *white*) vertices.

The sets of vertices are the following (see Fig. 3): X_1 and X_2 are defined as the black and white endpoints of $\mu - a - b$ arbitrary black-white edges of M. Sets X_3 and X_4 are obtained by splitting the b black vertices outside of M into two sets of equal sizes (rounding if necessary). Sets X_5 and X_6 are obtained by splitting the $n - 2\mu - b$ white vertices outside of M into two sets of equal sizes. Finally, X_9 and X_{10} are obtained by dividing the remaining b vertices of the matching into sets of equal sizes. We define $x_i = |X_i|$ for each set X_i.

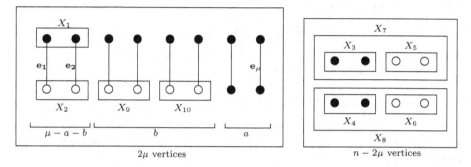

Fig. 3. Notations for the vertex sets

We first show

$$\bar{m} \geq \binom{x_1}{2} + \binom{x_2}{2} + \binom{x_7}{2} + \binom{x_8}{2} + x_2 x_9 + x_2 x_5 + x_2 x_{10} + x_2 x_6 \quad (3)$$

Note that each set X_i except X_1 is stable, because it either contains only white vertices or only vertices outside M. This explains the second, third and fourth terms in the above sum. For each term of the form $x_i x_j$ in the sum, both X_i and X_j contain only white vertices, and therefore share no edge, since any set of white vertices in G is stable. Note that no anti-edge is counted twice, since our anti-edges involve vertices taken in and between disjoint vertex sets.

Concerning the additional number of $\binom{x_1}{2}$ anti-edges required, we use Lemma 1 to prove that every edge inside X_1 is compensated for by an anti-edge between a vertex in X_1 and a vertex outside X_1. For each $v_j \in X_1$, we have:

$$d_{X_1}^<(v_j) \leq d_{X_1}(v_j)$$
$$= d(v_j) - d_{V \setminus X_1}(v_j).$$

Applying Lemma 1 yields:

$$d_{X_1}^<(v_j) \leq |T| - d_{V \setminus X_1}(v_j).$$

Using $|T| = \mu + a + b$ and $\mu \leq n/2$ yields

$$d_{X_1}^<(v_j) \leq n - (\mu - a - b) - d_{V \setminus X_1}(v_j).$$

Finally, since $|V\backslash X_1| = n - (\mu - a - b)$, the definition of the anti-degree yields

$$d^{\leq}_{\bar{X}_1}(v_j) \leq \bar{d}_{V\backslash X_1}(v_j).$$

We now take sums over the elements of X_1:

$$\sum_{v_j \in X_1} d^{\leq}_{\bar{X}_1}(v_j) \leq \sum_{v_j \in X_1} \bar{d}_{V\backslash X_1}(v_j).$$

Since the sets $N^{\leq}_{\bar{X}_1}(v_j)$ corresponding to the values $d^{\leq}_{\bar{X}_1}(v_j)$ in the above sum form a partition of the edges of $G[X_1]$, we have

$$m(G[X_1]) \leq \sum_{v_j \in X_1} \bar{d}_{V\backslash X_1}(v_j).$$

From the definition of $\bar{m}$, we have:

$$\binom{x_1}{2} \leq \bar{m}(G[X_1]) + \sum_{v_j \in X_1} \bar{d}_{V\backslash X_1}(v_j).$$

The above relation thus implies the existence of at least $\binom{x_1}{2}$ anti-edges involving vertices of X_1.

There remains to show that bound 3 is greater than $2\binom{n/2-a-b}{2}$. Plugging $x_2 = x_1$, $x_9 = x_3$, and $x_{10} = x_4$ into 3 yields:

$$\bar{m} \geq \binom{x_1}{2} + \binom{x_2}{2} + \binom{x_7}{2} + \binom{x_8}{2} + x_1 x_3 + x_1 x_5 + x_2 x_4 + x_2 x_6.$$

and therefore

$$\bar{m} \geq \binom{x_1}{2} + \binom{x_2}{2} + \binom{x_7}{2} + \binom{x_8}{2} + x_1 x_7 + x_2 x_8.$$

The desired result follows from repeated applications of the relation $\binom{x+y}{2} = \binom{x}{2} + \binom{y}{2} + xy$:

$$\bar{m} \geq \binom{x_1}{2} + \binom{x_2}{2} + \binom{x_7}{2} + \binom{x_8}{2} + x_1(x_7) + x_2(x_8)$$

$$= \binom{|X_1 \cup X_7|}{2} + \binom{|X_2 \cup X_8|}{2}$$

$$= \binom{x_1 + x_7}{2} + \binom{x_2 + x_8}{2}$$

$$= \binom{\lfloor n/2 - a - b \rfloor}{2} + \binom{\lceil n/2 - a - b \rceil}{2}$$

$$=^* \begin{cases} 2\binom{n/2-a-b}{2} & \text{if } n \text{ is even} \\ 2\binom{n/2-a-b}{2} + 1/4 & \text{otherwise.} \end{cases}$$

$$\geq 2\binom{n/2-a-b}{2}.$$

$\square$

Theorem 1 is essentially a consequence of this upper bound on the number of edges.

Theorem 1. *The approximation ratio of the greedy heuristic in ϵ-dense graphs with n vertices is at most*

$$
\begin{cases}
2 & \text{if } \epsilon \leq \frac{1}{2} + \frac{1}{n-1} \\
\left[1 - \frac{1}{2n} - \sqrt{\frac{1}{4n^2} + \left(1 - \frac{1}{n}\right)\frac{(1-\epsilon)}{2}}\right]^{-1} & \text{otherwise.}
\end{cases}
$$

$$
\xrightarrow{n \to \infty}
\begin{cases}
2 & \text{if } \epsilon \leq \frac{1}{2} \\
\left[1 - \sqrt{\frac{1-\epsilon}{2}}\right]^{-1} & \text{otherwise.}
\end{cases}
$$

Proof. We know from Lemma 2 that $\bar{m} \geq 2\binom{n/2-a-b}{2}$. Simple algebra using $\beta = \mu/|OPT|$, $2|OPT| = \mu + a + b$ and $\mu \leq n/2$ implies

$$
a + b \leq \frac{n}{2}\left[\frac{2-\beta}{\beta}\right].
$$

and therefore

$$
\bar{m} \geq 2\binom{n/2 - \frac{n}{2}\left[\frac{2-\beta}{\beta}\right]}{2} = 2\binom{n\left(\frac{\beta-1}{\beta}\right)}{2}. \tag{4}
$$

Let $x = (\beta - 1)/\beta$. We would like to express the above inequality as an upper bound on β, i.e. on x. The inequality can now be written as

$$
f(x) = n^2x^2 - nx - \bar{m} \leq 0.
$$

Differentiating f with respect to x shows that f decreases when $x < \frac{1}{2n}$ and increases when $x > \frac{1}{2n}$. The value of $f(x)$ can therefore only be negative when $x^- \leq x \leq x^+$, where x^- and x^+ are the roots of $f(x)$. Solving the second-order equation $f(x) = 0$ yields

$$
x^- = \frac{1}{2n} - \sqrt{\frac{1}{4n^2} + \frac{\bar{m}}{n^2}}
$$

and

$$
x^+ = \frac{1}{2n} + \sqrt{\frac{1}{4n^2} + \frac{\bar{m}}{n^2}}.
$$

The value of x^- is always negative and thus $x^- \leq x$ brings us no additional knowledge on the ratio. Rewriting inequality $x \leq x^+$ yields

$$
\frac{\beta - 1}{\beta} \leq \frac{1}{2n} + \sqrt{\frac{1}{4n^2} + \frac{\bar{m}}{n^2}}
$$

and

$$
\beta \leq \left[1 - \frac{1}{2n} - \sqrt{\frac{1}{4n^2} + \frac{\bar{m}}{n^2}}\right]^{-1}.
$$

Reverting to m and setting $m \geq \epsilon\binom{n}{2}$ yields the desired result

$$\beta \leq \left[1 - \frac{1}{2n} - \sqrt{\frac{1}{4n^2} + \left(1 - \frac{1}{n}\right)\frac{(1-\epsilon)}{2}}\right]^{-1}$$

$$= \left[1 - O\left(\frac{1}{n}\right) - \sqrt{\frac{1-\epsilon}{2} + O\left(\frac{1}{n}\right)}\right]^{-1}.$$

Direct algebraic manipulations show that

$$\left[1 - \frac{1}{2n} - \sqrt{\frac{1}{4n^2} + (1-\epsilon)\left(\frac{1}{2} - \frac{1}{2n}\right)}\right]^{-1} < 2$$

$$\iff \epsilon > \frac{1}{2}\left(\frac{n}{n-1}\right)$$

$$\iff \epsilon > \frac{1}{2} + \frac{1}{n-1}.$$

□

Tightness

The case $\epsilon \geq 7/9$. Let $\zeta_{n,k} = K_{n-2k} \times K_{k,k}$, where K_{n-2k} is a complete graph with $n - 2k$ vertices and $K_{k,k}$ a complete bipartite graph with two stable sets of size k (see Fig. 4(b) for an example). Such a graph can be compared with the *complete split graph* $\Psi_{n,k}$ (see Fig. 4(a)), which is defined as the join of a clique of size $n - k$ and an independent set of size k and is a tight example for the simpler greedy algorithm analyzed in [1].

Algorithm 1 always finds a perfect matching in $\zeta_{n,k}$. On the other hand, the following matching is clearly maximal: match k vertices of the clique with k vertices of one independent set, and match the remaining vertices of the clique among themselves. This is always possible when k and n are even and $k \leq n/3$ and yields a matching of size $(n - k)/2$. Therefore we have the following bound on the approximation ratio: $\beta = \mu/|OPT| \geq n/(n - k)$.

The number of edges of $\zeta_{n,k}$ is given by $m = \binom{n}{2} - 2\binom{k}{2}$ and therefore $k = \left(1 + \sqrt{1 + 4\left[\binom{n}{2} - m\right]}\right)/2$. We denote by ϵ the ratio $m/\binom{n}{2}$, i.e. the density of $\zeta_{n,k}$. From the above equality, we have $k = \left(1 + \sqrt{1 + 4\binom{n}{2}(1-\epsilon)}\right)/2$.

Plugging this equation into the inequality for β above yields

$$\beta \geq \left[1 - \frac{1}{2n} - \sqrt{\frac{1}{4n^2} + \left(1 - \frac{1}{n}\right)\frac{(1-\epsilon)}{2}}\right]^{-1},$$

which matches the upper bound on the ratio obtained in Theorem 1. Plugging the condition $k \leq n/3$ into $m = \binom{n}{2} - 2\binom{k}{2}$ yields $\epsilon \geq 7/9 + O(1/n)$. The graphs $\zeta_{n,k}$ with n and k even are thus a collection of tight examples for our bound when $\epsilon \geq 7/9$.

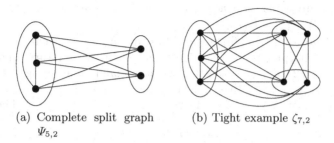

(a) Complete split graph
$\Psi_{5,2}$

(b) Tight example $\zeta_{7,2}$

Fig. 4. Tight examples

The general case. A slightly more intricate family of graphs can be built, which provide a collection of asymptotically tight examples for our ratio for any $\epsilon \geq 1/2$. We first describe the special case when $\epsilon = 1/2$ and $\beta \to 2$. The graph is the following (see Fig. 5) :

$$B \equiv M_{k/2} \times I_k \times K_1$$

where $M_{k/2}$ is a matching of $k/2$ edges, I_k an empty graph with k vertices, and K_1 is an isolated vertex.

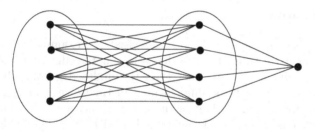

Fig. 5. Tight example $A_{9,4}$

It is easy to see that at each step of Algorithm 1 there exists an edge between the matching and the stable set that has maximum degree. Therefore the algorithm might choose k of these edges thus obtaining a cover of size k. On the other hand, taking all the edges of $M_{k/2}$ and an additional edge incident to K_1 yields a cover of size $k/2 + 1$. Therefore $\beta \geq k/(k/2 + 1)$ which tends to 2 as k tends to infinity. It is further straightforward to check that the density of this graph is $1/2 + O(1/n)$.

For other values of ϵ, we generalize the above example by joining it to a clique, i.e. we build the following general family:

$$A_{n,k} \equiv (M_{k/2} \times I_k \times K_1) \times K_{n-2k-1}$$

Note that $A_{n,k}$ is well-defined for any odd n and even $0 \leq k < n/2$ and that the limiting values of k correspond respectively to B and to K_n. The density of $A_{n,k}$ therefore spans the whole range $]1/2, 1]$. It is further easy to check that $m(A_n, k) = \binom{n}{2} - \binom{k}{2} + O(n)$.

In $A_{n,k}$, Algorithm 1 will first empty the clique K_{n-2k-1}, leaving us again with a subgraph isomorphic to B. The algorithm can therefore return a matching of size $(n-2k-1)/2+k = (n-1)/2$. On the other hand, taking a perfect matching in the clique K_{n-2k-1} together with the same $k/2+1$ edges described above for B yields a cover of size $(n - 2k - 1)/2 + k/2 + 1 = (n - k + 1)/2$. Therefore we have

$$\beta \geq \frac{(n-1)/2}{(n-k+1)/2} = \frac{n-1}{n-k+1}$$

Setting $m(A_n, k) = \binom{n}{2} - \binom{k}{2} + O(n)$ and $\epsilon = m/\binom{n}{2}$ as for $\zeta_{n,k}$ yields

$$\frac{n-1}{n-k+1} \rightarrow \left[1 - \sqrt{\frac{1-\epsilon}{2}}\right]^{-1} \quad \text{as } n \to \infty$$

Our lower bound on the ratio thus asymptotically matches the upper bound of Theorem 1. Note that we have made no special assumption on k as we had done for $\zeta_{n,k}$. The graphs $A_{n,k}$ with odd n and even $0 \leq k < n/2$ are therefore a family of asymptotically tight graphs for $\epsilon \geq 1/2$. Asymptotically tight examples for even n may be obtained by slight adaptation of the above graphs.

3   Conclusion

Several variants to Algorithm 1 could be devised. For example, one could decide to slightly alter Algorithm 1 by each time selecting the edge that has the highest degree in the original graph rather than the updated graph. This variant is interesting as it can easily be implemented in time $O(n + m)$ using counting sort. Another interesting variant is the one in which one does not select the highest degree edge, but rather the edge defined by the highest degree vertex and its highest degree neighbor. We claim that Theorem 1 remains valid for these two variants. One should first notice that the only place in our analysis where explicit use is made of the strategy for choosing an edge is in Lemma 1. It is almost straightforward to adapt its proof for both variants.

Further, it can be checked that the asymptotic bound of $1/\epsilon$ for graphs with minimum degree at least ϵn obtained for the maximal matching heuristic in [1] is also tight for Algorithm 1, with the same tight examples as those described in section 2.

Finally, Algorithm 1 also provides a 2-approximation for MINIMUM VERTEX COVER by taking the endpoints of the maximal matching returned by the algorithm. The ratio obtained in Theorem 1 is also valid for this problem by slight adaptations to the proofs. The analytical form of our asymptotic result compares interestingly with that of both the simplest [1] and the best known approximation algorithm for MINIMUM VERTEX COVER in ϵ-dense graphs [10] : $1/(1 - \sqrt{(1-\epsilon)/2})$ against $1/(1 - \sqrt{(1-\epsilon)})$ and $1/(1 - \sqrt{(1-\epsilon)/4})$. Fig. 6 compares these ratios.

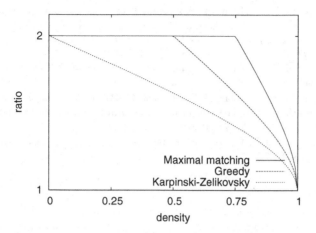

Fig. 6. A comparison of the ratios provided by the maximal matching heuristic, the greedy algorithm and Karpinski and Zelikovsky's algorithm

Acknowledgments. The authors wish to thank Martine Labbé, with whom this research was initiated, and Hadrien Mélot, author of the GraPHedron software [11], which was used to formulate the initial conjectures.

References

1. J. Cardinal, M. Labbé, S. Langerman, E. Levy, and H. Mélot. A tight analysis of the maximal matching heuristic. In *Proc. of The Eleventh International Computing and Combinatorics Conference (COCOON)*, LNCS, pages 701–709. Springer–Verlag, 2005.

2. R. Carr, T. Fujito, G. Konjevod, and O. Parekh. A 2 1/10-approximation algorithm for a generalization of the weighted edge-dominating set problem. *Journal of Combinatorial Optimization*, 5:317–326, 2001.

3. M. Chlebík and J. Chlebíková. Approximation hardness of edge dominating set problems. *Journal of Combinatorial Optimization*, 11(3):279–290, 2006.

4. A.V. Eremeev. On some approximation algorithms for dense vertex cover problem. In *Proc. of SOR*, pages 58–62. Springer–Verlag, 1999.

5. T. Fujito and H. Nagamochi. A 2-approximation algorithm for the minimum weight edge dominating set problem. *Discrete Appl. Math.*, 118:199–207, 2002.

6. M. R. Garey and D. S. Johnson. *Computers and intractability. A guide to the theory of NP-completeness.* Freeman and Company, 1979.

7. E. Halperin. Improved approximation algorithms for the vertex cover problem in graphs and hypergraphs. *Siam Journal on Computing*, 31:1608–1623, 2002.

8. J.D. Horton and K. Kilakos. Minimum edge dominating sets. *SIAM J. Discrete Math.*, 6:375–387, 1993.

9. T. Imamura and K. Iwama. Approximating vertex cover on dense graphs. In *Proc. of the 16th ACM-SIAM Symposium on Discrete Algorithms (SODA)*, pages 582–589, 2005.

10. M. Karpinski and A. Zelikovsky. Approximating dense cases of covering problems. In P. Pardalos and D. Du, editors, *Proc. of the DIMACS Workshop on Network Design: Connectivity and Facilites Location*, volume 40 of *DIMACS series in Disc. Math. and Theor. Comp. Sci.*, pages 169–178, 1997.
11. H. Mélot. Facets Defining Inequalities among Graph Invariants: the system GraPHedron. Submitted, 2005.
12. A. Srinivasan, K. Madhukar, P. Navagamsi, C. Pandu Rangan, and M.-S. Chang. Edge domination on bipartite permutation graphs and cotriangulated graphs. *Inf.Proc. Letters*, 56:165–171, 1995.
13. M. Yannakakis and F. Gavril. Edge dominating sets in graphs. *SIAM J. Appl. Math.*, 38(3):364–372, 1980.

A Randomized Algorithm for Online Unit Clustering*

Timothy M. Chan and Hamid Zarrabi-Zadeh

School of Computer Science, University of Waterloo
Waterloo, Ontario, Canada, N2L 3G1
{tmchan,hzarrabi}@uwaterloo.ca

Abstract. In this paper, we consider the online version of the following problem: partition a set of input points into subsets, each enclosable by a unit ball, so as to minimize the number of subsets used. In the one-dimensional case, we show that surprisingly the naïve upper bound of 2 on the competitive ratio can be beaten: we present a new randomized 15/8-competitive online algorithm. We also provide some lower bounds and an extension to higher dimensions.

1   Introduction

Clustering problems—dividing a set of points into groups to optimize various objective functions—are fundamental and arise in a wide variety of applications such as information retrieval, data mining, and facility location. We mention two of the most basic and popular versions of clustering:

Problem 1 (k-Center). *Given a set of n points and a parameter k, cover the set by k congruent balls, so as to minimize the radius of the balls.*

Problem 2 (Unit Covering). *Given a set of n points, cover the set by balls of unit radius, so as to minimize the number of balls used.*

Both problems are NP-hard in the Euclidean plane [10,19]. In fact, it is NP-hard to approximate the two-dimensional k-center problem to within a factor smaller than 2 [9]. Factor-2 algorithms are known for the k-center problem [9,11] in any dimension, while polynomial-time approximation schemes are known for the unit covering problem [14] in fixed dimensions.

Recently, many researchers have considered clustering problems in more practical settings, for example, in the online and data stream models [4,5,12], where the input is given as a sequence of points over time. In the online model, the solution must be constructed as points arrive and decisions made cannot be subsequently revoked; for example, in the unit covering problem, after a ball is opened to cover an incoming point, the ball cannot be removed later. In the related streaming model, the main concern is the amount of working space; as

---

* Work of the first author has been supported in part by NSERC.

T. Erlebach and C. Kaklamanis (Eds.): WAOA 2006, LNCS 4368, pp. 121–131, 2006.
© Springer-Verlag Berlin Heidelberg 2006

points arrive, we must decide which point should be kept in memory. We focus on the online setting in this paper.

The online version of the unit covering problem is one of the problems addressed in the paper by Charikar et al. [4]. They have given an upper bound of $O(2^d d \log d)$ and a lower bound of $\Omega(\frac{\log d}{\log \log \log d})$ on the competitive ratio of deterministic online algorithms in d dimensions; for $d = 1$ and 2, the lower bounds are 2 and 4 respectively.

In this paper, we address the online version of the following variant:

Problem 3 (Unit Clustering). *Given a set of n points, partition the set into clusters (subsets), each of radius at most one, so as to minimize the number of clusters used. Here, the* radius *of a cluster refers to the radius of its smallest enclosing ball.*

At first glance, Problem 3 might look eerily similar to Problem 2; in fact, in the usual offline setting, they are identical. However, in the on-line setting, there is one important difference: as a point p arrives, the unit clustering problem only requires us to decide on the choice of the cluster containing p, not the ball covering the cluster; the point cannot subsequently be reassigned to another cluster, but the position of the ball may be shifted.

We show that it is possible to get better results for Problem 3 than Problem 2. Interestingly we show that even in one dimension, the unit clustering problem admits a nontrivial algorithm with competitive ratio better than 2, albeit by using randomization. In contrast, such a result is not possible for unit covering. To be precise, we present an online algorithm for one-dimensional unit clustering that achieves expected competitive ratio 15/8 against oblivious adversaries. Our algorithm is not complicated but does require a combination of ideas and a careful case analysis. We contrast the result with a lower bound of 4/3 and also extend our algorithm for the problem in higher dimensions under the L_∞ metric.

We believe that the one-dimensional unit clustering problem itself is theoretically appealing because of its utter simplicity and its connection to well-known problems. For example, in the exact offline setting, one-dimensional unit clustering/covering is known to be equivalent to the dual problem of finding a largest subset of disjoint intervals among a given set of unit intervals—i.e., finding maximum independent sets in unit interval graphs. Higher-dimensional generalizations of this dual independent set problem have been explored in the map labeling and computational geometry literature [2,3,8], and online algorithms for various problems about geometric intersection graphs have been considered (such as [18]). The one-dimensional independent set problem can also be viewed as a simple scheduling problem (dubbed "activity selection" by Cormen et al. [6]), and various online algorithms about intervals and interval graphs (such as [1,7,16,17]) have been addressed in the literature on scheduling and resource allocation. In the online setting, one-dimensional unit clustering is equivalent to clique partitioning in unit interval graphs, and thus, equivalent to coloring in unit co-interval graphs. It is known that general co-interval graphs can be colored with competitive ratio at most 2 [13], and that, no online deterministic algorithm can beat this 2 bound [15]. To the best of our knowledge, however, online coloring of unit co-interval graphs has not been studied before.

2   Naïve Algorithms

In this section, we begin our study of the unit clustering problem in one dimension by pointing out the deficiencies of some natural strategies.

Recall that the goal is to assign points to clusters so that each cluster has length at most 1, where the *length* of a cluster refers to the length of its smallest enclosing interval. (Note that we have switched to using lengths instead of radii in one dimension; all intervals are closed.) We say that a point *lies* in a cluster if inserting it to the cluster would not increase the length of the cluster. We say that a point *fits* in a cluster if inserting it to the cluster would not cause the length to exceed 1. The following are three simple online algorithms, all easily provable to have competitive ratio at most 2:

Algorithm 1 (CENTERED). *For each new point p, if it is covered by an existing interval, put p in the corresponding cluster, else open a new cluster for the unit interval centered at p.*

Algorithm 2 (GRID). *Build a uniform unit grid on the line (where cells are intervals of the form $[i, i+1)$). For each new point p, if the grid cell containing p is nonempty, put p in the corresponding cluster, else open a new cluster for the grid cell.*

Algorithm 3 (GREEDY). *For each new point p, if p fits in some existing cluster, put p in such a cluster, else open a new cluster for p.*

The first two algorithms actually solve the stronger unit covering problem (Problem 2). No such algorithms can break the 2 bound, as we can easily prove:

Theorem 1. *There is a lower bound of 2 on the competitive ratio of any randomized (and deterministic) algorithm for the online unit covering problem in one dimension.*

Proof. To show the lower bound for randomized algorithms, we use Yao's technique and provide a probability distribution on the input sequences such that the resulting expected competitive ratio for any deterministic online algorithm is at least 2. The adversary provides a sequence of 3 points at position 1, x, and $1 + x$, where x is uniformly distributed in $[0, 1]$. The probability that a deterministic algorithm produces the optimal solution (of size 1 instead of 2 or more) is 0. Thus, the expected value of the competitive ratio is at least 2.   □

The 2 bound on the competitive ratio is also tight for Algorithm 3: just consider the sequence $\langle \frac{1}{2}, \frac{3}{2}, \ldots, 2k - \frac{1}{2} \rangle$ followed by $\langle 0, 2, \ldots, 2k \rangle$ (where the greedy algorithm uses $2k + 1$ clusters and the optimal solution needs only $k + 1$ clusters). No random combination of Algorithms 1–3 can lead to a better competitive ratio, as we can easily see by the same bad example. New ideas are needed to beat 2.

3   The New Algorithm

In this section, we present a new randomized algorithm for the online unit clustering problem. While the competitive ratio of this algorithm is not necessarily less than 2, the algorithm is carefully designed so that when combined with Algorithm 2 we get a competitive ratio strictly less than 2.

Our algorithm builds upon the simple grid strategy (Algorithm 2). To guard against a bad example like $\langle \frac{1}{2}, \frac{3}{2}, \ldots \rangle$, the idea is to allow two points in different grid cells to be put in a common cluster "occasionally" (as controlled by randomization). Doing so might actually hurt, not help, in many cases, but fortunately we can still show that there is a net benefit (in expectation), at least in the most critical case.

To implement this idea, we form *windows* each consisting of two grid cells and permit clusters crossing the two cells within a window but try to "discourage" clusters crossing two windows. The details of the algorithm are delicate and are described below. Note that only one random bit is used at the beginning.

Algorithm 4 (RANDWINDOW). *Group each two consecutive grid cells into a window of the form* $[2i, 2i+2)$. *With probability* $1/2$, *shift all windows one unit to the right. For each new point* p, *find the window* w *and the grid cell* c *containing* p, *and do the following:*

1:    **if** w is empty **then** open a new cluster for p
2:    **else if** p lies in a cluster **then** put p in that cluster
3:    **else if** p fits in a cluster entirely inside c **then** put p in that cluster
4:    **else if** p fits in a cluster intersecting w **then** put p in that cluster
5:    **else if** p fits in a cluster entirely inside a neighboring window w' and
6:         w' intersects > 1 clusters **then** put p in that cluster
7:    **else** open a new cluster for p

To summarize: the algorithm is greedy-like and opens a new cluster only if no existing cluster fits. The main exception is when the new point is the first point in a window (line 1); another exception arises from the (seemly mysterious) condition in line 6. When more than one cluster fits, the preference is towards clusters entirely inside a grid cell, and against clusters from neighboring windows. These exceptional cases and preference rules are vital to the analysis.

4   Analysis

For a grid cell (or a group of cells) x, the *cost* of x denoted by $\mu(x)$ is defined to be the number of clusters fully contained in x plus half the number of clusters crossing the boundaries of x, in the solution produced by our algorithm. Observe that μ is additive, i.e., for two adjacent groups of cells x and y, $\mu(x \cup y) = \mu(x) + \mu(y)$. This definition of cost will be useful for accounting purposes.

To prepare for the analysis, we first make several observations concerning the behavior of the RANDWINDOW algorithm. In the following, we refer to a cluster as a *crossing cluster* if it intersects two adjacent grid cells, or as a *whole cluster* if it is contained completely in a grid cell.

Observation 1

(i) *The enclosing intervals of the clusters are disjoint.*
(ii) *No grid cell contains two whole clusters.*
(iii) *If a grid cell c intersects a crossing cluster u_1 and a whole cluster u_2, then u_2 must be opened after u_1 has been opened, and after u_1 has become a crossing cluster.*

Proof. (i) holds because of line 2. (ii) holds because line 3 precedes line 7.

For (iii), let p_1 be the first point of u_1 in c and p'_1 be the first point of u_1 in a cell adjacent to c. Let p_2 be the first point of u_2. Among these three points, p_1 cannot be the last to arrive: otherwise, p_1 would be assigned to the whole cluster u_2 instead of u_1, because line 3 precedes lines 4–7. Furthermore, p'_1 cannot be the last to arrive: otherwise, p_1 would be assigned to u_2 instead, again because line 3 precedes lines 4–7. So, p_2 must be the last to arrive.   □

For example, according to Observation 1(ii), every grid cell c must have $\mu(c) \leq 1 + \frac{1}{2} + \frac{1}{2} = 2$.

Let σ be the input sequence and $\mathsf{opt}(\sigma)$ be an optimal covering of σ by unit intervals, with the property that the intervals are disjoint. (This property is satisfied by some optimal solution, simply by repeatedly shifting the intervals to the right.) We partition the grid cells into blocks, where each *block* is a maximal set of consecutive grid cells interconnected by the intervals from $\mathsf{opt}(\sigma)$ (see Fig. 1). Our approach is to analyze the cost of the solution produced by our algorithm within each block separately.

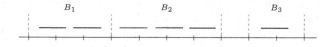

Fig. 1. Three blocks of sizes 2, 3, and 1

A block of size $k \geq 2$ contains exactly $k-1$ intervals from $\mathsf{opt}(\sigma)$. Define $\rho(k)$ to be the competitive ratio of the RANDWINDOW algorithm within a block of size k, i.e., $\rho(k)$ upper-bounds the expected value of $\mu(B)/(k-1)$ over all blocks B of size k. The required case analysis is delicate and is described in detail below. The main case to watch out for is $k = 2$: any bound for $\rho(2)$ strictly smaller than 2 will lead to a competitive ratio strictly smaller than 2 for the final algorithm (as we will see in Section 5), although bounds for $\rho(3), \rho(4), \ldots$ will affect the final constant.

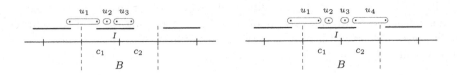

Fig. 2. Impossibility of Subcase 1.1 (left) and Subsubcase 1.3.2 (right)

Theorem 2. $\rho(2) = 7/4$, $\rho(3) = 9/4$, $\rho(4) \leq 7/3$, and $\rho(k) \leq 2k/(k-1)$ for all $k \geq 5$.

Proof. We first analyze $\rho(2)$. Consider a block B of size 2, consisting of cells c_1 and c_2 from left to right. Let I be the single unit interval in B in $\mathsf{opt}(\sigma)$. There are two possibilities:

- LUCKY CASE: B falls completely in one window w. After a cluster u has been opened for the new point (by line 1), all subsequent points in I are put in the same cluster u (by lines 3 and 4). Note that the condition put in line 6 prevents points from the neighboring windows to join u and make crossing clusters. So, u is the only cluster in B, and hence, $\mu(B) = 1$.
- UNLUCKY CASE: B is split between two neighboring windows. We first rule out some subcases:
  - SUBCASE 1.1: $\mu(c_1) = 2$. Here, c_1 intersects three clusters $\langle u_1, u_2, u_3 \rangle$ (from left to right), where u_1 and u_3 are crossing clusters and u_2 is a whole cluster (see Fig. 2, left). By Observation 1(iii), u_2 is opened after u_3 has become a crossing cluster, but then the points of u_2 would be assigned to u_3 instead (because line 4 precedes line 7 and $u_2 \cup u_3 \subset I$ has length at most 1): a contradiction.
  - SUBCASE 1.2: $\mu(c_2) = 2$. Similarly impossible.
  - SUBCASE 1.3: $\mu(c_1) = \mu(c_2) = 3/2$. We have only two scenarios:
    * SUBSUBCASE 1.3.1: B intersects three clusters $\langle u_1, u_2, u_3 \rangle$, where u_2 is a crossing cluster, and u_1 and u_3 are whole clusters. By Observation 1(iii), u_1 is opened after u_2 has become a crossing cluster, but then the points of u_1 would be assigned to u_2 instead (because of line 4 and $u_1 \cup u_2 \subset I$): a contradiction.
    * SUBSUBCASE 1.3.2: B intersects four clusters $\langle u_1, u_2, u_3, u_4 \rangle$, where u_1 and u_4 are crossing clusters and u_2 and u_3 are whole clusters (see Fig. 2, right). W.l.o.g., say u_2 is opened after u_3. By Observation 1(iii), u_2 is the last to be opened after u_1, u_3, u_4, but then u_2 would not be opened as points in u_2 may be assigned to u_3 (because lines 5–6 precedes line 7, $u_2 \cup u_3 \subset I$, and c_2 intersects more than one cluster): a contradiction.

In all remaining subcases, $\mu(B) = \mu(c_1) + \mu(c_2) \leq \frac{3}{2} + 1 = \frac{5}{2}$.

Since the lucky case occurs with probability exactly $1/2$, we conclude that $\rho(2) \leq \frac{1}{2}(1) + \frac{1}{2}(\frac{5}{2}) = \frac{7}{4}$. (This bound is tight.)

Fig. 3. Impossibility of Cases 2.1 (left) and 2.2 (right)

Next, we analyze $\rho(3)$. Consider a block B of size 3, consisting of cells c_1, c_2, c_3 from left to right. (It will not matter below whether c_1 and c_2 fall in the same window, or c_2 and c_3 instead.) Let I_1, I_2 be the two unit intervals in B in $\mathrm{opt}(\sigma)$ from left to right.

- CASE 2.1: $\mu(c_2) = 2$. Here, c_2 intersects three clusters $\langle u_1, u_2, u_3 \rangle$ (from left to right), where u_1 and u_3 are crossing clusters and u_2 is a whole cluster (see Fig. 3, left). By Observation 1(iii), u_2 is opened after u_1 and u_3 have become crossing clusters, but then the points of u_2 would be assigned to u_1 or u_3 instead (because of line 4 and $u_1 \cup u_2 \cup u_3 \subset I_1 \cup I_2$): a contradiction.
- CASE 2.2: $\mu(c_1) = \mu(c_3) = 2$. Here, c_1 intersects three clusters $\langle u_1, u_2, u_3 \rangle$ and c_3 intersects three clusters $\langle u_4, u_5, u_6 \rangle$ (from left to right), where u_1, u_3, u_4, u_6 are crossing clusters and u_2, u_5 are whole clusters (see Fig. 3, right). Then u_3 cannot be entirely contained in I_1: otherwise, by Observation 1(iii), u_2 is opened after u_1 and u_3 have become crossing clusters, but then the points of u_2 would be assigned to u_3 instead. Similarly, u_4 cannot be entirely contained in I_2. However, this implies that the enclosing intervals of u_3 and u_4 overlap: a contradiction.
- CASE 2.3: $\mu(c_1) = 2$ and $\mu(c_2) = \mu(c_3) = 3/2$. Here, B intersects six clusters $\langle u_1, \ldots, u_6 \rangle$ (from left to right), where u_1, u_3, u_6 are crossing clusters and u_2, u_4, u_5 are whole clusters. As in Case 2.2, u_3 cannot be entirely contained in I_1. This implies that $u_4 \cup u_5 \subset I_2$. We now proceed as in Subcase 1.3.2. Say u_4 is opened after u_5 (the other scenario is symmetric). By Observation 1(iii), u_4 is the last to be opened after u_3, u_5, u_6, but then u_4 would not be opened as points in u_4 may be assigned to u_5: a contradiction.
- CASE 2.4: $\mu(c_1) = \mu(c_2) = 3/2$ and $\mu(c_3) = 2$. Similarly impossible.

In all remaining subcases, $\mu(B) = \mu(c_1) + \mu(c_2) + \mu(c_3)$ is at most $2 + \frac{3}{2} + 1 = \frac{9}{2}$ or $\frac{3}{2} + \frac{3}{2} + \frac{3}{2} = \frac{9}{2}$. We conclude that $\rho(3) \le 9/4$. (This bound is tight.)

Now, we analyze $\rho(4)$. Consider a block B of size 4, consisting of cells $c_1, \ldots, c_4$ from left to right. Let I_1, I_2, I_3 be the three unit intervals in B in $\mathrm{opt}(\sigma)$ from left to right.

- CASE 3.1: $\mu(c_1) = \mu(c_3) = 2$. Here, c_1 intersects three clusters $\langle u_1, u_2, u_3 \rangle$ and c_3 intersects three clusters $\langle u_4, u_5, u_6 \rangle$ (from left to right), where u_1, u_3, u_4, u_6 are crossing clusters and u_2, u_5 are whole clusters. As in Case 2.2, u_3 cannot be entirely contained in I_1. Thus, $u_4 \cup u_5 \cup u_6 \subset I_2 \cup I_3$. We now proceed as in Case 2.1. By Observation 1(iii), u_5 is opened after u_4 and u_6

have become crossing clusters, but then the points of u_5 would be assigned to u_4 or u_6 instead: a contradiction.

- CASE 3.2: $\mu(c_2) = \mu(c_4) = 2$. Similarly impossible.

In all remaining subcases, $\mu(B) = (\mu(c_1) + \mu(c_3)) + (\mu(c_2) + \mu(c_4)) \leq (2 + \frac{3}{2}) + (2 + \frac{3}{2}) \leq 7$. We conclude that $\rho(4) \leq 7/3$.

For $k \geq 5$, we use a rather loose upper bound. Consider a block B of size k. As each cell c has $\mu(c) \leq 2$, we have $\mu(B) \leq 2k$, and hence $\rho(k) \leq 2k/(k-1)$. □

5   The Combined Algorithm

We can now combine the RANDWINDOW algorithm (Algorithm 4) with the GRID algorithm (Algorithm 2) to obtain a randomized online algorithm with competitive ratio strictly less than 2. Note that only two random bits in total are used at the beginning.

Algorithm 5 (COMBO). *With probability* 1/2, *run* RANDWINDOW, *else run* GRID.

Theorem 3. COMBO *is* 15/8-*competitive (against oblivious adversaries).*

Proof. The GRID algorithm uses exactly k clusters on a block of size k. Therefore, the competitive ratio of this algorithm within a block of size k is $k/(k-1)$.

The following table shows the competitive ratio of the RANDWINDOW, GRID, and COMBO algorithms, for all possible block sizes.

Table 1. The competitive ratio of the algorithms within a block

Block Size	2	3	4	$k \geq 5$
GRID	2	3/2	4/3	$k/(k-1)$
RANDWINDOW	7/4	9/4	$\leq 7/3$	$\leq 2k/(k-1)$
COMBO	15/8	15/8	$\leq 11/6$	$\leq 3/2 \cdot k/(k-1)$

As we can see, the competitive ratio of COMBO within a block is always at most 15/8. By summing over all blocks and exploiting the additivity of our cost function μ, we see that expected total cost of the solution produced by COMBO is at most 15/8 times the size of $\mathrm{opt}(\sigma)$ for every input sequence σ. □

We complement the above result with a quick lower bound argument:

Theorem 4. *There is a lower bound of* 4/3 *on the competitive ratio of any randomized algorithm for the online unit clustering problem in one dimension (against oblivious adversaries).*

Proof. We use Yao's technique. Consider two point sequences $P_1 = \langle 1, 2, \frac{1}{2}, \frac{5}{2} \rangle$ and $P_2 = \langle 1, 2, \frac{3}{2}, \frac{3}{2} \rangle$. With probability 2/3 the adversary provides P_1, and with probability 1/3 it provides P_2. Consider a deterministic algorithm $\mathcal{A}$. Regardless of which point sequence is selected by the adversary, the first two points provided to $\mathcal{A}$ are the same. If $\mathcal{A}$ clusters the first two points into one cluster, then it uses 3 clusters for P_1 and 1 cluster for P_2, giving the expected competitive ratio of $\frac{2}{3}(\frac{3}{2}) + \frac{1}{3}(1) = \frac{4}{3}$. If $\mathcal{A}$ clusters the first two points into two distinct clusters, then no more clusters are needed to cover the other two points of P_1 and P_2. Thus, the expected competitive ratio of $\mathcal{A}$ in this case is $\frac{2}{3} \cdot (1) + \frac{1}{3} \cdot (2) = \frac{4}{3}$ as well.   □

6   Beyond One Dimension

In the two-dimensional L_∞-metric case, we want to partition the given point set into subsets, each of L_∞-diameter at most 1 (i.e., each enclosable by a unit square), so as to minimize the number of subsets used. (See Fig. 4.)

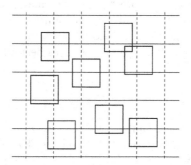

Fig. 4. Unit clustering in the L_∞ plane

All the naïve algorithms mentioned in Section 2, when extended to two dimensions, provide 4-competitive solutions to the optimal solution. Theorem 1 can be generalized to a deterministic lower bound of 4 on the competitive ratio for the unit covering problem. We show how to extend Theorem 3 to obtain a competitive ratio strictly less than 4 for unit clustering.

Theorem 5. *There is a 15/4-competitive algorithm for the online unit clustering problem in the L_∞ plane.*

Proof. Our online algorithm is simple: just use COMBO to find a unit clustering C_i for the points inside each horizontal strip $i \leq y < i + 1$. (Computing each C_i is indeed a one-dimensional problem.)

Let σ be the input sequence. We denote by σ_i the set of points from σ that lie in the strip $i \leq y < i + 1$. Let Z_i be an optimal unit covering for σ_i. Let O be an optimal unit covering for σ, and O_i be the set of unit squares in O that intersect the grid line $y = i$. Since all squares in O_i lie in the strip $i-1 \leq y < i+1$, we have $|Z_i| \leq |O_{i-1}| + |O_i|$. Therefore $\sum_i |Z_i| \leq 2|O|$, so $\sum_i |C_i| \leq \frac{15}{8} \sum_i |Z_i| \leq \frac{15}{4}|O|$. □

The above theorem can easily be extended to dimension $d > 2$, with ratio $2^d \cdot 15/16$.

7   Closing Remarks

We have shown that determining the best competitive ratio for the online unit clustering problem is nontrivial even in the simplest one-dimensional case. The obvious open problem is to close the gap between the 15/8 upper bound and 4/3 lower bound. An intriguing possibility that we haven't ruled out is whether a nontrivial result can be obtained without randomization at all. There is an obvious 3/2 deterministic lower bound, but we do not see any simple argument that achieves a lower bound of 2.

We wonder if ideas that are more "geometric" may lead to still better results than Theorem 5. Our work certainly raises countless questions concerning the best competitive ratio in higher-dimensional cases, for other metrics besides L_∞, and for other geometric measures of cluster sizes besides radius or diameter.

References

1. U. Adamy and T. Erlebach. Online coloring of intervals with bandwidth. In *Proc. 1st Workshop Approx. Online Algorithms*, volume 2909 of *Lecture Notes Comput. Sci.*, pages 1–12, 2003.
2. P. K. Agarwal, M. van Kreveld, and S. Suri. Label placement by maximum independent set in rectangles. *Comput. Geom. Theory Appl.*, 11:209–218, 1998.
3. T. M. Chan. Polynomial-time approximation schemes for packing and piercing fat objects. *J. Algorithms*, 46:178–189, 2003.
4. M. Charikar, C. Chekuri, T. Feder, and R. Motwani. Incremental clustering and dynamic information retrieval. *SIAM J. Comput.*, 33(6):1417–1440, 2004.
5. M. Charikar, L. O'Callaghan, and R. Panigrahy. Better streaming algorithms for clustering problems. In *Proc. 35th ACM Sympos. Theory Comput.*, pages 30–39, 2003.
6. T. H. Cormen, C. E. Leiserson, R. L. Rivest, and C. Stein. *Introduction to Algorithms*. MIT Press, Cambridge, MA, 2nd edition, 2001.
7. L. Epstein and M. Levy. Online interval coloring and variants. In *Proc. 32nd International Colloquium on Automata, Languages, and Programming (ICALP)*, volume 3580 of *Lecture Notes Comput. Sci.*, pages 602–613, 2005.
8. T. Erlebach, K. Jansen, and E. Seidel. Polynomial-time approximation schemes for geometric intersection graphs. *SIAM J. Comput.*, 34:1302–1323, 2005.
9. T. Feder and D. H. Greene. Optimal algorithms for approximate clustering. In *Proc. 20th ACM Sympos. Theory Comput.*, pages 434–444, 1988.
10. R. J. Fowler, M. S. Paterson, and S. L. Tanimoto. Optimal packing and covering in the plane are NP-complete. *Inform. Process. Lett.*, 12(3):133–137, 1981.

11. T. Gonzalez. Covering a set of points in multidimensional space. *Inform. Process. Lett.*, 40:181–188, 1991.
12. S. Guha, N. Mishra, R. Motwani, and L. O'Callaghan. Clustering data streams. In *Proc. 41st IEEE Sympos. Found. Comput. Sci.*, pages 359–366, 2000.
13. A. Gyárfás and J. Lehel. On-line and First-Fit colorings of graphs. *J. Graph Theory*, 12:217–227, 1988.
14. D. S. Hochbaum and W. Maas. Approximation schemes for covering and packing problems in image processing and VLSI. *J. ACM*, 32:130–136, 1985.
15. H. A. Kierstead and J. Qin. Coloring interval graphs with First-Fit. *SIAM J. Discrete Math.*, 8:47–57, 1995.
16. H. A. Kierstead and W. A. Trotter. An extremal problem in recursive combinatorics. *Congressus Numerantium*, 33:143–153, 1981.
17. R. J. Lipton and A. Tomkins. Online interval scheduling. In *Proc. 5th Sympos. Discrete Algorithms*, pages 302–311, 1994.
18. M. V. Marathe, H. Breu, H. B. Hunt III, S. S. Ravi, and D. J. Rosenkrantz. Simple heuristics for unit disk graphs. *Networks*, 25:59–68, 1995.
19. N. Megiddo and K. J. Supowit. On the complexity of some common geometric location problems. *SIAM J. Comput.*, 13(1):182–196, 1984.

On Hierarchical Diameter-Clustering, and the Supplier Problem

Aparna Das and Claire Kenyon

Brown University, Providence RI 02918, USA

Abstract. Given a data set in metric space, we study the problem of hierarchical clustering to minimize the maximum cluster diameter, and the hierarchical k-supplier problem with customers arriving online.

We prove that two previously known algorithms for hierarchical clustering, one (offline) due to Dasgupta and Long and the other (online) due to Charikar, Chekuri, Feder and Motwani, are essentially the same algorithm when points are considered in the same order. We show that the analysis of both algorithms are tight and exhibit a new lower bound for hierarchical clustering. Finally we present the first constant factor approximation algorithm for the online hierarchical k-supplier problem.

1 Introduction

Clustering is the partitioning of data points into disjoint clusters (or groups) according to similarity [1,10]. For example if the data points are books, a two clustering might consist of the clusters fiction, and non-fiction. In this way clustering can provide a concise view of large amounts of data. In many application domains it is useful to build a partitioning of the data that starts with broad categories which are gradually refined thus allowing the the data to be viewed simultaneously at different levels of conciseness. This calls for a *hierarchical* or nested clustering of the data where clusters have subclusters, these have sub-subclusters, and so on. For example a hierarchical clustering might first separate the books into clusters fiction and non-fiction, then separate the fiction cluster into classics and non-classics and the non-fiction cluster into math, science and history, and so on. More formally, a *hierarchical clustering* of n data points is a recursive partitioning of the points into $1, 2, 3, 4, \ldots, n$ clusters such that the $(k+1)$th clustering is obtained by dividing one of the clusters of the kth clustering into two parts, thus making the clustering gradually more fine-grained ([5], Section 10.9). This framework has long been popular among statisticians, biologists (particularly taxonomists) and social scientists [11].

A criteria commonly used to measure of the quality of a clustering is the maximum cluster diameter, where the diameter of a cluster is the distance between the two farthest points in the cluster. The goal is to find clusterings which minimize the maximum cluster diameter, thus similar points are placed in the same cluster while dissimilar points are separated. In this paper, we focus on the hierarchical diameter-clustering problem: finding a hierarchical clustering where the value of the clustering is the maximum cluster diameter.

T. Erlebach and C. Kaklamanis (Eds.): WAOA 2006, LNCS 4368, pp. 132–145, 2006.

Every associated k-clustering of the hierarchical clustering should be close to the optimal k-clustering, where the optimal k-clustering is the one that minimizes the maximum cluster diameter. The competitive ratio of a hierarchical clustering algorithm A is the supremum, over n and over input sets S of size n, of the quantity $\max_{k \in [1,n]} A_k(S)/\mathrm{OPT}_k(S)$, where $\mathrm{OPT}_k(S)$ is the value of the optimal k-clustering of S and $A_k(S)$ is the value[1] of the k-clustering constructed by algorithm A. Thus a hierarchical clustering algorithm with a small competitive ratio, produces k-clusterings which are close to the optimal for all $1 \leq k \leq n$.

The hierarchical diameter-clustering problem was studied in work by Dasgupta and Long [4] and by Charikar, Chekuri, Feder and Motwani [2]. A simple and commonly used algorithm for this problem is the greedy "agglomerative" algorithm [5], which starts with n singletons clusters and repeatedly merges the two clusters whose union has smallest diameter. However, it is proved in [4] that this algorithm has competitive ratio $\Omega(\log k)$. The authors then propose a better, constant-factor algorithm, inspired by the "divisive" k-clustering algorithm of Gonzales [7]. The algorithm proposed in [2] is instead "coalescent" and may be partially inspired by a k-clustering algorithm by Hochbaum and Shmoys [9]. Superficially the two papers look quite different. Quoting [4]: "the earlier work of [2] uses similar techniques for a loosely related problem, and achieves the same bounds". Indeed, both papers present a 8 competitive deterministic algorithm and a 2e competitive randomized variant. Additionally, the algorithm from [2] focuses on online clustering, where points arrive one by one in an arbitrary sequence. We refer to the algorithm from [2] as the *tree-doubling algorithm* and to the algorithm from [4] as the *farthest algorithm*. Here are the main results from [4,2].

Theorem 1. *For the hierarchical diameter-clustering problem,*
The farthest algorithm is 8-competitive, in its deterministic form and 2e-competitive in its randomized form [4].
The tree doubling algorithm is 8-competitive, in its deterministic form and 2e-competitive in its randomized form [2].

Our first contribution is to formally relate the two algorithms. Their specification contains some non-deterministic choices: the farthest algorithm starts from an arbitrary point, and the tree-doubling algorithm considers the points in arbitrary order. Assuming some conditions which remove the non-determinism, we prove that both in the deterministic and in the randomized cases the clustering produced by the farthest algorithm is always a refinement of the clustering produced by the tree-doubling algorithm, where refinement is defined as follows:

Definition 1. *A partition $F_1, F_2, \ldots F_l$ is a refinement of a partition $D_1, D_2, \ldots D_k$ iff $\forall i \leq l, \exists j \leq k$ such that $F_i \subseteq D_j$.*

Interestingly, both algorithms could actually be viewed as a coarser version of the greedy agglomerative algorithm used in practice.

---

[1] If A is randomized, then $A_k(S)$ should be replaced by $E(A_k(S))$.

Theorem 2 (Refinement). *Assume that the first two points labeled by the farthest algorithm have distance equal to the diameter of the input. Also assume that the tree-doubling algorithm considers points in the order in which they were labeled by the farthest algorithm. Moreover, in the randomized setting, assume that the two algorithms choose the same random value r.*

Then, for every k, the k-clustering produced by the farthest (deterministic or randomized) algorithm is a refinement of the k-clustering produced by the tree-doubling (deterministic or randomized) algorithm.

With this interpretation, we see that the competitive ratio of the farthest algorithm can be seen as a corollary of the competitive ratio of the tree-doubling algorithm. Could it be that the farthest algorithm is actually better? We answer this question in the negative by proving that the analysis of the farthest algorithm in [4] is tight.

Theorem 3 (Tightness). *The competitive ratio of the deterministic farthest algorithm is at least 8.*

This means that the 8 competitive ratio upper bound for the farthest algorithm is tight, and, by the refinement theorem, the 8 competitive ratio upper bound for the tree-doubling algorithm is also tight. Proving tightness of the randomized variants are open.

Can the competitive ratio be improved? We turn to the question of what is the best competitive ratio achievable for any hierarchical clustering algorithm with no computational restrictions. In other words, what is the best we can expect from a hierarchical clustering algorithm if it is allowed to have non-polynomial running time. We prove that no deterministic algorithm can achieve a competitive ratio better than 2, and no randomized algorithm can achieve competitive ratio better than $(3/2)$. (Note that the lower bounds proved in [2] apply to the online model only and thus are incomparable to our lower bounds.)

Theorem 4 (Hierarchical lower bound). *No deterministic (respectively randomized) hierarchical clustering algorithm can have competitive ratio better than 2 (respectively better than 3/2), even with unbounded computational power.*

How general are these techniques? In our final contribution, we extend the tree-doubling algorithm to design the first constant factor approximation algorithm for the *online hierarchical supplier problem*.

In the standard (offline, non hierarchical) k-*supplier problem*, we are given a set S of suppliers and a set C of customers, with customer-supplier distances. We wish to select a set S_k of k suppliers and an assignment of each customer c to a supplier $f(c)$ in S_k so as to minimize the maximum distance from any customer to its supplier, $\max_{c \in C} d(c, f(c))$. For example, the suppliers are the fixed database templates against which we are comparing the data (customers) and which we use for classification. A 3-approximation algorithm for the k-supplier problem is mentioned in [8].

In the more difficult *online hierarchical* setting, the set S of suppliers is known in advance but new customers arrive as time goes on, so C is a *sequence* of customers. When a new customer arrives, it is either assigned to one of the existing open suppliers, or it opens a new supplier. If opening a new supplier results in more than k open suppliers then two existing open suppliers merge their customer lists, and one of them closes. This requirement ensures that the hierarchical condition is satisfied, i.e that $S_{i-1} \subseteq S_i$ and that for each supplier $s \in S_i \setminus S_{i-1}$, all the customers assigned to s are assigned to the same supplier in S_{i-1}. For example, suppose customers arrive over time to use resources and we would like to dynamically increase/decrease the total number of resources allocated without having to do extensive recomputation. Using the hierarchical supplier solution, this only requires splitting/merging the customers currently assigned to one of the resources. The online hierarchical model is an increasingly important framework for clustering problems, when large amounts of data are gathered over time and needs to be categorized on the fly (see [13] for example).

Using the tree-doubling algorithm as a subroutine, we obtain a constant-factor approximation algorithm for the online hierarchical supplier problem. (Note that in the offline case, we could equivalently have used the farthest algorithm as a subroutine. In fact, we conjecture that in the offline case, a similar result may also be obtainable using methods from [3,12].)

Theorem 5 (Online hierarchical supplier). *For the online hierarchical k-supplier problem, there exist a deterministic 17-approximation algorithm and a randomized $(1 + 4e) = 11.87$-approximation algorithm.*

2   Proof of the Refinement Theorem

2.1   Review of the Farthest Algorithm from [4]

The input is a set of n points $\{x_1, \ldots x_n\}$ with associated distance metric d. The algorithm has three main steps:

Labeling the points. Take an arbitrary point and label it 1. Give label i for $i \in \{2, \ldots, n\}$, to the point which is farthest away from the previously labeled points. Let d_i denote the distance from i to the previous $i - 1$ labeled points, i.e $d_i = \min_{1 \leq j \leq i-1} d(, so these are the directions : i, j)$. Thus $d_2 = d(1, 2)$.

Assigning levels to labelled points. For labelled point 1, set level(1) $= 0$. For labelled point $i \in \{2, \ldots, n\}$, set level(i) $= \lfloor \log_2(d_2/d_i) \rfloor + 1$.

Organizing labelled points into a tree. Organize the points into a tree referred to as the Π'-tree. Place point 1 as the root Π'-tree. For each point $i \in \{2, \ldots, n\}$, define its parent, $\pi'(i)$, to be the point closest to i among the points with level strictly less than level(i). Insert points $i > 1$ into Π'-tree in order of increasing levels connecting each point i with an edge to its parent $\pi'(i)$.

The hierarchical clustering is represented implicitly in the Π'-tree. To obtain the k-clustering (of the hierarchical clustering) remove edges $(i, \pi'(i))$, for $i \in \{2, \ldots, k\}$ from the Π'-tree. Deleting these $k - 1$ edges splits the Π'-tree into k

connected components such that points $\{1, \ldots, k\}$ are in separate components. The components are returned as the k clusters.

It is easy to verify that this defines a hierarchical clustering, and [4] proves that it satisfies the following properties. The distances $(d_i)_i$ are a monotone non-increasing sequence, and the levels $(\text{level}(i))_i$ are a monotone non-decreasing sequence. The definition of levels imply the following bounds on d_i.

$$d_2/2^{\text{level}(i)} < d_i \le d_2/2^{\text{level}(i)-1} . \tag{1}$$

In addition [4] proves that:

$$d(i, \pi'(i)) \le d_2/2^{\text{level}(i)-1} . \tag{2}$$

[4] also present a randomized variant of the farthest algorithm, where the only difference is in the definition of levels. A value r is chosen uniformly at random from the interval $[0, 1]$, and the levels are now defined by: $\text{level}(1) = 0$ and $\text{level}(i) = \lfloor \ln(d_2/d_i) + r \rfloor + 1$. The monotonicity properties are unchanged; and the two inequalities are replaced by the following.

$$e^r d_2/e^{\text{level}(i)} < d_i \le e^r d_2/e^{\text{level}(i)-1} \quad \text{and} \quad d(i, \pi'(i)) \le e^r d_2/e^{\text{level}(i)-1} . \tag{3}$$

2.2   Review of the Tree-Doubling Algorithm from [2]

Here the input consists of a *sequence* of n points $\{x_1, \ldots x_n\}$ with associated distance metric d. Let Δ denote the diameter of the points. The algorithm considers the points one by one in an online fashion and maintains a certain infinite rooted tree which we refer to as the T^+ tree. Each node in T^+ is associated to a point, and the set of nodes associated to the same point forms an infinite path in the tree. The first point is placed at depth 0 as the root of T^+, and a copy of this point is placed at each depth $d > 0$ along with a parent edge to the copy at depth $d - 1$. When a new point p arrives it is inserted at a depth d_p, as defined by the insertion rule given below. A copy of p is placed at each depth $d > d_p$ with a parent edge to the copy of p at depth $d - 1$.

(Insertion rule). *Find the largest depth d with a point q such that $dist(p, q) \le \Delta/2^d$. Point p is inserted into depth $d_p = d + 1$ with a parent edge to q.*

To obtain a k-clustering, find the maximum depth d in T^+ which has at most k nodes. Delete all tree nodes at depth less than d. This leaves $\le k$ subtrees rooted at the points at depth d. Delete all multiple copies of points from the subtrees and return these as the clusters.

By [2], the following properties are maintained as nodes are added to T^+ :

Property 1 (**Close-parent property**). Points at depth d are at distance at most $\Delta/2^{d-1}$ from their parents.

Property 2 (**Far-cousins property**). Points at depth d are at distance greater than $\Delta/2^d$ from one another.

[2] also presents a randomized variant, where the only difference is in the insertion rule. A value r is chosen uniformly at random from the interval $[0,1]$, and the insertion rule is now: *Find the largest depth d that contains a point q such that $dist(p,q) \leq e^r \Delta / e^d$. Point p is inserted into depth $d_p = d+1$ with a parent edge to q.*

The properties are replaced by the following.

Property 3. Points at depth d are at distance at most $e^r \Delta / e^{d-1}$ from their parents, and at distance greater than $e^r \Delta / e^d$ from one another.

2.3   Proof of the Refinement Theorem, Deterministic Version

To relate the farthest and tree doubling algorithms we first make some assumptions about their nondeterministic choices. The farthest algorithm starts its labelling at an arbitrary point. We will assume the first point labelled by the farthest algorithm is at distance Δ from the second point labelled by the algorithm and thus $d_2 = \Delta$. The tree doubling algorithm receives its input points in an arbitrary order. We assume that points arrive to the tree-doubling algorithm in the order they are labeled by the farthest algorithm. Lastly we assume that ties are broken in the same way by the two algorithms. Specifically if points q, q' are tied to be a parent, the point with the larger label (in the farthest algorithm) or the point which arrived first (in the tree doubling algorithm) is favored.

Our proof is based on the alternative construction of the tree doubling T^+ tree, given in algorithm 1, which builds a tree T based on the farthest algorithm's Π' tree. We prove that our construction is consistent with the tree doubling algorithm's insertion rule and hence T could have legitimately been constructed by the tree-doubling algorithm. Finally we argue that the k-clustering defined by the Π' tree is a refinement of the k-clustering defined by T. Given the Π' tree of the farthest algorithm, the following algorithm constructs a T^+ tree. Fig. 1 shows the Π' tree and the corresponding tree T constructed by algorithm 1.

Algorithm 1: Given Π' construct T^+

(1)	Let T be an empty tree
(2)	Let ℓ = the maximum level of the points in Π'
(3)	**For** each level $i = 0, \ldots \ell$
(4)	Let L_i denote the points with level i
(5)	Let $S = L_0 \cup L_1 \cup \ldots \cup L_i$
(6)	Insert each $p \in S$ at depth i of T with an edge to:
(7)	The copy of $\pi'(p)$ at depth $i - 1$, if $level(p) = i$ or
(8)	The copy of p at depth $i - 1$, if $level(p) \neq i$
(9)	**Return** T

Let T be the tree constructed by Algorithm 1.

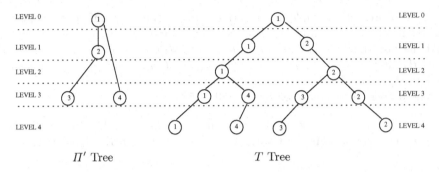

Π' Tree T Tree

Fig. 1. Π'-Tree to C^+ Tree

Lemma 1. T *satisfies the close-parent property.*

Proof. Let p be any point at depth d in T.

If $d \neq \text{level}(p)$, then by step (8) of algorithm 1, the parent of p in T is the copy of p at depth $d - 1$. Thus the close parent property follows trivially in this case since $d(p, p) = 0$.

Otherwise, $d = \text{level}(p)$. By step (7) of algorithm 1 $\text{parent}(p) = \pi'(p)$. Applying equation 2 with $d_2 = \Delta$, we have that, $d(p, \pi'(p)) \leq \Delta/2^{\text{level}(p)-1}$. Substituting $\text{level}(p) = d$ and $\pi'(p) = \text{parent}(p)$, we get: $d(p, \text{parent}(p)) \leq \Delta/2^{d-1}$.

Lemma 2. T *satisfies the insertion rule.*

Proof. By steps (3-6) of algorithm 1 if $\text{level}(p) = d$, then p appears in T for the first time at depth d and its parent q is $\pi'(p)$.

Since $\text{level}(\pi'(p)) < \text{level}(p)$, a copy of $\pi'(p)$ must be at depth $d - 1$ in T. Since T satisfies the close-parent property, $d(p, \pi'(p)) \leq \Delta/2^{d-1}$. Thus $\pi'(p)$ is qualified (distance-wise) to be the parent of p according to the insertion rule.

To show that insertion rule is satisfied we need to show that when p first arrives, there was no other point at a depth higher than $d - 1$ which was close enough to p to be its parent. Let q' be any point at depth $j > d - 1$, which arrived before p. Note that $q' \in \{1, \ldots, p-1\}$ since by assumption points arrive in the order they are labelled by the farthest algorithm. We need to show that $d(p, q') > \Delta/2^j$. Note that by definition of d_p, $d(p, q') \geq \min_{j \in [1, p-1]} d(p, j) = d_p$. Using the fact that $\text{level}(p) = d$ and equation 1 with $d_2 = \Delta$ we get

$$d_p > \Delta/2^{\text{level}(p)} = \Delta/2^d .$$

Combining the two statements above we have that

$$d(p, q') \geq d_p > \Delta/2^d \geq \Delta/2^j ,$$

where the last inequality follows since $j > d - 1 \Rightarrow j \geq d$. Since $d(p, q') > \Delta/2^j$ point q' cannot be parent of p.

We have shown that T satisfies the insertion rule and thus it can be constructed by the tree-doubling algorithm when the assumptions of Theorem 2 hold. Thus for the rest of the proof assume that the tree doubling algorithm constructs T.

Given k, the farthest algorithm removes exactly $k-1$ edges from the Π' tree and returns a clustering $F(k)$ with exactly k clusters. The tree-doubling algorithm looks for the deepest level of the T tree with at most k nodes and thus returns a clustering $D(k)$ with $\leq k$ clusters. We first show the two clusterings $D(k)$ and $F(k)$ are equivalent when they both have exactly k clusters. The refinement property then follows easily.

Lemma 3. *Let k be such that the tree doubling tree T has a depth d with exactly k vertices, then the clusterings $F(k)$ and $D(k)$ are the same.*

Proof. Let $F_1, \ldots F_k$ be the clusters returned by the farthest algorithm, where F_i contains point i. Let $D_1, \ldots D_k$ be the clusters returned by the tree-doubling algorithm, where the cluster are defined by the k points at depth d in T. Since depth d contains exactly k vertices, the monotonicity of $(\text{level}(i))_i$ implies that these points must be exactly the points $1, \ldots, k$. We will show that for any $1 \leq i \leq k$ if a point, x, is in D_i then $x \in F_i$. Since the k-clustering is a partition of the points, this immediately implies that $D_i = F_i$ for all $1 \leq i \leq k$.

Let x be a point in D_i. Since D_i contains the points in the subtree under i, there is a i-to-x path $P = (i = p_1, p_2, \ldots p_l = x)$ in T. Let $S = (i = s_1, s_2, \ldots s_m = x)$ be the sequence of points obtained by deleting all repetitions of points from P. By the construction of T we have that $s_j = \pi'(s_{j+1})$ for all $1 \leq j \leq m$ which implies that S is a valid i-to-x path in the Π'-tree. Since depth d of T contains all points in $\{1, \ldots, k\}$, only point i can appear in sequence S. Thus none of the points $\{1, \ldots, k\}$ except i are in the i-to-x path in Π' tree. This implies that $x \in F_i$.

Corollary 1. *Let k_1 be an input such that $D(k_1)$ has strictly less than k_1 clusters. Let k_2 be the minimum input such that $k_2 > k_1$ and $D(k_2)$ has exactly k_2 clusters. Then $D(k_1) \preceq F(k_1) \preceq D(k_2)$, where $A \preceq B$ stands for "B is a refinement of A".*

Proof. Let $k < k_1$ be the number of clusters in $D(k_1)$. Thus $D(k_1) = D(k)$ and T has a level with exactly k vertices. By Lemma 3, $F(k) = D(k)$. By definition of a hierarchical clustering, $F(k) \preceq F(k_1)$ as $k < k_1$. Thus we have $D(k_1) = D(k) = F(k) \preceq F(k_1)$.

Similarly, on input k_2, the tree-doubling algorithm produces a clustering with exactly k_2 clusters which implies that T has a level with exactly k_2 vertices. By Lemma 3, $F(k_2) = D(k_2)$. By definition of a hierarchical clustering, $F(k_1) \preceq F(k_2)$ as $k_1 < k_2$. Thus we have $F(k_1) \preceq F(k_2) = D(k_2)$.

2.4   Proof of the Refinement Theorem, Randomized Version

Suppose the random parameter r in the randomized versions of the farthest and the tree-doubling algorithms are chosen to be the same value. Then Lemma 3 and Corollary 1 also apply to the randomized algorithms. The only change to the analysis is to use inequalities 3 instead of inequalities 2 and 1 in the proof of correctness for algorithm 1.

2.5 Nondeterministic Choices

To prove the refinement theorem, we made some assumptions about the nondeterministic choices of the two algorithms. But how much do these choices affect the performance of the algorithms?

The first point chosen by the farthest algorithm determines the value of d_2 and this in turn determines the level threshold of the Π' tree, i.e level one contains the points which are at distance $[d_2, d_2/2)$ from previously labelled points and level two contains points which are at distance $[d_2/2, d_2/2^2)$ from previously labelled points and so on. The initial point can affect the performance of the farthest algorithm by a factor up to 8 as demonstrated on the example we present in Section 3, Fig. 2. On this example when the farthest algorithm chooses initial point p_1 it outputs a 5-clustering which has cost arbitrarily close to 8OPT. However the optimal 5-clustering can be obtained if p_4 is chosen as the initial point.

Points arrive to the tree doubling algorithm in an arbitrary order. How much can the ordering of points affect the performance of the tree doubling algorithm? By the refinement theorem, if points arrive in the order labelled by the farthest algorithm, there is always a way to break ties so that the tree doubling clustering is no better than the farthest clustering. However the arrival order of point can help the tree doubling algorithm perform better than the farthest algorithm. We demonstrate this on the tight example presented in 3, Fig. 2. If the points arrive as labelled by the farthest algorithm, the tree doubling and the farthest 5-clustering have cost 8OPT, while if the order starts with p_2, p_5, p_5', then tree doubling can construct the cost 2OPT, 5-clustering:

$$\left\{ (p_2), (p_3), (p_3') (p_1, p_4, p_5, q_1 \ldots q_n), (p_1', p_4', p_5', q_1' \ldots q_n') \right\} .$$

Combining these observations, we see that the farthest algorithm can produce clusterings which are 8 times better than the tree doubling algorithm clusterings if the farthest algorithm starts with the best possible initial point and the tree doubling is given its points in the worst possible ordering. On the other hand the tree doubling clusterings can be 4 times better than the farthest clusterings when its points are ordered favorably and the farthest algorithm starts at the worst possible initial point.

3 Proof of the Tightness Theorem

We will prove that, for any $\epsilon > 0$, there exists an input on which the farthest algorithm produces a hierarchical clustering where the $k = 5$ clustering is worse than the optimal 5-clustering by a factor of at least $8 - 4\epsilon$.

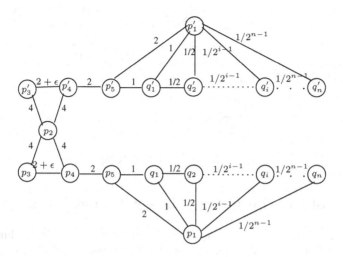

Fig. 2. Graph for Tight Example

Choose any $\epsilon > 0$ and let $n = 2\log(1/\epsilon)$. The input set S will have $2n + 9$ points; nine standard points, $p_1, p'_1, p_2, p_3, p'_3, p_4, p'_4, p_5, p'_5$, and $2n$ additional points $q_1, q'_1, q_2, q'_2, \ldots q_n, q'_n$ with distance as shown in Fig. 2. Note that the distance from q_i to q_{i+1} and the distance from p_1 to q_i for $i \in [1, n-1]$ is $1/2^i$ and the same holds for the distance from q'_i to q'_{i+1} and the distance from p'_1 to q'_i. It is easy to verify that the optimal 5-clustering of S is:

$$\left\{ (p_1, p_5, q_1, q_2 \ldots q_n), (p'_1, p'_5, q'_1, q'_2, \ldots, q'_n), (p_2), (p_3, p_4), (p'_3, p'_4) \right\}$$

where clusters (p_3, p_4) and (p'_3, p'_4) have the largest diameter of $2 + \epsilon = \text{OPT}(5)$.

We carry out the steps of farthest algorithm and show that its 5-clustering can have a cluster of diameter $16 - (2/2^{n-1})$. The algorithm starts with point p_1 and obtains the ordering: $p_1, p'_1, p_2, p_3, p'_3, p_4, p'_4, p_5, p'_5, q_1, q'_1, \ldots q_n, q'_n$. Thus $d(p_1, p'_1) = 16 = \Delta$ is used to define the levels for the points. The algorithm connects point $p \neq p_1$ to its parent $\pi'(p)$, the closest point to p at a strictly lower level. The resulting Π'-tree is shown in Fig. 3. To obtain a 5-clustering, the algorithm removes edges $(p_i, \pi'(p_i))$ for $p_i \in \{p'_1, p_2, p_3, p'_3\}$ which yields the clustering:

$$\left\{ (p_1), (p'_1), (p_3)(p'_3), (p_4, p_5, q_1 \ldots q_n, \ p_2, \ p'_4, p'_5, q'_1 \ldots q'_n,) \right\}$$

The diameter of the last cluster is the distance from q_n to q'_n which is $16 - 2/2^{n-1} = 16 - 4\epsilon^2 = (8 - 4\epsilon)\text{OPT}(5)$. This proves the Tightness theorem.

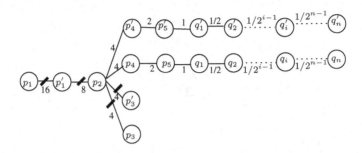

Fig. 3. Π-Tree for Tight Example

4   Proof of the Hierarchical Lower Bound Theorem

We demonstrate an input set S on which every deterministic algorithm obtains a competitive ratio at least 2 and every randomized algorithm obtains a competitive ratio at least 1.5. S has points p_{ij} for $i, j \in [1, 4]$ and $i \neq j$ with distances, $d(p_{ij}, p_{ji}) = 1$ and $d(p_{ij}, p_{ik}) = 2$ as shown in Fig. 4. (This resembles but is not the same as the example in [2], where the authors focus on the online setting). It is easy to verify that the optimal 6-clustering consists of the six pairs $p_{ij}p_{ji}$ each of diameter 1. Let $B_i = \{p_{ij} | j \in [1, 4], i \neq j\}$. Observe that B_i for $i \in [1, 4]$ is the optimal 4-clustering with each cluster having diameter 2.

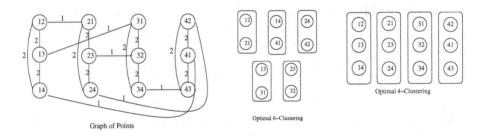

Fig. 4. Lower Bound Example

4.1   The Deterministic Lower Bound

Let A be any deterministic hierarchical clustering algorithm.

Case 1: Suppose A produces the optimal 6-clustering. Then A's clusters must be the 6 pairs $p_{ij}p_{ji}$. Since A is a hierarchical clustering algorithm, it must merge some of these pairs to obtain the 4-clustering. Merging any two of these pairs results in a cluster of diameter 4, giving us a competitive ratio of at least $A(4)/\text{OPT}(4) = 4/2 = 2$.

Case 2: Suppose A does not produce the optimal 6-clustering. Then some cluster in A's 6-clustering consist of points other than some pair $p_{ij}p_{ji}$. This cluster must have diameter ≥ 2. Thus the competitive ratio for A is at least $A(6)/OPT(6) \geq 2/1 = 2$.

4.2   The Randomized Lower Bound

Let B be any randomized hierarchical clustering algorithm. Let p be the probability that B outputs the optimal 6-clustering. Thus the maximum diameter is 1 with probability p, and at least 2 with probability $1 - p$. (See analysis for the deterministic scenario). We compute the expected competitive ratio of B for $k = 4$ and $k = 6$, and by definition, the expected competitive ratio of B over all values of k is at least the maximum of these two values.

For $k = 4$ the competitive ratio is

$$E(B_4(S))/\mathrm{OPT}_4(S) \geq (4p + 2(1 - p))/2 = 1 + p \ .$$

where the first inequality follows from the fact that when B chooses the optimal 6-clustering, its 4-clustering will have a cluster of diameter ≥ 4.

For $k = 6$ the competitive ratio is

$$E(B_6(S))/\mathrm{OPT}_6(S) \geq (p + 2(1 - p))/1 = 2 - p \ .$$

The expected competitive ratio is $\max(1 + p, 2 - p) \geq 1.5$.

5   Proof of the Online Hierarchical Supplier Theorem

Our online algorithm for k-supplier will use the (online) tree-doubling algorithm as a subroutine. Note that we could equivalently have used the farthest algorithm if we were designing an off-line algorithm.

5.1   The Algorithm

We denote a supplier as *active* if it is the closest supplier to one of the current customers. Throughout the algorithm, we will maintain a hierarchical clustering of the active suppliers by inserting them into the (deterministic or randomized) tree-doubling algorithm tree T^+.

When a new customer c arrives, we find the supplier s who is closest to c. If s is not yet in T^+, we mark s as an active supplier and add s to T^+ (using the deterministic or randomized tree-doubling algorithm).

To obtain a hierarchical k-supplier solution, find the largest depth d in T^+ which contains $k' \leq k$ active suppliers $s_1, s_2, \ldots s_{k'}$ and output these suppliers. For each customer c with closest supplier s_0, assign c to s_i for $i \in [1, k']$, if $s_0 = s_i$ or if s_0 is in the subtree below s_i in depth d of T^+.

5.2   The Deterministic Analysis

Suppose d is the largest depth containing at most k active suppliers. Let s (at depth d) be the supplier that customer c was assigned to and s_0 be the active supplier that c is closest to. Then there is a s_0-to-s path in T^+. Let $s_0, s_1, \ldots s_p$

be the sequence of the suppliers on the s_0-to-s path, where $s_p = s$. By the triangular inequality, the distance from c to s can be bounded as:

$$d(c,s) \leq d(c, s_0) + \sum_{i=0}^{p-1} d(s_i, s_{i+1}) \ . \tag{4}$$

Let Δ be the maximum distance between any two suppliers. By the close-parent property of the tree-doubling algorithm, the distance from s_i to s_{i+1} for $i \in [0, p-1]$ is at most $\Delta/2^{depth(s_i)-1}$. Since the depths of suppliers on the s_0-to-s path are strictly decreasing, and s_{p-1} is on level $d+1$, we have that,

$$\sum_{i=0}^{p-1} d(s_i, s_{i+1}) \leq \frac{\Delta}{2^{depth(s_{p-1})-1}}(1 + 1/2 + 1/4 + \ldots) \leq 2\frac{\Delta}{2^d} \ . \tag{5}$$

Now we derive two lower bounds for OPT_k. First, since s_0 is the closest supplier to c, we have that $\text{OPT}_k \geq d(c, s_0)$. Next, since d is the largest depth in T^+ with at most k active suppliers, depth $d+1$ contains at least $k+1$ active suppliers, $s_1, s_2, \ldots, s_{k+1}$. Using Lemma 4, we have $\text{OPT}_k \geq \delta/4$ where $\delta = \min_{1 \leq i < j \leq k+1} d(s_i, s_j)$. By the Far-Cousins property of T^+, δ is at least $\Delta/2^{d+1}$. Applying these bounds we obtain

$$d(c,s) \leq d(c, s_0) + \frac{2\Delta}{2^d} \leq \text{OPT}_k + 4\delta \leq 17 \, \text{OPT}_k \ .$$

Lemma 4. *Let d be the largest depth in T^+ with at most k active suppliers and let $s_1, s_2, \ldots, s_{k+1}$ be active suppliers at depth $d+1$. Let $\delta = \min_{1 \leq i < j \leq k+1} d(s_i, s_j)$, and OPT_k be the maximum distance from a customer to a supplier in the optimal k-supplier solution. Then $\delta \leq 4OPT_k$.*

Proof. Since suppliers $s_1, s_2, \ldots, s_{k+1}$ are active, each of them is the closest supplier to some customer c_i. The solution OPT_k uses at most k suppliers, so it will have to assign two of those customers, c_i and c_j, to the some supplier s^*. Thus,

$$\text{OPT}_k \geq \max(d(c_i, s^*), d(c_j, s^*)) \geq (d(c_i, s^*) + d(c_j, s^*))/2 \ .$$

Applying the triangle inequality on $d(s_i, s_j)$ we have that:

$$d(s_i, s_j) \leq d(s_i, c_i) + d(c_i, s^*) + d(s^*, c_j) + d(c_j, s_j)$$

Using the fact that s_i is the closest supplier to c_i and s_j is closest for c_j, we obtain

$$\delta \leq d(s_i, s_j) \leq 2(d(c_i, s^*) + d(c_j, s^*)) \leq 4OPT_k \ .$$

5.3   The Randomized Analysis

Equation 4 still holds. Instead of Equation 5 we now have:

$$\sum_{i=0}^{p-1} d(s_i, s_{i+1}) \leq \frac{e^r \Delta}{e^{depth(s_{p-1})-1}}(1 + 1/e + 1/e^2 + \ldots) \leq \frac{e}{e-1} \frac{e^r \Delta}{e^d} \ .$$

Now, by Property 3 the minimum distance δ between $s_1, \ldots, s_{k+1}$ satisfies $e^r \Delta / e^{d+1} < \delta \le e^r \Delta / e^d$. Write $\delta = e^\epsilon e^r \Delta / e^{d+1}$, where ϵ is distributed uniformly in $[0, 1)$. In expectation we have

$$E(e^r \Delta / e^{d+1}) = \delta \int_0^1 e^{-\epsilon} d\epsilon = \delta \frac{e-1}{e} \ .$$

Lemma 4 still holds, so we finally get:

$$E(d(c, s)) \le d(c, s_0) + \frac{e}{e-1} E(\frac{e^r \Delta}{e^d}) \le \mathrm{OPT}_k + \frac{e}{e-1} e\delta \frac{e-1}{e} \le (1 + 4e)\, \mathrm{OPT}_k \ .$$

References

1. P. Arabie, L. J. Hubert and G. De Soete, editors. *Clustering and Classification.* World Scientific, River Edge, NJ, 1998.
2. M. Charikar, C. Chekuri, T. Feder and R. Motwani. Incremental clustering and dynamic information retrieval. In *Proceedings of the 29th Annual ACM Symposium on the Theory of Computing,* pages 626–635, 1997.
3. Marek Chrobak, Claire Kenyon, John Noga and Neal E. Young. Online Medians via Online Bribery. *Lecture Notes in Computer Science* 3887:311-322 (2006); Latin American Theoretical Informatics, 2006.
4. S. Dasgupta, P. Long. Performance guarantees for hierarchical clustering. *Journal of Computer and System Sciences,* 70(4):555-569, 2005.
5. R. O. Duda, P. E. Hart, and D. G. Sork. *Pattern Classification.* Wiley and Sons, 2001.
6. M. E. Dyer and A. M. Frieze. A simple heuristic for the p-center problem. *Operations Research Letters.,* 3:285–288, 1985.
7. T. F. Gonzalez. Clustering to Minimize the Maximum Intercluster Distance. In *Proceedings of the 17th Annual ACM Symposium on the Theory of Computing,* 38:293-306, 1985.
8. D.S Hochbaum. Various Notions of Approximations: Good, Better, Best and More. In D.S Hochbaum, editor, *Approximation Algorithms for NP-Hard Problems.* PWS Publishing Company. 1996.
9. Dorit S. Hochbaum and David B. Shmoys. A best possible heuristic for the k-center problem. *Mathematics of Operations Research,* 10:180–184, 1985.
10. A. K. Jain and R. C. Dubes. *Algorithms for Clustering Data.* Prentice Hall, Englewood Cliffs, NJ, 1988.
11. L. Kaufman and Peter J. Rousseeuw, *Finding Groups in Data: An Introduction to Cluster Analysis,* Wiley, NY, 1990.
12. Guolong Lin, Chandrashekhar Nagarajan, Rajmohan Rajamaran, and David P. Williamson. A general approach for incremental approximation and hierarchical clustering. In *Proceedings of the seventeenth annual ACM-SIAM Symposium on Discrete algorithm (SODA),* pages 1147-1156, 2006.
13. NSF Workshop Report on *Emerging Issues in Aerosol Particle Science and Technology* (NAST), UCLA, 2003, Chapter 1, Section 18, "Improved and rapid data analysis tools (Chemical Characterization)". Available at http://www.nano.gov/html/res/NSFAerosolParteport.pdf.

Bin Packing with Rejection Revisited

Leah Epstein

Department of Mathematics, University of Haifa, 31905 Haifa, Israel
lea@math.haifa.ac.il

Abstract. We consider the following generalization of bin packing. Each item is associated with a size bounded by 1, as well as a rejection cost, that an algorithm must pay if it chooses not to pack this item. The cost of an algorithm is the sum of all rejection costs of rejected items plus the number of unit sized bins used for packing all other items.

We first study the offline version of the problem and design an AP-TAS for it. This is a non-trivial generalization of the APTAS given by Fernandez de la Vega and Lueker for the standard bin packing problem. We further give an approximation algorithm of absolute approximation ratio $\frac{3}{2}$, this value is best possible unless $\mathbf{P} = \mathbf{NP}$.

Finally, we study an online version of the problem. For the bounded space variant, where only a constant number of bins can be open simultaneously, we design a sequence an algorithms whose competitive ratios tend to the best possible asymptotic competitive ratio. We show that our algorithms have the same asymptotic competitive ratios as these known for the standard problem, whose ratios tend to $\Pi_\infty \approx 1.691$. Furthermore, we introduce an unbounded space algorithm which achieves a much smaller asymptotic competitive ratio. All our results improve upon previous results of Dósa and He.

1 Introduction

In the classical bin packing problem [17,4,3], a set (or sequence) of items, which are positive numbers no larger than 1, are to be packed into unit sized bins. The sum of items packed into one bin cannot exceed its size and the existing supply of such bins is unbounded. Each item must be packed into exactly one bin, minimizing the number of bins used. However, in many applications, it is possible to *refuse* to pack an item. This rejection needs to be compensated, and costs some given amount for each item, which is called the "rejection cost" of the item. In an application where bins are disks and items are files to be saved on these disks, the rejection cost of a file is the cost of transferring it to be saved on alternative media. In another application, where bins are storage units, a rejection cost is paid to a disappointed customer whose goods cannot be stored.

We call the packing problem studied in this paper BIN PACKING WITH REJECTION. In this problem, an item has both a size and a rejection cost associated with it. Each item must be either assigned to a bin or rejected. A bin is *empty* if no item is assigned to it, otherwise it is *used*. Unlike the standard problem where the goal is to minimize the number of used bins, the target function in the problem with rejection is the sum of the following two amounts. The first one

T. Erlebach and C. Kaklamanis (Eds.): WAOA 2006, LNCS 4368, pp. 146–159, 2006.
© Springer-Verlag Berlin Heidelberg 2006

is the sum of all rejection costs of rejected items. The second one is the number of bins used to pack the accepted items, i.e., items which are not rejected. The goal is to minimize this sum. Clearly, standard bin packing is a special case of bin packing with rejection, where all rejection costs are larger than 1.

We denote the set of items by I. For an item $i \in I$, we denote its size by p_i and its rejection cost by r_i. In this paper we study both offline and online algorithms for bin packing with rejection. In online environments of the bin packing problem, we receive the items as a sequence σ. Every element in the sequence is a pair, giving the size and rejection cost of this element. Thus, we get a sequence $(p_1, r_1), (p_2, r_2) \ldots (p_n, r_n)$, and the set I contains the same elements as σ. The elements arrive one by one. Upon arrival, an item must be either assigned or rejected. Such a decision is irrevocable.

The bin packing problem with rejection was introduced and studied by Dósa and He [6]. They suggested an interesting application for the offline version of the problem which is related to caching. Items are files which would need to be used in a local system. Each file would be needed exactly once at a later time. A file can be downloaded in advance to this local system, and stored on local web servers. The process of downloading a file from a local server (when it is actually needed) is fast, but stored files consume space on the servers. In this case the incurred cost results from the cost of local servers. The second option is to download a file only when it is actually needed, without storing it first. In the last case, a rejection cost occurs which is associated with the communication cost of downloading the file from an external server. An algorithm would need to have a cost as low as possible with respect to the two types of costs.

For an algorithm $\mathcal{A}$, we denote its cost by $\mathcal{A}$ as well. The cost of an optimal offline algorithm that knows the complete sequence of vertices is denoted by OPT. In this paper we mostly consider the asymptotic competitive ratio and the asymptotic approximation ratio criteria. When we discuss the performance guarantees of algorithms, we use the term "competitive" for online algorithms and the term "approximation" for offline algorithms. The asymptotic measures are standard measures of algorithm quality for bin packing problems. For a given input σ, let $\mathcal{A}(\sigma)$ be cost of algorithm $\mathcal{A}$ on σ. Let $\mathrm{OPT}(\sigma)$ be the minimum possible cost of serving all items in σ (i.e., the cost of packing a subset of the items plus the cost of rejecting all other items). The *asymptotic approximation ratio* (or *asymptotic competitive ratio*) for an algorithm $\mathcal{A}$ is defined to be $\mathcal{R}_\mathcal{A} = \limsup\limits_{n \to \infty} \sup\limits_{\sigma} \left\{ \frac{\mathcal{A}(\sigma)}{\mathrm{OPT}(\sigma)} \middle| \mathrm{OPT}(\sigma) = n \right\}$. We also consider the absolute approximation ratio in this paper. The absolute approximation ratio (or competitive ratio) of $\mathcal{A}$ is the infimum $\mathcal{R}$ such that for any input, $\mathcal{A} \leq \mathcal{R} \cdot \mathrm{OPT}$. If the approximation (competitive) ratio of a polynomial time offline (online) algorithm is at most $\mathcal{R}$, we say that it is a $\mathcal{R}$-approximation ($\mathcal{R}$-competitive), this applies to both types of approximation and competitive ratios.

Previous work. In [6], Dósa and He study four variants of bin packing with rejection. These are offline and online bin packing with respect to the absolute and the asymptotic measures. For the offline problem, the approximation ratios of the algorithms shown in the paper are 2 and $\frac{3}{2}$, where the latter applies only to the asymptotic measure. Moreover, it is mentioned that unless $\mathbf{P} = \mathbf{NP}$, no algorithm can have absolute approximation ratio of less than $\frac{3}{2}$ (due to a simple reduction from the PARTITION problem). Note that this holds already for standard bin packing.

For the online problem, they design an algorithm of absolute competitive ratio 2.618 and an algorithm of asymptotic competitive ratio $1.75 + \varepsilon$. They show a lower bound of 2.343 for the first online variant, and mention that the lower bound of 1.5401 for the standard online bin packing problem, due to Van Vliet [18] is the best lower bound known for the second variant.

As the standard bin packing problem is a special case of the problem with rejection, we next compare the above results with these known for the standard bin packing problem. The offline bin packing problem admits an APTAS (Asymptotic Polynomial Time Approximation Scheme), as was shown by Fernandez de la Vega and Lueker [5]. This scheme returns for every given value $\varepsilon > 0$ an algorithm with asymptotic approximation ratio $1 + \varepsilon$. The algorithm has polynomial running time if ε is seen as a constant. Karmarkar and Karp [12] designed an AFPTAS (Asymptotic Fully Polynomial Time Approximation Scheme) for the problem. They use a similar (but much more complex) algorithm, to achieve a running time which also depends on $\frac{1}{\varepsilon}$ polynomially.

As stated above, the absolute approximation ratio of any algorithm cannot be expected to be better than $\frac{3}{2}$. Several algorithms are known to achieve this bound. Specifically, the simple First-Fit-Decreasing (FFD) algorithm, which sorts the items according to non-increasing size, and applies First Fit (each item is packed to the earliest bin where it fits), is one of these algorithms. This result is implied by bounds on the performance of FFD, which are given e.g. by [20] and also proved directly using a simple proof in [16]. Several other algorithms with the same approximation ratio are known, (see e.g. [22]).

As for the online problem, the currently best known upper bound on the asymptotic competitive ratio is 1.58889 due to Seiden [15]. This problem has been extensively studied. The online bin packing problem was first investigated by Ullman [17]. He showed that the FIRST FIT algorithm has performance ratio $\frac{17}{10}$. This result was then published in [10]. Johnson [11] showed that the NEXT FIT algorithm has performance ratio 2. Yao [19] designed an algorithm called REVISED FIRST FIT and showed that it has performance ratio $\frac{5}{3}$.

Lee and Lee [13] developed the REFINED HARMONIC algorithm, which they showed to have a performance ratio of $\frac{273}{228} < 1.63597$. The next improvements were MODIFIED HARMONIC and MODIFIED HARMONIC 2. [14] showed that the first algorithm has competitive ratio of at most $\frac{538}{333} < 1.61562$ and claimed that the second algorithm has competitive ratio of at most $\frac{239091}{148304} < 1.61217$.

There is less study of the absolute competitive ratio, and the existent study focuses on the performance of First Fit. An upper bound of 1.75 was proved by Simchi-Levi [16]. A lower bound of $\frac{5}{3}$ was given by Zhang [21].

An important version of online bin packing (which is not studied in [6]) is the bounded space model. Bounded space algorithms can only have a constant number of bins available to accept items at any point during processing. The available bins are also called "open bins". The bounded space assumption is a quite natural one. Essentially the bounded space restriction guarantees that output of packed bins is steady, and that the packer does not accumulate an enormous backlog of bins which are only output at the end of processing. For the classical bin packing problem, Lee and Lee [13] presented an algorithm called HARMONIC, which partitions items into $m > 1$ classes and uses bounded space of at most $m - 1$ open bins. For any $\varepsilon > 0$, there is a number m such that the HARMONIC algorithm that uses m classes has a performance ratio of at most $(1 + \varepsilon)\Pi_\infty$ [13], where $\Pi_\infty \approx 1.69103$ is the sum of series (see Section 3.2). They also showed there is no bounded space algorithm with a performance ratio below Π_∞. The algorithms mentioned above REFINED HARMONIC, MODIFIED HARMONIC and MODIFIED HARMONIC 2 are all unbounded space adaptations of HARMONIC. Note that the 1.75 upper bound of Dósa and He [6] does not use bounded space, as it is based on First Fit. It is achieved by a sequence of algorithms, whose sequence of competitive ratios tends to 1.75 from above.

There has been a fair amount of research on variants of well known problem, where a notion of rejection is introduced. Such studies include research on variants of various important scheduling problems (see [1,9,7]). Since scheduling is strongly related to bin packing, this gives another motivation to the study of the bin packing problem with rejection.

Our results. We first study the offline problem. We design an APTAS for bin packing with rejection problem which uses techniques from [5] but also from [8] and [2]. For a given value of ε, the APTAS has cost of at most $(1 + \varepsilon)\text{OPT} + 1$.

Next, we design an algorithm with absolute approximation ratio $\frac{3}{2}$. To do that, we use the APTAS using a constant value of ε, combined with adaptations of the APTAS and and additional arguments for cases where the value OPT is small. Note that here the costs do not always take integer values unlike in standard bin packing. Our $(1 + \varepsilon)$-approximation (in the asymptotic case) and $\frac{3}{2}$-approximation (in the absolute case) improve the previous results of [6] for the two measures which are $\frac{3}{2}$ and 2 respectively.

We continue with a study of the online problem. To be able to prove upper bounds for online algorithms, we generalize the notion of weighting [17,15] to algorithms which allow rejection. We establish the best asymptotic competitive ratio for bounded space algorithms, and show it is the same as for the problem without rejection. For this, we adapt the HARMONIC algorithm of Lee and Lee [13] to be able to handle the notion of rejection. We show that the adapted algorithms still have the same asymptotic competitive ratios, and thus, achieve the best possible performance. Finally we show an improved unbounded space algorithm which is a modification of MODIFIED HARMONIC which can handle rejections. Both our algorithms, the rejective variants of HARMONIC and MODIFIED HARMONIC achieve better asymptotic competitive ratios than the algorithm of [6]. Their ratios are approximately 1.69103 and 1.61562, whereas the algorithm of [6] has a competitive ratio $1.75 + \varepsilon$.

Proofs that were omitted from this version can be found in the full version.

2   Offline Bin Packing with Rejection

2.1   An APTAS

To design an APTAS, we use methods similar to the well known APTAS for the classical bin packing problem, given by Fernandez de la Vega and Lueker [5]. The adaptation we design here has some similarities with [2], however there are many differences due to the different natures of the problems. In order to be able to deal with rejection costs, we also use methods similar to ones used for scheduling, as in [8].

We assume that without loss of generality, each rejection cost r_i satisfies $r_i < 1$. We can make this assumption since an item of rejection cost at least 1, that is rejected in some solution, can be placed in a bin of its own instead, and the solution cost does not increase. We also assume OPT ≥ 1. In order to be able to assume this, note that if OPT < 1 this means that all jobs are rejected, since any solution which uses at least one bin has cost of at least 1. Therefore, we can compute the sum of all rejection costs. If this sum is smaller than 1 we output this solution and otherwise, we run the APTAS. We can always check the solution which rejects all jobs and output it if it turns out to be better than the result of the APTAS. This will be useful to get a better approximation for small values of OPT which is done later.

As in [5], a first partition is done into "large items" and "small items". Let δ be a function of ε defined later. We require δ to be an inverse of an integer. An item j is considered to be large if both $r_j \geq \delta$ and $p_j \geq \delta$. All other items are small. We denote the multiset of large items by L and the multiset of small items by M. We have $I = L \cup M$.

The first step is to construct a set of possible packings of the large items. For each such packing of large items only, we add the other items in a near optimal way. The number of packings of large items would be polynomially bounded, yet, packings are enumerated in a way that a packing, which is close enough for our purposes to an optimal packing (restricted to large items only), is tested.

Let N be the number of large items in the input ($N = |L|$). If the number of large items is relatively small, that is $N < \frac{1}{\delta^4}$, we simply enumerate all possible solutions for these large items (these are partial packings of the large items where the unpacked items are rejected) into at most N bins. Since a packing contains at most N bins, and each item can be either placed into one of these bins or rejected, there are at most $(N+1)^N \leq (\frac{1}{\delta^4})^{\frac{1}{\delta^4}}$ possible packings. Note that in this process with opened bins but possibly some of them remained empty. The set of bins remaining empty after this process are removed from the packing. We would like to add empty bins later and to test all possible amounts of empty bins, such bins are added to the packing to accommodate small items.

For the case where $N \geq \frac{1}{\delta^4}$, we perform a rounding of the rejection costs of all items in L. We define intervals $[\delta + i\delta^2, \delta + (i+1)\delta^2)$ for $i = 0, \ldots, \Delta = \frac{1}{\delta^2} - \frac{1}{\delta} - 1$. For every item $j \in L$, we define r'_j to be the left endpoint of the interval to which r_j belongs (i.e., it is the value of r_j, rounded down to the closest value $\delta + i\delta^2$). Let I' be the adapted input. Let $A(I')$ be the cost of a solution of an algorithm

A for the rounded input, and let $A'(I')$ be the cost of the same solution on the original items. Then we can show the following.

Lemma 1. $A'(I') \le (1+\delta)A(I')$ and $OPT(I') \le OPT(I)$

For $0 \le i \le \Delta$, let N_i be the number of items with rounded rejection cost $\delta + i\delta^2$, and let $a_{i,1} \ge \ldots \ge a_{i,N_i}$ be (the sizes of) these items. Note that $N = \sum\limits_{i=0}^{\Delta} N_i$.

We can consider only the sizes of items for each i, since they all have the same rejection cost $\delta + i\delta^2$. Therefore, in this case we can identify between items and their sizes. For a given $1 \le i \le \Delta$, denote the multi-set of item sizes by B_i.

We perform a linear grouping on each one of the multi-sets of large items $B_i = \{a_{i,1}, \ldots, a_{i,N_i}\}$. Let $m = \frac{1}{\delta^2}$. We partition the sorted set of large items into m consecutive sequences $S_{i,j}$ $(j = 1, \ldots, m)$ of $k_i = \lceil \frac{N_i}{m} \rceil = \lceil N_i\delta^2 \rceil$ items each (to make the last sequence be of the same cardinality, we define $a_{i,t} = 0$ for $t > N_i$). I.e., $S_{i,j} = \{a_{i,(j-1)k_i+1}, \ldots, a_{i,(j-1)k_i+k_i}\}$ for $j = 1, 2, \ldots, m$. For $j \ge 2$, we define a modified sequence $\hat{S}_{i,j}$ which is based on the sequence S_j as follows. $\hat{S}_{i,j}$ is a multiset which contains exactly k_i items of size $a_{i,(j-1)k_i+1}$, i.e., all items are rounded up to the size of the largest element of $S_{i,j}$. The set $S_{i,1}$ is not rounded and therefore $\hat{S}_{i,1} = S_{i,1}$. Let L'_i be the union of all multisets $\hat{S}_{i,j}$ $(L'_i = \bigcup\limits_{j=1}^{m} \hat{S}_{i,j})$ and $L' = \bigcup\limits_{i=0}^{\Delta} L'_i$ and let $L''_i = \bigcup\limits_{j=2}^{m} \hat{S}_{i,j}$, $L'' = \bigcup\limits_{i=0}^{\Delta} L''_i$.

We find solutions for the two sets $L_1 = \bigcup\limits_{i=0}^{\Delta} S_{i,1} = L' - L''$ and L'' separately.
The items of L_1 are packed each in a separate bin. The input L'' is treated as follows. This input contains at most $T = (m-1)(\Delta+1) < \frac{1}{\delta^4}$ different type of items (where two items are of the same type if they are of the same size and have the same rounded rejection cost).

We enumerate all possible packings of the L'' items into i bins, where $0 \le i \le N$. The input L'' contains at most T distinct sizes of elements. We are interested in computing all solutions of a bin packing instance with a constant number of distinct large types. Let $(b_1, \rho_1), \ldots, (b_T, \rho_T)$ be the set of types, where $\delta \le b_j \le 1$ is the size of items of type (b_j, ρ_j) and $\delta \le \rho_j \le 1$ is its (rounded) rejection cost. We represent a multiset of items by a vector $J = (u_1, \ldots, u_T)$, where u_j is the number of items of type (b_j, ρ_j). Let $\hat{N} = (n_1, \ldots, n_T)$ denote an input. A *pattern* is a vector of non-negative integers such that the multiset of items represented by it can fit in a single bin, i.e. q is a pattern if $\sum\limits_{i=j}^{T} q_j b_j \le 1$.

Let Q be the set of all patterns. A packing can be described by specifying for every $q \in Q$, the number of bins y_q that are packed using pattern q.

As noted above, we remove empty bins from the packing, therefore an empty pattern (for which $q_i = 0$ for $1 \le i \le T$), may be considered to be a legal pattern, but is useless. The difference between n_j and the number of items of type (b_j, ρ_j) that are packed in the packing are rejected items.

We now argue that $|Q| \le (T+1)^{\frac{1}{\delta}}$. A bin can contain at most $\frac{1}{\delta}$ items. To show the bound, we can represent each bin by a list of length $\frac{1}{\delta}$. In this list we first provide an complete enumeration of all items of this bin, if any slots remain empty, we fill them with "null". There are $T + 1$ options for each item in the list, since an item can be absent as well as of any size among the T possible sizes. This gives an upper bound of $(T + 1)^{\frac{1}{\delta}}$ on the number of patterns $|Q|$.

A vector $y \in \mathcal{N}_0^Q$ specifies a valid packing of an input $\hat{N}$ into ℓ bins if and only if the following constraints hold.

$$\sum_{q \in Q} y_q = \ell, \quad \text{and for all} \quad 1 \leq j \leq T, \quad \sum_{q \in Q} q_j y_q \leq n_j \quad (1)$$

Since for each $1 \leq j \leq T$, there are $n_j - \sum_{q \in Q} q_j y_q$ items of this type which remain unpacked. The rejection cost of each of them is ρ_j and thus the cost of the entire packing including rejection costs of rejected items is $\ell + \sum_{j=1}^{T} \rho_j(n_j - \sum_{q \in Q} q_j y_q)$.

Since $\ell \leq N$, we are only interested in vectors y where each component is in the set $\{0, \ldots, N\}$. Thus, the number of possible vectors y is polynomial.

For every packing, constructed for large items, we do the following. Consider all non-empty bins packed with large items. If the packing was created for the original items (in the case where N is small), the packing is not changed.

Otherwise, keep the bins of L_1 items unchanged. Note that a vector y defines a packing of the L'' items completely, these are linearly grouped items, and not the input items. After the process of packing is completed, including the packing of small items that are packed in the next step, we can replace the items of $\hat{S}_{i,j}$ in the packing by items of $S_{i,j}$. Clearly, the items of $S_{i,j}$ are never larger than the items of $\hat{S}_{i,j}$, and so the resulting packing is feasible.

Let ℓ be the number of bins in the packing. Since the final packing cannot contain more than n non-empty bins, we perform the following for all the following values of d, $d = \ell, \ldots, n$. Thus, d will be the number of used bins in the resulting packing. For each bin, which is already packed with some large items, compute the empty space in it (that is 1 minus the sum of sizes of all items assigned to it). Denote the empty spaces in bins $z = 1, \ldots, d$ by x_z. We define $x_z = 1$ for $\ell < z \leq d$. To assign the small items (all items of M), construct the following integer program. Let $n' = n - N$ be the number of small items, and $\{(c_1, r_1), \ldots, (c_{n'}, r_{n'})\}$ be pairs of sizes and rejection costs of these items. For $1 \leq z \leq d + 1$ and $1 \leq j \leq n'$, let $X_{j,z}$ be an indicator variable. If $z \leq d$, the value of $X_{j,z}$ is 1 if item j is assigned to bin z and 0 otherwise. If $z = d + 1$ the value of $X_{j,z}$ is 1 if item j is rejected and 0 otherwise.

We apply the upper bounds on sum of sizes of items in the bins as follows. For each $1 \leq z \leq d$, $\sum_{j=1}^{n'} c_j \cdot X_{j,z} \leq x_z$. We clearly have $\sum_{z=1}^{d+1} X_{j,z} \geq 1$ for all $1 \leq j \leq n'$, since each item must be either assigned to at least one bin or rejected. If it is assigned to more than one bin, one of its occurences can be removed without violating the other constraints. If it is both assigned and rejected, it is again removed from any bin it is assigned to.

The linear goal function is to minimize the expression $\sum_{j=1}^{n'} r_j \cdot X_{j,d+1}$. This is the sum of rejected items, and since the number of used bins is d, the cost of an algorithm is d plus the sum of rejection costs.

We relax the integrality constraint, and replace it with $X_{j,z} \geq 0$. We are left with a linear program which clearly has a solution if the original integer program does. Solving the linear program we can find a basic solution. This basic solution has at most $d+n'$ non-zero variables (as the number of constraints). Clearly, each item j has at least one non-zero variable $X_{j,z}$ and thus we get that the number of items that are not assigned completely to a bin or completely rejected (i.e., that have more than one non-zero variable associated with them) is at most d. These items are not assigned according to the solution found by the linear program. Since these items are small, for each item, either the rejection cost is at most δ, or the size it at most δ (or both). Therefore, out of the (at most) d items we still need to assign, we reject all items with rejection cost of at most δ, and pack the other items into bins, so that each bin packed in this way, (possibly except for the last one) contains exactly $\frac{1}{\delta}$ items. Out of the d small items that participate in this process, let d_1 be the number of rejected small items and $d - d_1$ the number of small items which are packed into bins.

Therefore, the additional cost for these items is at most $\delta d_1 + \lceil \delta(d - d_1) \rceil \leq \delta d + 1$. As an output, it is possible to choose the solution with smallest cost out of all resulting solutions.

We next analyze the performance guarantee of the above algorithm. We make use of the following definitions and lemma.

For two multisets A, B, whose elements are pairs of sizes and rejection costs of items. We say that A is *dominated* by B and denote $A \leq B$ if there exists an injection $f : A \to B$ with the following properties. Let $a = (p_a, r_a) \in A$, and let $f(a) = b = (p_b, r_b) \in B$, then $p_b \geq p_a$ and $r_b \geq r_a$.

Lemma 2. *If A and B are multisets such that $A \leq B$, then $OPT(A) \leq OPT(B)$.*

We would like to analyze the minimum cost of any solution we get. To upper bound the cost of this minimal solution, we actually upper bound the cost of one specific solution, defined later. In the full version of the paper, we prove the following theorem.

Theorem 1. *Algorithm FL is an APTAS.*

As mentioned earlier, the bin packing problem with rejection, if analyzed by the absolute approximation ratio, cannot have an approximation algorithm with approximation ratio smaller than $\frac{3}{2}$ (unless $\mathbf{P} = \mathbf{NP}$). In the full version of the paper we design an algorithm with this (probably best possible) absolute approximation ratio, and prove the following.

Theorem 2. *There exists a polynomial offline approximation algorithm, whose absolute approximation ratio is $\frac{3}{2}$.*

3   Online Bin Packing with Rejection

3.1   Analysis of Online Bin Packing Algorithms with Rejection

In this section we develop a scheme which is useful for analyzing bin packing algorithms with rejection. It is possible to apply the method both to offline and online algorithms, however, in this paper we only use it for online algorithms. The method is based on weighting, and is similar to the method used already by Ullman [17] (see also [13,15]). We describe the basic method briefly as our method generalizes it.

The essence of this method is to assign weights to items. The weights must be assigned so that the cost of the algorithm, i.e. the number of used bins is roughly the sum of weights. A small deviation is allowed when dealing with the asymptotic competitive ratio, thus an additive constant does not degrade the performance of an algorithm. As the next step, the problem of upper bounding the asymptotic competitive ratio is reduced into that of finding the maximum sum of weights of items which can fit into a single bin. In some cases, a constant number k of distinct weighting functions are defined to handle several major behaviors of the algorithm (resulting from specific inputs). For each outcome of the algorithm, at least one of the k weighting functions needs to have the above property regarding the cost of the algorithm. In this case, an upper bound on the competitive ratio is the maximum between the k maximum sums of weights in a single bin for the k weight functions. The method in [15] is more complex and generalizes the above method.

Surprisingly, the method can be generalized to deal with weights which are not related only to the cost of packing items (i.e., numbers of bins) but to rejection costs as well.

Let $\mathcal{A}$ be an online algorithm and let $\mathcal{C}$ be a desired competitive ratio. Let $w^1, \ldots, w^k$ be a set of functions $w^i : (0,1] \to R_0^+$ (where R_0^+ denotes the set of non-negative real numbers). For an item j, we denote its weight with respect to weight function w^i by w_j^i.

Theorem 3. *A value $\mathcal{C}$ is an upper bound on the asymptotic competitive ratio of algorithm $\mathcal{A}$ if the following conditions hold.*
1. For every item j, and for every weight function w^i, we have that $w_j^i \leq \mathcal{C} r_j$, that is, for every weight function, the weight assigned to each item is no larger than $\mathcal{C}$ times its rejection cost. 2. There exists a constant μ, such that for every input, there exists a value $1 \leq i \leq k$ such that $\mathcal{A} \leq \sum_{j=1}^{n} w_j^i + \mu$. 3. For every set of items J such that $\sum_{j \in J} p_j \leq 1$, and every $1 \leq i \leq k$, we have $\sum_{j \in J} w_j^i \leq \mathcal{C}$.

3.2   Algorithm REJECTIVE HARMONIC

We now define our adaptation of the HARMONIC$_k$ algorithm of Lee and Lee [13]. The algorithm is called REJECTIVE HARMONIC$_k$ (REJH$_k$). The fundamental idea of "harmonic-based" algorithms is to first classify items by size, and then pack an item according to its class (as opposed to letting the exact size influence packing decisions). We use a similar classification, but after classification is applied, we further use a decision rule (based on a threshold) to identify whether the item should be packed or rejected.

For the classification of items, we partition the interval $(0, 1]$ into sub-intervals. We use $k - 1$ sub-intervals of the form $(\frac{1}{i+1}, \frac{1}{i}]$ for $i = 1, \ldots, k - 1$ (intervals $1, \ldots, k - 1$) and one final sub-interval $(0, \frac{1}{k}]$ (interval k). Each packed bin will contain only items from one sub-interval. Items in sub-interval i that are not rejected, are packed i to a bin for $i = 1, \ldots, k - 1$ (except for possibly the very last bin dedicated to this interval). The items in interval k that are not rejected are packed using the greedy algorithm NEXT FIT. This algorithm keeps a single open bin and packs items of interval k that are not rejected to this bin until some item does not fit. Then a new bin is opened for interval k, and the previous bin is never used again. For $1 \leq i \leq k - 1$, a bin which received the full amount of items (according to its type) is closed, therefore a total of at most $k - 1$ bins are open or active simultaneously (one per interval, except for $(\frac{1}{2}, 1]$ which does not need an active bin).

We next define the thresholds for acceptance or rejection of a new item. Given an item $a \in I$, let $\frac{1}{s_a}$ be the right endpoint of the sub-interval $1 \leq s_a \leq k$ to which p_a belongs. If $s_a < k$, item a is rejected if $r_a \leq \frac{1}{s_a}$, and otherwise a is accepted and packed according to the algorithm above. If $s_a = k$, item a is rejected if $r_a \leq \frac{k}{k-1} p_a$, and otherwise a is accepted.

As a first step of analyzing the algorithm, we assign weights to items. We will use the method introduced in the previous section for the analysis. The assignment is similar to the proof of [13], however, unlike the proof in [13], our weights are a function of both the sizes and rejection costs. We use a single weight function w, and the weight of item $a \in I$ is denoted w_a.

In order to use the method, we need to assign the weights so that the three conditions in Theorem 3 hold. We do the assignment so that the cost of the algorithm satisfies $RejH_k \leq \sum_{a \in I} w_a + k - 1$.

An item a which is rejected by the algorithm gets weight r_a. An item a which is accepted gets weight $\frac{1}{s_a}$, if $s_a < k$ and $\frac{k}{k-1} p_a$, if $s_a = k$. Thus each item of sub-intervals $1, \ldots, k - 1$ gets weight $\min\{r_a, \frac{1}{s_a}\}$ and each item of sub-interval k gets weight $\min\{r_a, \frac{k}{k-1} p_a\}$.

For the analysis, we use the following well known sequence π_i, $i \geq 1$, which often occurs in bin packing. Let $\pi_1 = 2, \pi_{i+1} = \pi_i(\pi_i - 1) + 1$ and let $\Pi_\infty = \sum_{i=1}^{\infty} \frac{1}{\pi_i - 1} \approx 1.69103$. This sequence is presented in [13]. It is not difficult to show that $1 - \sum_{i=1}^{t} \frac{1}{\pi_i} = \frac{1}{\pi_{i+1} - 1}$. It is shown in [13] that the sequence of asymptotic competitive ratios of the algorithms HARMONIC$_k$ tends to Π_∞ as k grows, and that no bounded space algorithm can have an asymptotic competitive ratio smaller than Π_∞. We can show that the generalization RejH$_k$ has the same properties.

Clearly, the lower bound for the problem with rejection follows from the lower bound on the special case without rejection.

Theorem 4. *The asymptotic competitive ratio of RejH$_k$ tends to Π_∞ as k grows. No algorithm can have a smaller asymptotic competitive ratio.*

Consider an optimal offline algorithm OPT. For this algorithm, denote by R_{OPT} the set of rejected items and by A_{OPT} the set of accepted items. Let B_{OPT} denote the number of used bins.

Then $\text{OPT} = B_{\text{OPT}} + \sum_{a \in R_{\text{OPT}}} r_a \geq B_{\text{OPT}} + \sum_{a \in R_{\text{OPT}}} w_a$. Therefore, in order to prove an asymptotic competitive ratio $\mathcal{C}$, it is enough to prove $\sum_{a \in A_{\text{OPT}}} w_a \leq \mathcal{C} \cdot B_{\text{OPT}}$. To prove this, it is enough to consider every bin of OPT separately, and to show that the sum of weights of items in this bin is at most $\mathcal{C}$. Finally, to show this, we upper bound the sum of weights of items that can fit in a single bin.

To summarize the technique used here, we assign weights to items, so that the cost of an algorithm is roughly the sum of weights. Then we reduce the problem into that of finding the maximum sum of weights of items in a single bin. This method is often used in bin packing problems, and was already used in [17]. Surprisingly, the method here is applied even though the weights are not related only to packing items but to rejection costs as well.

3.3 Algorithm REJECTIVE MODIFIED HARMONIC

In this section we show how to design improved algorithms which are unbounded space. As an example, we adapt one of the best algorithms known for online bin packing to allow rejection. This algorithm MODIFIED HARMONIC was introduced by Ramanan et al. [14]. We give a short description of this algorithm.

As HARMONIC, MODIFIED HARMONIC also classifies items by size, and packs items according to classes. A disadvantage of HARMONIC is in the packing of items of the sub-interval $I_1 = (\frac{1}{2}, 1]$. These items are packed one per bin, possibly wasting a lot of space in each single bin. To avoid this large waste of space, MODIFIED HARMONIC and other later algorithms (see [15]) use two extra interval endpoints, of the form $\frac{1}{2} < \Delta < 1$ and $1 - \Delta$. Then, some small items can be combined in one bin together with an item of size in $(\frac{1}{2}, \Delta]$. Items larger than Δ (i.e., in the interval I_1^1) are still packed one per bin as in HARMONIC. These algorithms furthermore use parameters α^i ($i = 2, \ldots, n-1$) which represent the fraction of items of intervals $I_i = (\frac{1}{i+1}, \frac{1}{i}]$ which are supposed to be combined with an item of size in $I_1^2 = (\frac{1}{2}, \Delta]$. For $i = 2$ α^2 is the fraction of items in the interval $I_2^2 = (\frac{1}{3}, 1 - \Delta]$. This fraction of items, when they arrive, is either immediately combined with such a large item (if this large item was not combined with items of different intervals yet), or else space is reserved for the larger item. Once such a large item arrives, it is inserted into a space reserved for it. The remaining bins with items of interval I_i (or I_2^2, for $i = 2$) still contain i items per bin. Moreover, items of the interval $I_2^1 = (1 - \Delta, \frac{1}{2}]$ are not combined with larger items and are packed in pairs. The items of the last interval $I_n = (0, \frac{1}{n}]$ are not combined with larger items and are packed using NEXT FIT.

MODIFIED HARMONIC (MH) is defined using four intervals of items in $(\frac{1}{3}, 1]$ as above, 35 intervals I_i for $i = 3, ..., 37$ and one last interval $I_{38} = (0, \frac{1}{38}]$. It uses $\Delta = \frac{419}{684}$.

$$\alpha^2 = \frac{1}{9}; \ \alpha^3 = \frac{1}{12}; \ \alpha^4 = \alpha^5 = 0; \ \alpha^i = \frac{37 - i}{37(i+1)}, \text{ for } 6 \le i \le 36 \text{ and } \alpha^{37} = 0.$$

The results of [14] imply that the asymptotic performance ratio of MODIFIED HARMONIC is at most $\frac{538}{333} < 1.61562$. (In the original definition, Δ was used to denote $1 - \Delta$.) Note that for every interval I_i (or I_2^2, for $i = 2$) for which smaller items that are possible to be combined with a larger item in a bin, we compute the maximum amount m_i of such items that can fit into the bin, leaving an empty space of size at least Δ. In this calculation, a maximum size of item is taken into account. Thus we get $m_2 = m_3 = 1$, $m_6 = m_7 = 2$, $m_8 = m_9 = m_{10} = 3$, $m_{11} = m_{12} = 4$, $m_{13} = m_{14} = m_{15} = 5$, $m_{16} = m_{17} = m_{18} = 6$, $m_{19} = m_{20} = 7$, $m_{21} = m_{22} = m_{23} = 8$, $m_{24} = m_{25} = 9$, $m_{26} = m_{27} = m_{28} = 10$, $m_{29} = m_{30} = 11$, $m_{31} = m_{32} = m_{33} = 12$, $m_{34} = m_{35} = m_{36} = 13$.

In the analysis we ignore incomplete bins which did not receive the full amount of items they are supposed to get. These are bins with items of size in $(0, \frac{1}{2}]$, that were not supposed to be combined with larger items, and bins with items of these sizes that are supposed to be combined with a larger item, but did not get m_i items. The number of such incomplete bins is bounded by a constant since we do not open a new bin until the previous one receives the full amount of items. However, a bin which received an item of size in I_1^2 but did not receive smaller items, or a bin which has space reserved for am item of size in I_1^2, that never arrived, cannot be ignored since their amount can be arbitrary. We note however, that after removing incomplete bins, there cannot be both types of bins mentioned above, and we either need to deal with "waiting" bins with an items of size in I_1^2, or "waiting" bins with space reserved for such an item.

We define a version of MODIFIED HARMONIC which allows rejection, and call it REJECTIVE MODIFIED HARMONIC (MHR). This algorithm has a decision rule for every interval. Upon arrival of an item, it is either rejected, or assigned by MH. Rejected items are simply ignored by this sub-routine that runs MH.

We therefore only need to define a rejection rule for every interval. Let x be an item, we consider all possible cases. If $x \in I_1^1 = (\Delta, 1]$, x is rejected if $r_j \le 1$ and otherwise accepted. If $x \in I_1^2 = (\frac{1}{2}, \Delta]$, x is rejected if $r_j \le \frac{2}{3}$ and otherwise accepted. If $x \in I_2^1 = (1 - \Delta, \frac{1}{2}]$, x is rejected if $r_j \le \frac{1}{2}$ and otherwise accepted. If $x \in I_1^2 = (\frac{1}{3}, \Delta]$, x is rejected if $r_j \le \frac{4}{9}$ and otherwise accepted. If $x \in I_3 = (\frac{1}{4}, \frac{1}{3}]$, x is rejected if $r_j \le \frac{11}{36}$ and otherwise accepted. If $x \in I_i = (\frac{1}{i+1}, \frac{1}{i}]$, for the following values of i; $i = 4, 5, 37$, x is rejected if $r_j \le \frac{1}{i}$ and otherwise accepted. If $x \in I_i = (\frac{1}{i+1}, \frac{1}{i}]$ for $6 \le i \le 36$, x is rejected if $r_j \le \frac{38}{37(i+1)}$ and otherwise accepted. If $x \in I_{38} = (0, \frac{1}{38}]$, x is rejected if $r_j \le \frac{38 \cdot p_j}{37}$ and otherwise accepted.

We assign two sets of weights w^1 and w^2 to items as follows. The proof is similar to the proof in [14], with differences resulting from rejections. A rejected item has $w_j^1 = w_j^2 = r_j$. An accepted item j of an interval I_i for $i = 4, 5, 37$ is assigned weight $w_j^1 = w_j^2 = \frac{1}{i}$. An item of interval I_1^1 gets weight $w_j^1 = w_j^2 = 1$.

An item of interval I_1^2 gets weight $w_j^1 = 1$, $w_j^2 = \frac{2}{3}$. An item of interval I_2^1 gets weight $w_j^1 = w_j^2 = \frac{1}{2}$. An item of interval I_2^2 gets weight $w_j^1 = \frac{4}{9}$, $w_j^2 = \frac{5}{9}$. An item of interval I_3 gets weight $w_j^1 = \frac{11}{36}$, $w_j^2 = \frac{7}{18}$. An item of interval I_i for $6 \leq i \leq 36$ gets weight $w_j^1 = \frac{38}{37(i+1)}$, $w_j^2 = w_j^1 + \frac{37-i}{37m_i(i+1)}$. An item of interval I_{38} gets weight $w_j^2 = w_j^1 = \frac{38 \cdot p_j}{37}$.

Weights are defined as in [14] except for rejected items and items in the interval I_1^2 for which we defined $w_j^2 = \frac{2}{3}$. Weights and rejection rules are defined so that for a rejected item j, its weight is never larger than the weight $\min\{w_j^1, w_j^2\}$ that it would have received if it had a larger rejection cost and were accepted.

In order to analyze the competitive ratio and show it is at most $C_1 = \frac{538}{333} < 1.61562$ (as for the original algorithm), we show that all conditions of Theorem 3 hold. The second condition holds due to the following. The proof of [14] shows that the condition holds in the case where no items are rejected, and for an item j in the interval I_1^2, the second weight function is defined by $w_j^2 = 0$. Since the weight of rejected items is exactly their rejection cost, and the weights we define are never smaller than the weights in [14], the condition follows.

To prove the first condition, note that for each item j, either its weight is equal to its rejection cost, or its rejection cost is at least its weight w_j^1. Thus we need to show for every item that $w_j^2 \leq C_1 w_j^1$. We only need to consider cases in which the two weights are not the same. For an item j in the interval I_1^2 we have $\frac{w_j^2}{w_j^1} = 1.5$. For an item j in the interval I_2^2 we have $\frac{w_j^2}{w_j^1} = 1.25$. For an item j in the interval I_3 we have $\frac{w_j^2}{w_j^1} = \frac{14}{11}$. For an item j in interval I_i for $6 \leq i \leq 36$ we have $\frac{w_j^2}{w_j^1} = 1 + \frac{37-i}{38m_i}$. This value is maximized for $i = 6$, since m_i is monotonically increasing. For $i = 6$ we have $m_i = 2$ and thus $\frac{w_j^2}{w_j^1} \leq 1 + \frac{31}{76} < \frac{3}{2}$.

To prove the last condition of Theorem 3, we note again that weights of rejected items are never larger than their weights according to each weight function, and thus we need to consider the weight functions as they are defined. Items for which the weights are defined as in [14], the proof follows from the result in that paper. Thus we need to consider only w^2 and only sets of items that can fit in a bin and which contain an item of size in I_1^2. Denote this large item by x. Such a bin can contain in addition only items smaller than $\frac{1}{2}$. In order to give an upper bound on the total weight of items in the bin, we find an upper bound on the ratio $\frac{w_j^2}{p_j}$ for items no larger than $\frac{1}{2}$. We can see that this ratio is no larger than $\frac{5}{3}$ for items in $(\frac{1}{6}, \frac{1}{2}]$, no larger than $\frac{38}{37}$ for items in $(0, \frac{1}{37}]$, and no larger than $\frac{38}{37} + \frac{37-i}{37m_i}$. This ratio is smaller than $\frac{3}{2}$. Thus an upper bound on the total weight in the bin with respect to w^2 is $\frac{2}{3} + \frac{1}{2} \cdot \frac{5}{3} = \frac{3}{2}$.

Since all conditions hold for $C_1 = \frac{538}{333}$ we establish the following theorem.

Theorem 5. *The competitive ratio of* MHR *is at most* $C_1 = \frac{538}{333}$.

References

1. Y. Bartal, S. Leonardi, A. Marchetti-Spaccamela, J. Sgall, and L. Stougie. Multiprocessor scheduling with rejection. *SIAM Journal on Discrete Mathematics*, 13(1):64–78, 2000.
2. A. Caprara, H. Kellerer, and U. Pferschy. Approximation schemes for ordered vector packing problems. *Naval Research Logistics*, 92:58–69, 2003.
3. E. G. Coffman, M. R. Garey, and D. S. Johnson. Approximation algorithms for bin packing: A survey. In D. Hochbaum, editor, *Approximation algorithms*. PWS Publishing Company, 1997.
4. J. Csirik and G. J. Woeginger. On-line packing and covering problems. In *A. Fiat and G. J. Woeginger, editors,* Online Algorithms: The State of the Art, pages 147–177, 1998.
5. W. Fernandez de la Vega and G. S. Lueker. Bin packing can be solved within $1 + \varepsilon$ in linear time. *Combinatorica*, 1:349–355, 1981.
6. G. Dósa and Y. He. Bin packing problems with rejection penalties and their dual problems. *Information and Computation*, 204(5):795–815, 2006.
7. D. W. Engels, D. R. Karger, S. G. Kolliopoulos, S. Sengupta, R. N. Uma, and J. Wein. Techniques for scheduling with rejection. *Journal of Algorithms*, 49(1):175–191, 2003.
8. D. S. Hochbaum and D. B. Shmoys. Using dual approximation algorithms for scheduling problems: theoretical and practical results. *Journal of the ACM*, 34(1):144–162, 1987.
9. H. Hoogeveen, M. Skutella, and G. J. Woeginger. Preemptive scheduling with rejection. In *Proc. of the 8th Annual European Symposium on Algorithms (ESA2000)*, pages 268–277, 2000.
10. D. S. Johnson, A. Demers, J. D. Ullman, Michael R. Garey, and Ronald L. Graham. Worst-case performance bounds for simple one-dimensional packing algorithms. *SIAM Journal on Computing*, 3:256–278, 1974.
11. David S. Johnson. Fast algorithms for bin packing. *Journal of Computer and System Sciences*, 8:272–314, 1974.
12. N. Karmarkar and R. M. Karp. An efficient approximation scheme for the one-dimensional bin-packing problem. In *Proceedings of the 23rd Annual Symposium on Foundations of Computer Science (FOCS'82*, pages 312–320, 1982.
13. C. C. Lee and D. T. Lee. A simple online bin packing algorithm. *J. ACM*, 32(3):562–572, 1985.
14. P. Ramanan, D. J. Brown, C. C. Lee, and D. T. Lee. Online bin packing in linear time. *Journal of Algorithms*, 10:305–326, 1989.
15. S. S. Seiden. On the online bin packing problem. *J. ACM*, 49(5):640–671, 2002.
16. D. Simchi-Levi. New worst-case results for the bin-packing problem. *Naval Res. Logist.*, 41(4):579–585, 1994.
17. J. D. Ullman. The performance of a memory allocation algorithm. Technical Report 100, Princeton University, Princeton, NJ, 1971.
18. A. van Vliet. An improved lower bound for online bin packing algorithms. *Information Processing Letters*, 43(5):277–284, 1992.
19. A. C. C. Yao. New algorithms for bin packing. *J. ACM*, 27:207–227, 1980.
20. M. Yue. A simple proof of the inequality $FFD(L) \leq (11/9)OPT(L) + 1, \forall L$, for the FFD bin-packing algorithm. *Acta. Math. Appl. Sinica*, 7:321–331, 1991.
21. G. Zhang. Private communication.
22. G. Zhang, X. Cai, and C.K. Wong. Linear time approximation algorithms for bin packing. *Operations Research Letters*, 26:217–222, 2000.

On Bin Packing with Conflicts

Leah Epstein[1] and Asaf Levin[2]

[1] Department of Mathematics, University of Haifa, 31905 Haifa, Israel
lea@math.haifa.ac.il
[2] Department of Statistics, The Hebrew University, Jerusalem, Israel
levinas@mscc.huji.ac.il

Abstract. We consider the offline and online versions of a bin packing problem called BIN PACKING WITH CONFLICTS. Given a set of items $V = \{1, 2, \ldots, n\}$ with sizes $s_1, s_2 \ldots, s_n \in [0, 1]$ and a conflict graph $G = (V, E)$, the goal is to find a partition of the items into independent sets of G, where the total size of each independent set is at most one, so that the number of independent sets in the partition is minimized. This problem is clearly a generalization of both the classical (one-dimensional) bin packing problem where $E = \emptyset$ and of the graph coloring problem where $s_i = 0$ for all $i = 1, 2, \ldots, n$. Since coloring problems on general graphs are hard to approximate, following previous work, we study the problem on specific graph classes. For the offline version we design improved approximation algorithms for perfect graphs and other special classes of graphs, these are a $\frac{5}{2} = 2.5$-approximation algorithm for perfect graphs, a $\frac{7}{3} \approx 2.33333$-approximation for a sub-class of perfect graphs, which contains interval graphs, and a $\frac{7}{4} = 1.75$-approximation for bipartite graphs. For the online problem on interval graphs, we design a 4.7-competitive algorithm and show a lower bound of $\frac{155}{36} \approx 4.30556$ on the competitive ratio of any algorithm. To derive the last lower bound, we introduce the first lower bound on the asymptotic competitive ratio of any online bin packing algorithm with known optimal value, which is $\frac{47}{36} \approx 1.30556$.

1 Introduction

We consider the following BIN PACKING WITH CONFLICTS problem (BPC) (see [15,3] and also the information on the bin packing problem given in [4]). Given a set of items $V = \{1, 2, \ldots, n\}$ with sizes $s_1, s_2 \ldots, s_n \in [0, 1]$ and a conflict graph $G = (V, E)$, the goal is to find a partition of the items into independent sets of G where the total size of each independent set is at most one, so that the number of independent sets in the partition is minimized. This problem is clearly a generalization of both the classical (one-dimensional) bin packing problem where $E = \emptyset$ and of the graph coloring problem where $s_i = 0$ for all $i = 1, 2, \ldots, n$. In an online environment, items arrive one by one to be packed immediately and irrevocably. A new item is introduced by its size, together with all its edges in the current conflict graph (i.e., edges which connect it to previously introduced items).

T. Erlebach and C. Kaklamanis (Eds.): WAOA 2006, LNCS 4368, pp. 160–173, 2006.

This problem arises in assigning processes or tasks to processors. In this case we are given a set of tasks, where some pairs of tasks are not allowed to execute on the same processor due to efficiency or fault tolerance reasons. The goal is to assign a minimum number of processors to this set of processes given that the makespan is bounded by some constant (see Jansen [14]). Other applications of this problem arise in the area of database replicas storage, school course time tables construction, scheduling communication systems (see de Werra [5]), and finally in load balancing, the parallel solution of partial differential equations by two dimensional domain decomposition (see Irani and Leung [13]). We follow earlier work and consider the BPC on sub-classes of perfect graphs. This restriction is motivated by the theoretical hardness of approximating graph coloring on general graphs.

In order to analyze our approximation and online algorithms we use common criteria which are the approximation ratio (also called performance guarantee) and competitive analysis. For an algorithm $\mathcal{A}$, we denote its cost by $\mathcal{A}$ as well. An optimal offline algorithm that knows the complete sequence of items is denoted by OPT. We consider the (absolute) approximation (competitive) ratio that is defined as follows. The (absolute) approximation (competitive) ratio of $\mathcal{A}$ is the infimum $\mathcal{R}$ such that for any input, $\mathcal{A} \leq \mathcal{R} \cdot \text{OPT}$. If the absolute approximation (competitive) ratio of an offline (online) algorithm is at most ρ we say that it is a ρ-approximation (ρ-competitive). For the offline problem, we restrict ourselves to algorithms that run in polynomial time. Our online algorithm is also a polynomial time algorithm (though this property is not always required in the competitive analysis literature). We focus on the absolute criteria and not on the criteria of asymptotic approximation ratio and asymptotic competitive ratio (these criteria are commonly used for bin packing problems) since a conflict graph can allow us to magnify small bad instances into large ones (with large enough values of OPT) with the same absolute ratio. So in general, we do not expect to have a better asymptotic approximation ratio than the corresponding absolute approximation ratio, even though this may be possible.

Since the BPC problem generalizes the classical coloring problem that is known to be extremely hard to approximate, we follow earlier studies and consider the BPC problem on the class of perfect graphs for which the coloring problem is polynomially solvable (see [25]). The best previously known approximation algorithm for BPC on perfect graphs is the algorithm of Jansen and Öhring [15] with an approximation ratio of 2.7. In Section 3.1 we improve this result and present our 2.5-approximation algorithm for BPC on perfect graphs.

Following Jansen and Öhring [15] we consider the class of graphs for which one can solve in polynomial time the PRECOLORING EXTENSION problem defined as follows. Given an undirected graph $G = (V, E)$ and k distinct vertices $v_1, v_2, \ldots, v_k$, the problem is to find a minimum coloring f of G such that $f(v_i) = i$ for $i = 1, 2, \ldots, k$. This problem is reviewed in [12,21], and it is known to be polynomially solvable for the following graph classes: interval graphs, forests, split graphs, complements of bipartite graphs, cographs, partial K-trees and complements of Meyniel graphs[1] (see [12] for a review of these results) and it is also polynomially solvable for chordal graphs as shown by Marx [22]. However, it is known to be NP-complete for bipartite graphs [12]. We denote by $\mathcal{C}$ the class of graphs G for which one can solve in polynomial time the precoloring extension problem for any induced subgraph of G (including G itself). I.e., $\mathcal{C}$ is closed under the operation of induced subgraph extraction. Jansen and Öhring [15] analyzed the following *algorithm with precoloring* for the case where G belongs to $\mathcal{C}$. Denote the set of large items by $L = \{j : s_j > \frac{1}{2}\}$, and denote by $\chi_I(G)$ the minimum number of colors used by an optimal solution for the precoloring extension problem defined by G. Finally, we define the set of precolored vertices to be L. Compute a feasible coloring of G using $\chi_I(G)$ colors, where for each pair of items in L they are assigned different colors. For each color class, apply a bin-packing heuristic such as the First-Fit-Decreasing algorithm. They proved that the resulting algorithm is a $\frac{5}{2}$-approximation algorithm. In Section 3.2 we improve this result by presenting a $\frac{7}{3}$-approximation algorithm.

For all $\varepsilon > 0$ Jansen and Öhring [15] also presented a $(2 + \varepsilon)$-approximation algorithm for BPC on cographs and partial K-trees. Furthermore, they presented a 2-approximation algorithm for bipartite graphs. A d-inductive graph has the property that the vertices can be assigned distinct numbers $1, \dots, n$ such that each vertex is adjacent to at most d lower numbered vertices. Jansen [14] showed an asymptotic fully polynomial time approximation scheme for BPC on d-inductive graphs where d is a constant. This result includes the cases of trees, grid graphs, planar graphs and graphs with constant treewidth. Oh and Son [24] and McCloskey and Shankar [23] considered BPC on graphs that are union of cliques, but their results are inferior to the 2.7-approximation algorithm of Jansen and Öhring [15].

The hardness of approximation of BPC follows from the hardness of standard offline bin packing (with respect to the absolute approximation ratio). It is not hard to see that unless $\mathbf{P} = \mathbf{NP}$, no algorithm can have absolute approximation ratio of less than $\frac{3}{2}$ (due to a simple reduction from the PARTITION problem, see problem SP12 in [8]). Since standard bin packing is a special case of BPC, where the conflict graph is an independent set, we get that for all graph classes studied in this paper, BPC is **APX**-hard, and unless $\mathbf{P} = \mathbf{NP}$, cannot be approximated within a factor smaller than $\frac{3}{2}$. Note that for bin packing, already the simple First-Fit-Decreasing algorithm is a $\frac{3}{2}$-approximation [27].

Our results. In Section 2 we describe the methods applied in this paper. We use weights for our analysis. The weights used throughout the paper have the unique and novel property that weights are assigned not only as a function of size of items, but also as a function of the location of items in an optimal solution or in an approximate solution. We think that this new technical approach can contribute to the analysis of algorithms for other problems as well.

We use these methods in Section 2 to give improved and tight bounds on two algorithms designed in [15]. We show that their algorithm for perfect graphs

---

[1] A graph is Meyniel if every cycle of odd length at least five has at least two chords.

has performance guarantee of approximately 2.691 and their algorithm with pre-coloring has performance guarantee of approximately 2.423. These tight results follow from our analysis together with bad examples for these algorithms given in [15]. Note that these bounds and their proofs resemble the analysis of the Harmonic algorithm [19] (the bounds are one unit higher than the upper bounds for Harmonic). However, neither the algorithms of [15] nor our algorithms use a partition into classes as is done in the Harmonic algorithm. Moreover, such a partition in our case would result in an arbitrarily high approximation ratios.

In Section 3 we present our improved new algorithms for the offline case of BPC. In Section 3.1 we design an improved algorithm for perfect graphs with performance guarantee of 2.5. Our algorithm is also a 2.5-approximation algorithm for BPC on all graph classes where one can solve the regular coloring problem (i.e., coloring the vertex set of a graph using a minimum number of colors) in polynomial time. In Section 3.2 we design an improved algorithm with precoloring with performance guarantee of $\frac{7}{3}$. In Section 3.3 we design a $\frac{7}{4}$-approximation algorithm for bipartite graphs.

In Section 4 we discuss *online* algorithms for BPC on interval graphs. We design a simple 4.7-competitive algorithm and show a lower bound of $\frac{155}{36} \approx 4.30556$ on the competitive ratio of any online algorithm. We derive the last lower bound by introducing the first non-trivial lower bound for online bin packing with known optimal value, which is $\frac{47}{36} \approx 1.30556$. We also show an $O(\log n)$ competitive algorithm for bipartite graphs, which is best possible. Both algorithms are adaptations of online algorithms for the standard coloring problem, see [18,20].

Proofs that were omitted from this version due to lack of space can be found in the full version of the paper.

2   Weighting Functions and the Performance of FFD Based Algorithms

In this section, we define weighting functions which are a major tool in the analysis of algorithms for bin packing. The weights defined in this section are later adapted and used for the analysis of our improved algorithms.

The idea of such weights is simple. An item receives a weight according to its size and its packing in some fixed solution. The weights are assigned in a way that the cost of an algorithm is close to the total sum of weights. In order to complete the analysis, it is usually necessary to consider the total weight that can be packed into a single bin of an optimal solution.

In this paper, we exploit this method in order to achieve improved algorithms for BPC. Though this method was not applied to BPC before, it was widely used for standard bin packing, and many variants on bin packing. This technique was used already in 1971 by Ullman [28] (see also [17,19,26]).

In this section, we define a set of weights which depends solely on the size of items. For an item x such that $s_x > \frac{1}{2}$ we define $weight(x) = 1$. We define the interval $\mathcal{I}_1$ by $\mathcal{I}_1 = (\frac{1}{2}, 1]$. For an item x such that $s_x \leq \frac{1}{2}$, let j be an integer such that $s_x \in \mathcal{I}_j = (\frac{1}{j+1}, \frac{1}{j}]$. We define $weight(x) = s_x + \frac{1}{j(j+1)}$. Note that even though this classification to intervals was used before, the weight function is non-standard. Typically either all items in an interval receive the same weight or are scaled by a common multiplicative factor (see e.g. [19,2]). We note that the weight function does not round up the size of an item to the next unit fraction.

We need to use this special weight function in order to make sure that the amount of weight is large enough, even if the input is partitioned into several classes, each of which is packed separately. On the other hand, we must make sure that the weights are not too large, so that the bound on the performance guarantee is not increased artificially. A similar (though different) weight function was used before by Galambos and Woeginger [6]. Their weight function can be used to prove Corollary 2 and Theorem 1 but not the other results of this paper. Therefore, we need to modify the weight function of [6] for our needs.

Given this set of weights, we note that for an item x of size $s_x \in \mathcal{I}_j$ $(j \geq 2)$, the ratio between its weight and its size is bounded as follows, $\frac{j+2}{j+1} \leq \frac{weight(x)}{s_x} < \frac{j+1}{j}$.

For a set of items X, we denote the sum of weights of all items in X by $W(X)$. I.e. $W(X) = \sum_{x \in X} weight(x)$. We next show that any algorithm which first partitions the input into μ classes, and then applies the algorithm First-Fit-Decreasing on each class separately, satisfies the following condition on its cost as a function of the total weight and μ.

Lemma 1. *Consider an algorithm $\mathcal{A}$ and a subset of items J which forms an independent set and is packed using First-Fit-Decreasing (FFD). Let Y be the number of bins used for this packing. Then we have $Y \leq W(J) + 1$.*

Proof. Note that for the above weight function, any bin which contains an item of size in $(\frac{1}{2}, 1]$ has total weight of items at least 1. Note also that the weight of an item in $\mathcal{I}_j = (\frac{1}{j+1}, \frac{1}{j}]$ is at least $\frac{1}{j+1} + \frac{1}{j(j+1)} = \frac{1}{j}$. Therefore, any bin which contains j items of size in the interval $(\frac{1}{j+1}, \frac{1}{j}]$ has total weight of at least 1. We can remove such bins from the packing and focus on all other bins called *transition bins* (if no bins are left after the removal, we are done).

A transition bin contains only items whose size is at most $\frac{1}{2}$. Note that the last bin ever opened may result in a transition bin, and it contains at least one item. Moreover, let a transition bin be of *type j* (for some $j \geq 2$), if the first item ever packed into it has size in $\mathcal{I}_j$. Next, we argue that there can be at most one transition bin of each type. Since the items are packed using FFD, transition bins are created in a sorted order, starting with the smallest type. If there are two bins of the same type j, this means that during the time between the packing the first items in these two bins, all packed items were also of size in interval $\mathcal{I}_j$. Therefore, the first bin must be assigned j such items before the second transition bin of this type is opened, and thus the first bin is not a transition bin. Let k be the largest type of any transition bin ever opened (i.e., the transition bin with the smallest item). Remove from the packing all items of size at most $\frac{1}{k+1}$. This removal may only decrease the total weight. As stated above, the weight of all remaining items in the transition bins is at least a multiplicative factor of $\frac{k+2}{k+1}$ their size.

Let α be the size of the first item in the last transition bin. Since the last transition bin is opened, all other bins have a total size of items which is more than $1 - \alpha$. Let $i_1 < \ldots < i_t < k$ be the sorted list of types of transition bins. We consider two cases which are $t \leq \lfloor \frac{k+2}{2} \rfloor$, and $t > \lfloor \frac{k+2}{2} \rfloor$. In both cases we need to show that the total weight in all transition bins is at least t (since there are $t + 1$ transition bins).

In the first case, if $t = 0$ we are done. Assume therefore $t \geq 1$. We get a total weight of at least $t(1 - \alpha)\frac{k+2}{k+1} + \alpha + \frac{1}{k(k+1)} = t\frac{k+2}{k+1} + \frac{1}{k(k+1)} - \alpha(t\frac{k+2}{k+1} - 1) \geq t\frac{k+2}{k+1} + \frac{1}{k(k+1)} - \frac{t\frac{k+2}{k+1}-1}{k} = t\frac{k+2}{k+1}\frac{k-1}{k} + \frac{k+2}{k(k+1)}$. The inequality holds since the coefficient multiplied by α is negative and $\alpha \leq \frac{1}{k}$. We need to show that the weight is at least t, i.e. that $t(\frac{k^2+k-2}{k^2+k} - 1) + \frac{k+2}{k(k+1)} = \frac{-2t}{k^2+k} + \frac{k+2}{k^2+k} \geq 0$. We get that this holds for $t \leq \frac{k+2}{2}$.

Consider the second case. The proof of the first case shows that it is enough to consider the first $f = t - \lfloor \frac{k+2}{2} \rfloor$ transition bins, and to show that the total weight of items in these bins is at least f. These bins are bins of types $i_1, \ldots, i_f$. Consider the bin of type i_{f+1}. Note that $i_{f+1} \leq k - \lfloor \frac{k+2}{2} \rfloor$ since no two transition bins are of the same type, and $i_{f+1} \geq 3$, since $i_f \geq 2$. Let β be the size of the first item in the bin of type i_{f+1}. Let $m = i_{f+1}$. Considering only items of sizes in $(\frac{1}{m}, \frac{1}{2}]$, we have that each bin out of the first f transition bins has total size of such items of at least $1 - \beta$. However, they also have a total size of items in $(\frac{1}{k+1}, \frac{1}{2}]$ of at least $1 - \alpha$. Therefore the weight of items in each such bin is at least $(1 - \beta)\frac{m+2}{m+1} + (\beta - \alpha)\frac{k+2}{k+1} = \frac{m+2}{m+1} + \beta(\frac{1}{k+1} - \frac{1}{m+1}) - \alpha\frac{k+2}{k+1}$. We will show that this amount is never smaller than 1. This expression is minimized for maximum values of α, β and thus we need to show, $(1 - \frac{1}{m}) \cdot \frac{m+2}{m+1} + (\frac{1}{m} - \frac{1}{k}) \cdot \frac{k+2}{k+1} - 1 \geq 0$, which is equivalent to $\frac{k-m}{km} \cdot \frac{k+2}{k+1} \geq \frac{2}{m(m+1)}$. Note that $k - m \geq \frac{k+1}{2}$, $m + 1 \geq 4$ and thus $\frac{k-m}{km}\frac{k+2}{k+1} \cdot \frac{m(m+1)}{2} \geq \frac{k+2}{k} > 1$. This completes the proof. □

In the sequel, we consider algorithms for an input I of the following structure. The set I is partitioned into ν independent sets. Out of these sets $\mu \leq \nu$ are packed using FFD. Each other independent set J is packed into a single bin and is assigned a total weight of at least 1.

Corollary 1. *An algorithm $\mathcal{B}$ as above satisfies $\mathcal{B} \leq W(I) + \mu$.*

We now give a tight analysis of the FFD based algorithm given in [15] for perfect graphs. That algorithm finds a coloring of all items with a minimum number of colors, and then uses FFD to pack each color class. It was shown in [15] that the performance guarantee of this algorithm is at most 2.7 and at least $1 + \Pi_\infty \approx 2.69103$. The value Π_∞ is the sum of a series and is computed using the well known sequence π_i, $i \geq 1$, which often occurs in bin packing. Let $\pi_1 = 2, \pi_{i+1} = \pi_i(\pi_i - 1) + 1$. Then $\Pi_\infty = \sum_{i=1}^{\infty} \frac{1}{\pi_i - 1}$. This sequence is presented e.g. in [2,19].

We are now ready to prove a matching upper bound of $1 + \Pi_\infty = 2.691$ for this algorithm. In order to do so, we need to find an upper bound on the total weight which can reside in one bin. The proof is similar to those of [2,19], however our weights are defined differently since these proofs do not hold in our case. We assume that the weight of an item s_x of size in $\mathcal{I}_1$ is $x + \frac{1}{2}$, which may only increase the total weight, since we assigned weight 1 to these items.

Lemma 2. *Consider a set of items J packed into one bin in* OPT. *Then* $W(J) \leq \Pi_\infty \approx 1.69103$.

Corollary 2. *The performance guarantee of the* FFD *based algorithm $\mathcal{A}$ of [15] for perfect graphs is $\Pi_\infty + 1 \approx 2.69103$.*

Note that Lemma 2 can be generalized as follows. For an integer $z \geq 1$, let $\pi_1(z) = z + 1, \pi_{i+1}(z) = \pi_i(z)(\pi_i(z) - 1) + 1$. Then $\Pi_\infty(z) = \sum_{i=1}^{\infty} \frac{1}{\pi_i(z) - 1}$.

Lemma 3. *Consider a set of items J which consists of the contents of a sub-bin of size $\frac{1}{z}$. Then $W(J) \leq \Pi_\infty(z)$.*

In order to analyze the *algorithm with precoloring*, we need to define a set of weights which does not give very high weights to items in $\mathcal{I}_1 = (\frac{1}{2}, 1]$. We define the weight for $s_x \in \mathcal{I}_1$ to be $weight(x) = s_x + \frac{1}{6}$. This unique definition it possible due to the special treatment of items in $\mathcal{I}_1$.

In order to establish a lemma regarding the sum of weights in an independent set, we modify the type of algorithms we allow to use. Once again, the set I is partitioned into ν independent sets. Each independent set has at most one item of size in $\mathcal{I}_1$. Out of these sets $\mu \leq \nu$ are packed using FFD. Each other independent set J is packed into a single bin, and is assigned a total weight of at least 1.

Lemma 4. *An algorithm $\mathcal{B}$ as above satisfies $\mathcal{B} \leq W(I) + \mu$.*

We can now show a tight analysis of the FFD based *algorithm with precoloring* given in [15]. That algorithm finds a coloring of all items with a minimum number of colors, with the restriction that items of size in $\mathcal{I}_1$ receive distinct colors, and then uses FFD to pack each color class. It was shown in [15] that the performance guarantee of this algorithm is at most 2.5 and at least $\Pi_\infty(3) + 2 \approx 2.4231$.

Theorem 1. *The performance guarantee of the* FFD *based algorithm with precoloring $\mathcal{B}$ of [15] is $\Pi_\infty(3) + 2 \approx 2.4231$.*

3   Improved Algorithms

In the previous section we showed better bounds for two variants of the problem, based on previously known algorithms from [15]. Though this already gives an improvement over the previously known bounds, the bounds we have shown are tight bounds, and thus further improvement is possible only using new algorithms, which we now design. To analyze these algorithms we use weighting in a more complex way.

3.1   Perfect Conflict Graphs

We design an algorithm which uses a preprocessing phase.

Algorithm Matching Preprocessing

1. Define the following bipartite graph. One set of vertices consists of all items of size in $\mathcal{I}_1$. The other set of vertices consists of all other items. An edge (a, b) between vertices of items of sizes $s_a > \frac{1}{2}$ and $s_b \leq \frac{1}{2}$ occurs if the two following conditions hold.
   (a) $s_a + s_b \leq 1$.
   (b) $(a, b) \notin E(G)$.
   That is, if these two items can be placed in a bin together. If this edge occurs, we give it the cost $c(a, b) = weight(b)$, where $weight(b)$ is defined as above to be $s_b + \frac{1}{j(j+1)}$, for the integer j such that $s_b \in (\frac{1}{j+1}, \frac{1}{j}]$.
2. Find a maximum cost matching in the bipartite graph.
3. Each pair of matched vertices are removed from G and packed into a bin together.
4. Let G' denote the induced subgraph over the items that were not packed in the preprocessing.
5. Compute a feasible coloring of G' using $\chi(G')$ colors.
6. For each color class, apply the First-Fit-Decreasing algorithm.

Theorem 2. *The above algorithm is a $\frac{5}{2} = 2.5$-approximation algorithm.*

Proof. The outline of the proof is as follows. We assign weights according to an optimal packing. Afterwards, we take the total weight and re-assign it to items so that the total weight does not grow and the conditions of Corollary 1 hold.

Fix an optimal packing, OPT. For a bin with no items of size in $\mathcal{I}_1$, weights are defined as before. For an item of size in $\mathcal{I}_1$ the weight is always 1. Given a bin with an item of size in $\mathcal{I}_1$ which contains additional items, pick an item of largest size in the bin among the items in the bin with size at most $\frac{1}{2}$, and give it weight zero. All other items in the bin receive weights as before. Note that the items which received zero weight, together with the items of size in $\mathcal{I}_1$ placed together with them in the same bins of OPT form a valid matching in the bipartite graph, whose cost is exactly the total reduction in the weights of items (compared to the weights used for perfect graphs in Section 2). We use the notation $weight_1$ for this reduced weight function, and $weight$ for the regular weight function (as used in the proofs for perfect graphs in Section 2). Let ω be the cost of the matching removed by the algorithm. Then by the optimality of the removed matching, we conclude that $\sum_{x \in I} (weight(x) - weight_1(x)) \leq \omega$. We re-assign weights to items so that an item of size in $(0, \frac{1}{2}]$ that was removed in the matching receives weight zero, and any other item receives a weight as usual (as defined by the function $weight$). This weight function (after the re-assignment) is called $weight_2$. We have $\sum_{x \in I} weight_2(x) + \omega = \sum_{x \in I} weight(x) \leq \omega + \sum_{x \in I} weight_1(x)$. Therefore, the total weight does not grow, and we may analyze the algorithm (but not OPT) using the weights $weight_2$. Clearly, each of the bins removed by the algorithm in the matching has weight of at least 1 since each of these contains an item of unit weight. Therefore, we can use Corollary 1 since the weights of items that are packed using FFD are the same as before.

Finally, we need to analyze the largest amount of weight that can be packed into a single bin of OPT. Using Theorem 1, we can see that if all item sizes are no larger than $\frac{1}{2}$, then this amount is smaller than $\frac{3}{2}$. We can use this as the weights of all the items considered here, are the same as in that proof. Consider now a bin with an item of size in $\mathcal{I}_1$. If this is the only item in the bin, then the total weight is 1. Otherwise let $x_1 \geq \ldots \geq x_t$ be the sorted list of other items in the bin, where x_1 is the item which was assigned a zero weight in $weight_1$. Let $j \geq 2$ be an integer such that $x_1 \in \mathcal{I}_j$. The total weight of the large item and items $x_1, \ldots, x_t$ is therefore at most $\frac{j+1}{j}(\sum_{i=2}^{t} s_{x_i}) + 1 \leq 1 + \frac{j+1}{j}(1 - \frac{1}{2} - \frac{1}{j+1}) = 1 + \frac{j+1}{2j} - \frac{1}{j} = 1 + \frac{j-1}{2j} < \frac{3}{2}$.   □

Proposition 1. *The approximation ratio of Algorithm Matching Preprocessing is at least 2.5.*

Remark 1. Algorithm Matching Preprocessing is a 2.5-approximation algorithm for BPC on any hereditary class of graphs for which one can find in polynomial time a coloring that uses a minimum number of colors.

3.2   Conflict Graphs That Belong to $\mathcal{C}$

In this section we study an approximation algorithm for the case where the conflict graph G belongs to $\mathcal{C}$. I.e., given an induced subgraph of G, $G' = (V', E')$ and a set of vertices $L' \subseteq V'$, we can find a coloring of G using a minimum number of colors such that each pair of vertices from L' are assigned distinct colors.

We analyze the following algorithm. The weight function $weight$ is defined as in Section 2 for items with size at most $\frac{1}{2}$ and for an item x such that $s_x \in \mathcal{I}_1$, $weight(x) = s_x + \frac{1}{6}$. We can use Lemma 4 since our algorithm will follow its conditions.

Algorithm Greedy Preprocessing

1. **While** there is a set of three items $\{a, b, c\}$ that can fit into one bin (i.e., $s_a + s_b + s_c \leq 1$ and $\{a, b, c\}$ is an independent set of G) such that $weight(a) + weight(b) + weight(c) > 1$ and $s_c \leq s_b \leq s_a \leq \frac{1}{2}$, or two items $\{a, b\}$ that can fit into one bin (i.e., $s_a + s_b \leq 1$ and $\{a, b\}$ is an independent set of G) such that $weight(a) + weight(b) > 1$ **do** as follows.

   Choose such a set A of maximal total weight. Delete A from G, and assign a new bin for the items of A that is dedicated to this set of items.

   Denote by $G' = (V', E')$ the resulting conflict graph induced by the remaining items.
2. Denote the set of large items by $L = \{j \in V' : s_j > \frac{1}{2}\}$, and denote by $\chi_I(G')$ the minimum number of colors used by the optimal solution for

the precoloring extension problem defined by G' and the set of precolored vertices L. Compute a feasible coloring of G' using $\chi_I(G')$ colors, where any two items in L are assigned different colors.

3. For each color class, apply the First-Fit-Decreasing algorithm.

Theorem 3. *The approximation ratio of the above algorithm is exactly* $\frac{7}{3} \approx 2.33333$.

3.3   Bipartite Graphs

In [15] it was shown that a simple algorithm which finds some coloring of the graph with two colors, and packs each color class using Next-Fit, is a 2-approximation. It was shown there that even if Next-Fit is replaced by FFD, still this algorithm does not have a better approximation ratio.

We design an algorithm which gives special treatment to some of the problematic cases and thus get a $\frac{7}{4}$-approximation.

We start with an analysis of the algorithm above (with FFD), which we call TWO-SET (TS), as a function of the value OPT. Let A and B denote the sets of items of the two colors. Let $\ell(A)$ and $\ell(B)$ denote the numbers of bins packed by FFD for each of the two sets, let $s(X)$ denote the sum of item sizes in a set X, i.e., $s(X) = \sum_{x \in X} s_x$, and let OPT$(X)$ denote the cost of an optimal solution for a set X. Clearly, we have $s(X) \leq \mathrm{OPT}(X) \leq \mathrm{OPT}$ for $X = A, B$, and also $\mathrm{OPT} \geq s(A) + s(B)$.

Simchi-Levi [27] proved that for any input Y, the solution of FFD on this output satisfies $\mathrm{FFD}(Y) \leq \frac{3}{2}\mathrm{OPT}(Y)$. Therefore, if the size of one of the sets (without loss of generality, the set A) is small enough, namely, this set fits into one bin $s(A) \leq 1$, we get $TS \leq \mathrm{FFD}(B) + 1 \leq \frac{3}{2}\mathrm{OPT} + 1$.

Otherwise, if for both sets, the output of FFD created at least one bin where the smallest item that opens a new bin is in the interval $(0, \frac{1}{3}]$. Then, for each set A and B, all bins but the last one are occupied by more than $\frac{2}{3}$, and the sum of items in the two last bins together is more than 1. We get for $X = A, B$, $s(X) > \frac{2}{3}(\ell(X) - 2) + 1$. Thus $TS \leq \ell(A) + \ell(B) < \frac{3}{2}\mathrm{OPT} + 1$.

Suppose next that both sets A and B do not have a bin opened by an item with size in the interval $(0, \frac{1}{3}]$. Then, we remove all items smaller than $\frac{1}{3}$ from the input. Clearly, the output does not change. Each bin contains an item of size in $(\frac{1}{2}, 1]$ (and possibly one smaller item as well) or two items in the interval $(\frac{1}{3}, \frac{1}{2}]$, except possibly the last bin for each set, that may contain a single item of this last interval. Let Z denote the number of items of size in $(\frac{1}{2}, 1]$ in $A \cup B$ and let V denote the number of items from $A \cup B$ with size in the interval $(\frac{1}{3}, \frac{1}{2}]$. Therefore, $TS \leq Z + \frac{V-2}{2} + 2 = Z + \frac{V}{2} + 1$. However, for any packing and thus for an optimal one we have that each bin contains at most one item with size larger than $\frac{1}{2}$, and at most two items with size larger than $\frac{1}{3}$, thus we have $\mathrm{OPT} \geq Z$ and $\mathrm{OPT} \geq \frac{Z+V}{2}$. We get $TS \leq \frac{Z+V}{2} + \frac{Z}{2} + 1 \leq \frac{3}{2}\mathrm{OPT} + 1$.

We are left with the case where (without loss of generality) the set A contains a bin opened by an item in $(0, \frac{1}{3}]$, and B does not. If A does not contain a bin opened by an item of size in $(0, \frac{1}{4}]$, we can remove all items smaller than

$\frac{1}{4}$ from the input, and get the same output. Let Z denote again the number of items in $(\frac{1}{2}, 1]$ and V denote the number of items in $(\frac{1}{4}, \frac{1}{2}]$. We now argue that $V \leq 3(\text{OPT} - Z) + Z = 3\text{OPT} - 2Z$. this last inequality holds, since a bin with an item larger than $\frac{1}{2}$, can contain at most one item larger than $\frac{1}{4}$, and any other bin can contain at most three such items. Therefore, $TS \leq Z + \frac{V-2}{2} + 2 \leq Z + \frac{3}{2}\text{OPT} - Z + 1 \leq \frac{3}{2}\text{OPT} + 1$.

Finally, we need to consider the case that A contains at least one bin opened by an item of size in $(0, \frac{1}{4}]$, and B does not have a bin opened by an item whose size is at most $\frac{1}{3}$. Thus all bins of A but the last one are occupied by more than $\frac{3}{4}$. We get $s(A) > \frac{3}{4}(\ell(A) - 2) + 1$ and $s(B) > \frac{1}{2}\ell(B)$. The last inequality holds for any Any-Fit type algorithm, and for FFD in particular. Moreover, note that the packing of B is an optimal one. This can be proved using simple exchange arguments (see [27]). Thus we have $\ell(B) \leq \text{OPT}$. We get $\text{OPT} \geq s(A) + s(B) > \frac{3}{4}\ell(A) + \frac{1}{2}\ell(B) - \frac{1}{2}$. Thus $SL < \frac{4}{3}\text{OPT} + \frac{2}{3} + \frac{1}{3}\text{OPT} = \frac{5}{3}\text{OPT} + \frac{2}{3}$. Since both OPT and SL are integers, we get $SL \leq \frac{5}{3}\text{OPT} + \frac{1}{3}$.

Lemma 5. *If* $\text{OPT} \geq 3$ *then the algorithm above satisfies* $SL \leq \frac{7}{4}\text{OPT}$ *and this bound is tight when* $\text{OPT} = 4$.

As we can see, the only case which is left is $\text{OPT} = 2$ which requires a special treatment. This case can be identified by a solution of cost 4. Clearly, such solutions can be achieved also for $\text{OPT} = 3$ and $\text{OPT} = 4$. We define an algorithm and prove that it succeeds if $\text{OPT} = 2$. Thus, if it fails, then $\text{OPT} \geq 3$ which means that the original solution already does not violate the approximation ratio $\frac{7}{4}$ which we would like to prove. We call this algorithm MODIFIED TWO-SET.

If $\text{OPT} = 2$, this means that it is possible to color the input using two colors, and pack each independent set into a single bin. If the conflict graph is connected, there is a unique way to color the items, and thus this optimal packing can be achieved. However, a bipartite disconnected graph has more than one possible coloring with two colors, since the roles of the two colors in each connected component can be swapped. As a first step, we color each connected component using two colors. Let z be the number of components, and denote the items of component i by V_i. For each $1 \leq i \leq z$, we get two sets A_i and B_i, such that $A_i \cup B_i = V_i$ and $A_i \cap B_i = \emptyset$. Each set contains the vertices of V_i that share a color. We define $p_i = |s(A_i) - s(B_i)|$. Let $q_i = \max\{s(A_i), s(B_i)\} - p_i$. The sizes p_i define a scheduling problem on two machines. We run LPT (Longest Processing Time First) on this input. This means that we initialize two empty sets, A and B. Sort the sizes p_i in non-increasing order. Then starting from the largest size, we assign each size to the set whose total sum is minimal. Graham [10] defined and analyzed this algorithm for an arbitrary number of machines (subsets). It is not difficult to see that when the algorithm terminates, we have $|s(A) - s(B)| \leq p_k$, where k is the last index of size assigned to the set with larger sum. For $1 \leq i \leq z$, we define a coloring using two colors (which are defined by the sets C and D) as follows. If $s(A_i) \geq s(B_i)$, and p_i is in A or if $s(A_i) < s(B_i)$, and p_i is in B, assign the items in A_i to C and the items in B_i to D. Otherwise assign the items in B_i to C and the items in A_i to D. This assignment means that $s(C) = \sum_{i \in A} p_i + \sum_{i=1}^{z} q_i$ and $s(D) = \sum_{i \in B} p_i + \sum_{i=1}^{z} q_i$. Thus we have $|s(C) - s(D)| \leq p_k$ as well. Assume (without loss of generality) that $s(C) \geq s(D)$. Since OPT $= 2$, $S(C) + S(D) \leq 2$. Thus $s(D) \leq 1$ and all the items assigned to D fit into a single bin. Now remove $p_k + q_k$ from $s(C)$. We get a total of less than $s(D) \leq 1$, and thus the remaining items of C fit into one bin. Finally the items of the larger set among $s(A_k)$ and $s(B_k)$ must be packed together in a solution with two bins only, and since OPT $= 2$, we get that these items also fit into one bin.

Theorem 4. *Algorithm* MODIFIED TWO-SET *has an approximation ratio of exactly* $\frac{7}{4}$.

Proof. We showed that if OPT $= 2$, the process above succeeds to pack the input into two bins. Otherwise, the theorem follows from Lemma 5.   □

4   Online Algorithms

In this section we discuss online algorithms for interval graphs. For many classes of graphs, the online problem is hard to approximate. The coloring problem is a special case of BPC, where all item sizes are zero.

Consider e.g. the problem on trees. Gyárfás and Lehel [11] proved a deterministic lower bound of $\Omega(\log n)$ on the online coloring of bipartite graphs on n vertices, which holds already for trees. Lovász, Saks and Trotter [20] showed an online coloring algorithm which colors such a graph (which is 2 colorable) using $O(\log n)$ colors. This immediately implies an online coloring algorithm for BPC on bipartite graphs, which is optimal up to a constant multiplicative factor on the competitive ratio. This algorithm $\mathcal{A}$ uses the algorithm of [20] to color the conflict graph using C colors. Then each color class is colored by some reasonable algorithm, e.g. Next-Fit. We get that for each color class i, which contains ℓ_i bins, the total size of items S_i is more than $\frac{\ell_i - 1}{2}$ (since no two consecutive bins can be combined). We get that $\mathcal{A} \leq \sum_{i=1}^{C} \ell_i < \sum_{i=1}^{C} (2S_i + 1) \leq 2\text{OPT} + C \leq O(\log n)\text{OPT}$.

Since the same can be applied for any graph class for which no constant competitive algorithm exists, we focus on a graph class for which such an algorithm exists, namely, interval graphs. Kierstead and Trotter [18] constructed an online coloring algorithm for interval graphs which uses at most $3\omega - 2$ colors where ω is the maximum clique size of the graph. They also presented a matching lower bound of $3\omega - 2$ on the number of colors in a coloring of an arbitrary online coloring algorithm. Note that the chromatic number of interval graphs equals to the size of a maximum clique, which is equivalent in the case of interval graphs to the largest number of intervals that intersect any point (see [16,9]). The technique above implies a 5-competitive algorithm. We can show that using First-Fit (FF) instead of Next-Fit for coloring each class slightly improves this bound.

Theorem 5. *The algorithm of [18] combined with* FF *for coloring each class has competitive ratio 4.7.*

We can show that an algorithm of much smaller competitive ratio does not exist.

Theorem 6. *The competitive ratio of any online algorithm for* BPC *on interval graphs has competitive ratio of at least* $\frac{155}{36} \approx 4.30556$.

In order to prove this theorem, we prove the following two lemmas.

Lemma 6. *Let c be a lower bound on the asymptotic competitive ratio of any online algorithm for standard bin packing, which knows the value* OPT *in advance. Then the competitive ratio for any online algorithm for* BPC *on interval graphs has competitive ratio of at least* $3 + c$.

Lemma 7. *Any online algorithm for standard bin packing, which knows the value* OPT *in advance, has competitive ratio of at least* $\frac{47}{36} \approx 1.30556$.

5   Conclusion

We have improved the upper bounds for BPC on perfect graphs, interval graphs (and a few related classes) and bipartite graphs. Most our results follow from adaptation of weighting systems to enable analysis of algorithms for BPC, and new algorithms which carefully remove small subgraphs of items which cause problematic instances. There is still a gap between the inapproximability which follows from bin packing, and the upper bounds. An open problem would be to close this gap.

Another open question is the following. As in [15], we used the absolute approximation ratio to analyze the performance of our algorithms. It can be seen that using the asymptotic approximation ratio, we can achieve a slightly better upper bound for bipartite graphs. It is unclear whether the same is true for other graph classes, i.e., whether the asymptotic approximation ratio for BPC is strictly lower than the absolute one for some cases.

References

1. E. Arkin and R. Hassin. On local search for weighted packing problems. *Mathematics of Operations Research*, 23:640–648, 1998.
2. B. S. Baker and E. G. Coffman, Jr. A tight asymptotic bound for next-fit-decreasing bin-packing. *SIAM J. on Algebraic and Discrete Methods*, 2(2):147–152, 1981.
3. E. G. Coffman, Jr., J. Csirik, and J. Leung. Variants of classical bin packing. In T. F. Gonzalez, editor, *Approximation algorithms and metaheuristics*. Chapman and Hall/CRC. To appear.
4. P. Crescenzi, V. Kann, M. M. Halldórsson, M. Karpinski, and G. J. Woeginger. A compendium of NP optimization problems. *http://www.nada.kth.se/ viggo/ problemlist/compendium.html*.
5. D. de Werra. An introduction to timetabling. *European Journal of Operational Research*, 19:151–162, 1985.

6. G. Galambos and G. J. Woeginger. Repacking helps in bounded space online bin packing. *Computing*, 49:329–338, 1993.
7. M. R. Garey, R. L. Graham, D. S. Johnson, and A. C. C. Yao. Resource constrained scheduling as generalized bin packing. *Journal of Combinatorial Theory (Series A)*, 21:257–298, 1976.
8. M. R. Garey and D. S. Johnson. *Computers and intractability*. W. H. Freeman and Company, New York, 1979.
9. M. C. Golumbic. *Algorithmic Graph Theory and Perfect Graphs*. Academic Press, 1980.
10. R. L. Graham. Bounds on multiprocessing timing anomalies. *SIAM Journal on Applied Mathematics*, 17:263–269, 1969.
11. A. Gyárfás and J. Lehel. On-line and first-fit colorings of graphs. *Journal of Graph Theory*, 12:217–227, 1988.
12. M. Hujter and Z. Tuza. Precoloring extension, III: Classes of perfect graphs. *Combinatorics, Probability and Computing*, 5:35–56, 1996.
13. S. Irani and V. J. Leung. Scheduling with conflicts, and applications to traffic signal control. In *Proc. of 7th Annual ACM-SIAM Symposium on Discrete Algorithms (SODA'96)*, pages 85–94, 1996.
14. K. Jansen. An approximation scheme for bin packing with conflicts. *Journal of Combinatorial Optimization*, 3(4):363–377, 1999.
15. K. Jansen and S. Öhring. Approximation algorithms for time constrained scheduling. *Information and Computation*, 132:85–108, 1997.
16. T. R. Jensen and B. Toft. *Graph coloring problems*. Wiley, 1995.
17. D. S. Johnson, A. Demers, J. D. Ullman, Michael R. Garey, and Ronald L. Graham. Worst-case performance bounds for simple one-dimensional packing algorithms. *SIAM Journal on Computing*, 3:256–278, 1974.
18. H. A. Kierstead and W. T. Trotter. An extremal problem in recursive combinatorics. *Congressus Numerantium*, 33:143–153, 1981.
19. C. C. Lee and D. T. Lee. A simple online bin packing algorithm. *Journal of the ACM*, 32(3):562–572, 1985.
20. L. Lovász, M. Saks, and W. T. Trotter. An on-line graph coloring algorithm with sublinear performance ratio. *Discrete Math.*, 75:319–325, 1989.
21. D. Marx. Precoloring extension. *http://www.cs.bme.hu/ dmarx/prext.html*.
22. D. Marx. Precoloring extension on chordal graphs. manuscript, 2004.
23. B. McCloskey and A. Shankar. Approaches to bin packing with clique-graph conflicts. Technical Report UCB/CSD-05-1378, EECS Department, University of California, Berkeley, 2005.
24. Y. Oh and S. H. Son. On a constrained bin-packing problem. Technical Report CS-95-14, Department of Computer Science, University of Virginia, 1995.
25. A. Schrijver. *Combinatorial optimization polyhedra and efficiency*. Springer-Verlag, 2003.
26. S. S. Seiden. On the online bin packing problem. *Journal of the ACM*, 49(5):640–671, 2002.
27. D. Simchi-Levi. New worst-case results for the bin-packing problem. *Naval Res. Logist.*, 41(4):579–585, 1994.
28. J. D. Ullman. The performance of a memory allocation algorithm. Technical Report 100, Princeton University, Princeton, NJ, 1971.
29. A. van Vliet. An improved lower bound for online bin packing algorithms. *Information Processing Letters*, 43(5):277–284, 1992.
30. A. C. C. Yao. New algorithms for bin packing. *Journal of the ACM*, 27:207–227, 1980.

Approximate Distance Queries in Disk Graphs

Martin Fürer and Shiva Prasad Kasiviswanathan

Computer Science and Engineering, Pennsylvania State University
{furer,kasivisw}@cse.psu.edu

Abstract. We present efficient algorithms for approximately answering distance queries in disk graphs. Let G be a disk graph with n vertices and m edges. For any fixed $\epsilon > 0$, we show that G can be preprocessed in $O(m\sqrt{n}\epsilon^{-1} + m\epsilon^{-2}\log S)$ time, constructing a data structure of size $O(n^{3/2}\epsilon^{-1} + n\epsilon^{-2}\log S)$, such that any subsequent distance query can be answered approximately in $O(\sqrt{n}\epsilon^{-1} + \epsilon^{-2}\log S)$ time. Here S is the ratio between the largest and smallest radius. The estimate produced is within an additive error which is only ϵ times the longest edge on some shortest path.

The algorithm uses an efficient subdivision of the plane to construct a sparse graph having many of the same distance properties as the input disk graph. Additionally, the sparse graph has a small separator decomposition, which is then used to answer distance queries. The algorithm extends naturally to the higher dimensional ball graphs.

1 Introduction

In this paper we consider the problem of preprocessing a graph such that subsequent distance queries can be answered quickly within a small error. This natural extension to the all pairs shortest path problem captures practical situations, where more often than not, we are interested in estimating the distance between two vertices quickly and accurately. In this framework, the goodness of an algorithm is typically measured in terms of the preprocessing time, query time, space complexity and approximation factor (if any).

A disk graph is an intersection graph of disks in the plane. We consider weighted disk graphs where the weight of an edge is the Euclidean distance between centers. We present a new method for answering distance queries in disk graphs within an additive error which is only ϵ times the longest edge on some shortest path. The results are also extended to their higher dimensional versions, the ball graphs. The *difficulty* in answering queries for the disk graph metric when compared to the metric induced by a complete Euclidean graph (where it is trivial) is that two points that are spatially close are not necessarily close under the graph metric.

The algorithm uses a hierarchical subdivision of the plane into tiles of different sizes to replace the (possibly dense) disk graph by a sparse graph having many of the same distance properties. The sparse graph has a small geometric separator decomposition, which is then exploited for answering distance queries. The input graph might have no small separator, it could even be complete.

T. Erlebach and C. Kaklamanis (Eds.): WAOA 2006, LNCS 4368, pp. 174–187, 2006.

Disk graphs have been used widely to model the communication between objects in VLSI [1] and recently in the context of wireless ad-hoc networks [2,3]. For wireless networks they model the fact that two wireless nodes can directly communicate with each other only if they are within a certain distance.

Distance queries are important in disk graphs as they are widely used to determine coverage in wireless sensor networks, and for routing protocols [4,5,6]. In most potential applications (like military) one would not only desire high accuracy of these estimates but also the actual path producing this estimate. Our approximation algorithms are designed keeping this in mind.

1.1   Related Work

Let $G = (V, E)$ be a weighted graph, and let $d_G(u, v)$ denote the length of a shortest path between vertices u and v in G. An estimate $\delta(u, v)$ of the path length $d_G(u, v)$ is said to be a c-stretch if it satisfies $d_G(u, v) \leq \delta(u, v) \leq c d_G(u, v)$. For general undirected graphs, Thorup and Zwick [7] show that for any $c \geq 1$, a graph with n vertices and m edges can be preprocessed in $O(cmn^{1/c})$ expected time, constructing a data structure of size $O(cn^{1+1/c})$, such that a $(2c - 1)$-stretch answer to any distance query can be produced in $O(c)$ time. Many other time-space trade-off results are also known (See Zwick's [8] survey on this subject).

For unit disk graphs, Gao and Zhang [6] gave a construction of a c-well-separated pair decomposition (introduced by Callahan and Kosaraju in [9]) with $O(n \log n)$ pairs for any constant $c \geq 1$. Using the well-separated pair decomposition and $O(n\sqrt{n} \log n \epsilon^{-3})$ time preprocessing, they show that an $(1 + \epsilon)$-stretch answer to any distance query can be produced in $O(1)$ time. They also show that for unit ball graphs in $\mathbb{R}^k$ at least $\Omega(n^{2-2/k})$ pairs are needed for the well-separated pair decomposition. However, one cannot hope to extend these results to general disks graphs, as general disk graphs do not have a sub-quadratic well-separated pair-decomposition. One such example is the star graph, formed by a big disk and $n - 1$ pairwise disjoint small disks intersecting the big disk.

In a very recent paper [10] the authors have used similar techniques as in this paper to devise a sub-quadratic time algorithm for constructing spanners of ball graphs. It is also shown that after $O(n^{2-1/k} \epsilon^{-k+1/2} + \epsilon^{-2k+1} \log S)$ time and space preprocessing an $(1 + \epsilon)$-stretch answer to any distance query can be produced in $O(n^{1-1/k} \epsilon^{-k+1/2} + \epsilon^{-2k+1} \log S)$ time and a $(2 + \epsilon)$-stretch answer in $O(\log n)$ time. The lower cost of preprocessing comes with a trade-off of worse error guarantees.

2   Preliminaries

Let $\mathcal{P}$ be a set of points in $\mathbb{R}^k$ for any fixed dimension k. Let $\mathcal{D}$ be a set of n balls such that (i) $D_u \in \mathcal{D}$ is centered at $u \in \mathcal{P}$, and (ii) D_u has radius of r_u. Balls D_u and D_v intersect if $d(u, v) \leq (r_u + r_v)$, where $d(.,.)$ denotes the Euclidean metric. The ball graph G is a weighted graph where an edge between u and v with weight $d(u, v)$ exists if D_u and D_v intersect. Let d_G denote the metric induced by the connected graph G on its vertices by shortest paths.

We use m to denote the number of edges in graph G. We require that the input to the algorithms is the set of balls, not only the corresponding intersection graph. We re-scale to ensure that in $\mathcal{D}$, for every ball D_u there exists at least one ball center outside D_u. We then re-scale the balls such that the largest radius equals one. The global scale factor (ratio between largest and smallest radius) of $\mathcal{D}$ is then defined as

$$\rho(\mathcal{D}) = 1/\min\{r_u \mid D_u \in \mathcal{D}\} .$$

For disk graphs our algorithms use a variant of quadtrees. For a node s, denote by $P(s)$ the parent of s in the tree. We use d_s to denote the depth of node s in the tree. A point (x, y) is *contained* in a node s representing a square with center (x_s, y_s) and length l_s iff $x_s - l_s/2 \leq x < x_s + l_s/2$ and $y_s - l_s/2 \leq y < y_s + l_s/2$. For a set of squares $\mathcal{S}$ in the quadtree a point is contained in $\mathcal{S}$ iff there exists $s \in \mathcal{S}$, such that point is contained in s. For two squares s and s', the distance $dist(s, s')$ is the Euclidean distance between their centers.

To avoid ambiguities, throughout the paper we refer to the vertices of a graph as *vertices* and vertices of a tree as *nodes*. We assume w.l.o.g. that ϵ^{-1} is a power of 2. Floors and ceilings are omitted throughout the paper, unless needed.

2.1   Separators and Separator Decomposition

A subset of vertices S of a graph G with n vertices is an $f(n)$-separator that α-splits ($\alpha < 1$) if $|S| \leq f(n)$ and the vertices of $G - S$ can be partitioned into two sets V_1 and V_2 such that there are no edges from V_1 to V_2, $max\{|V_1|, |V_2|\} \leq \alpha n$, where f is a function. An $f(K)$-separator decomposition of G is a recursive decomposition of G using separators, where subgraphs of size K have separators of size $O(f(K))$.

We use a rooted binary tree T_G to represent a separator decomposition of a graph $G = (V, E)$. For a set V' of vertices in G, we use $Ne(V')$ to denote the neighborhood of V'. Each node $t \in T_G$ is labeled by two subsets of vertices $V(t) \subseteq V$ and $S(t) \subseteq V(t)$. Let $G(t) = (V(t), E(t))$ denote the subgraph induced by $V(t)$. Then $S(t)$ is the separator in $G(t)$. The root $r \in T_G$ has $V(r) = V$ and $S(r)$ is a separator in G. For any $t \in T_G$, the labels of its children t_0, t_1 are defined as follows: let $V_1 \subset V(t)$ and $V_2 \subset V(t)$ be the components separated by $S(t)$ in $G(t)$. Then $V(t_0) = V_1 \cup (S(t) \cap Ne(V_1))$, $V(t_1) = V_2 \cup (S(t) \cap Ne(V_2))$.

2.2   Our Contributions

In this paper we use a quadtree like partitioning scheme to construct a new sparse graph. Given an input ball (disk) graph $G = (V, E)$ a weighted graph $G' = (V', E')$ is constructed such that: (i) V' is a subset of V, (ii) every vertex in G is close to some vertex in G' under the d_G metric, and (iii) the distance between any two vertices present in G' is not much larger than the distance between them in G. We refer to the elements of V' as the *representative vertices*. We call G' the *cluster graph* of G.

For any fixed $\epsilon > 0$, an estimate $\delta(u, v)$ of the distance between u and v is said to be $(1 + \epsilon)$-approximate if $d_G(u, v) \leq \delta(u, v) \leq d_G(u, v) + \epsilon d_G(u, v)$. We define a stronger notion called *strong* $(1 + \epsilon)$-*approximation*. An $(1 + \epsilon)$-approximate estimate $\delta(u, v)$ is said to be strong if $d_G(u, v) \leq \delta(u, v) \leq d_G(u, v) + \epsilon \ell(u, v)$, where

$$\ell(u, v) = max\{\ell \mid \exists \text{ a shortest path in } G \text{ between } u \text{ and } v \text{ of edge length } \ell\} .$$

Let G be a ball graph defined on $\mathcal{D}$, we show that after $O(mn^{1-1/k} \epsilon^{-k+1} + m\epsilon^{-k} \log \rho(\mathcal{D}))$ time and $O(n^{2-1/k} \epsilon^{-k+1} + n\epsilon^{-k} \log \rho(\mathcal{D}))$ space preprocessing, a strong $(1 + \epsilon)$-approximate estimate for the distance between any two vertices can be obtained in $O(n^{1-1/k} \epsilon^{-k+1} + \epsilon^{-k} \log \rho(\mathcal{D}))$ time. We can also output a corresponding short path between the query vertices in $O(L)$ time, where L is the number of edges of the reported path. In all our cases $\ell(u, v)$ is strictly less than $d_G(u, v)$. Therefore our approximation is strictly better than the standard $(1 + \epsilon)$-approximation.

To illustrate the main ideas in our algorithm we start by considering the easier case where all the disks have *almost* the same radius, i.e., when every radius is $\Theta(1)$.

3   Distance Queries in Almost Unit Disk Graphs

We first describe the construction of the cluster graph G' and then the algorithm for finding a separator decomposition.

Imposing the Grid: The input to our algorithm is a set of disks in $\mathbb{R}^2$. Let $\mathcal{P}$ be the set of their centers. The *bounding box* of $\mathcal{P}$ is the smallest rectangle enclosing $\mathcal{P}$. We assume the left bottom corner as the origin. An ϵ-grid is defined by horizontal and vertical line segments drawn at $y \in \epsilon\mathbb{Z}$ and $x \in \epsilon\mathbb{Z}$ within the bounding box.

Constructing G': Let *Roots* represent the non-empty squares of the ϵ-grid. For every square $s \in Roots$, the vertices contained in s form a clique in G. One such vertex is chosen as the representative, R_s. Define the *neighborhood* $N(s)$, as the set of all $s' \in Roots$ which are within a distance of 2 from s. Note that the choice of distance 2 here is for the case where all the disks have the same radius. In general one can upper bound this distance as twice the radius of largest disk in $\mathcal{D}$. The vertices of G' are the representatives of the squares in *Roots* and the edge $(R_s, R_{s'})$ is added to G' if $s' \in N(s)$ (see Figure 1). For every square s, $|N(s)| = O(\epsilon^{-2})$. Since the graph G' has at most n vertices, it can be constructed in $O(n\epsilon^{-2})$ time.

Separator in G': A $\sqrt{n}\epsilon^{-1}$-vertex separator in G' with 2/3-split can be found in $O(n \log n)$ by noting that G' is an $O(\epsilon^{-1})$-overlap graph as defined by Miller *et al.* [11] and by using their results for geometric separators on these graphs. In Appendix A, we provide a simpler algorithm having the same running time, but guaranteeing a superior split ratio.

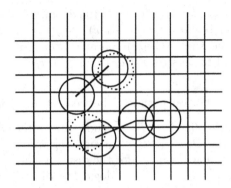

Fig. 1. The grid with $\epsilon = 1/2$. The centers of solid disks are the representative vertices used in G'. The centers of dotted disks are ignored. The edges of G' are also shown.

Theorem 1. *Let G be an (almost) unit disk graph and G' be the cluster graph constructed from G as described above. An $O(\sqrt{n}\epsilon^{-1})$-separator decomposition with $1/2$-split of G' can be found in $O(n \log n)$ time.*

Strong $(1 + \epsilon)$-approximate answers to distance queries: The query algorithm is similar to the one used by Arikati *et al.* [12]. We discuss the procedure for a single node $t \in T_{G'}$. The preprocessing phase involves the following steps: (i) Compute $T_{G'}$ the separator decomposition tree for G'. (ii) Let $H(t)$ denote the graph induced by the set of vertices

$$\{v \mid \exists s \in Roots \; \exists u \in V(t) \text{ such that } u = R_s \text{ and } v \text{ is contained in } s\}$$

on G. Intuitively the graph $H(t)$ is the induced graph of vertices belonging to either $V(t)$ or contained in the same square as a vertex in $V(t)$. From each node in $S(t)$ do a single source shortest path (SSSP) computation on $H(t)$.

The query procedure for finding an approximately shortest path between the vertices u and v of G consists of the following steps: (i) If there is an edge between u and v, set $\delta(u, v)$ to $d(u, v)$. (ii) Otherwise, (a) Initialize $\delta(u, v)$ to ∞. (b) Compute $s(u)$ and $s(v)$ as nodes in $Roots$ with u contained in $s(u)$ and v in $s(v)$. (c) Find the least common ancestor of $R_{s(u)}$ and $R_{s(v)}$ in $T_{G'}$, say t'. (d) Estimate $\delta(u, v)$ as $min\{\delta(u, v), min_{z \in S(t')}\{d_G(u, z) + d_G(z, v)\}\}$. (e) If t' is not the root of $T_{G'}$, set $t' = P(t')$ and repeat step (d).

Using the algorithm of Schieber and Vishkin [13], the least common ancestor queries can be answered in $O(1)$ time after linear time preprocessing. The proof of correctness of the algorithm follows as in [12] and is omitted in this extended abstract.

Theorem 2. *Let G be an (almost) unit disk graph with $m = \Omega(n \log n)$ edges. The graph G can be preprocessed in $O(m\sqrt{n}\epsilon^{-1})$ time, producing a data structure of size $O(n^{3/2}\epsilon^{-1})$, such that subsequent distance queries can be answered approximately, in $O(\sqrt{n}\epsilon^{-1})$ time. The outputs produced are strong $(1+\epsilon)$-approximate distance estimates.*

Proof. If there is an edge between u and v in G, then the actual distance is the estimate. Otherwise, we know $d_G(u, v) > 2$. Consider a shortest path $P_s = (u_1, u_2, \ldots, u_k)$ between $u_1 = u$ and $u_k = v$ with no shortcuts, i.e., no edge from u_{j-1} to u_{j+1} for any j between 2 and $k - 1$. Such a path P_s exists due to the triangle inequality. Since $d(u_{j-1}, u_{j+1}) > 2$, $max\{d(u_{j-1}, u_j), d(u_j, u_{j+1})\} \geq 1$. Let $(s(u_1), s(u_2), \ldots, s(u_k))$ be the sequence of squares in $Roots$ such that $s(u_i)$ contains u_i. We know the path $(R_{s(u_1)}, \ldots, R_{s(u_k)})$ exists in G'. Let t be a minimal depth node in $T_{G'}$ where this path gets separated. This implies that among the vertices $\{R_{s(u_1)}, \ldots, R_{s(u_k)}\}$ at least one is in $S(t)$, say $R_{s(u_i)}$.

We compute single source shortest path from the vertex $R_{s(u_i)}$, which is at most $\sqrt{2}\epsilon$ distance away from u_i. There is also an edge between u_i and $R_{s(u_i)}$ in G. This proves that the estimate is only $O(\epsilon)$ greater than actual shortest path. It is also a strong $(1 + \epsilon)$-approximate estimate, because there exists an edge in the path of P_s with length at least 1.

The running time of the preprocessing is dominated by the time for running Sssp from all the separator nodes. Using Dijkstra for Sssp gives a running time of $O(m\sqrt{n}\epsilon^{-1})$ for the preprocessing phase. The storage needed for all the results of Sssp computations is $O(n^{3/2}\epsilon^{-1})$. The query time is dominated by Steps (d) and (e) of the query algorithm and can be bounded by $O(\sqrt{n}\epsilon^{-1})$.   ☐

4   Distance Queries in Ball Graphs

In this section we extend the results to arbitrary ball graphs. Again we describe the algorithm for the case of disk graphs and then state the extensions to higher dimensions. Let G denote the input disk graph. Each disk is associated with a level, a disk D_u is of level l if: $2^{-l} \leq r_u < 2^{-l+1}$. Let l_{min} denote the level of the smallest disk in $\mathcal{D}$, i.e., $l_{min} = \lceil \log \rho(\mathcal{D}) \rceil$. The level l_{min} is the deepest level in Γ.

Imposing the Grid: We impose an ϵ-grid as in the case of unit disk graphs. Additionally, we recursively subdivide each non empty square in the ϵ-grid using a simple variant of quadtrees. We view this subdivision as a 4-ary forest with the root nodes as the non-empty squares in the ϵ-grid. Each square is partitioned into four equal squares, which form its children. We continue partitioning the non-empty squares until the size of the squares becomes $\epsilon 2^{-l_{min}}$. This construction differs from the standard quadtrees in: (a) when we stop partitioning it is not necessary that each square contains only one point, and (b) unlike in quadtrees we don't stop the partitioning as soon as the square has only one point inside it. We call this procedure *dissection* of the ϵ-grid.

Constructing G′: Let Γ denote the forest from the dissection of the ϵ-grid. Let $Roots$ initially be the collection of non-empty squares of the ϵ-grid. Γ is a collection of disjoint trees, each of which is rooted at a node of $Roots$. We introduce a disk of level l only at depth l in Γ. For a node $s \in \Gamma$, the set $C(s)$ only consists of disks which are of level d_s or less, and whose centers are contained in s. We also add to $Roots$ any node $s \in \Gamma$ satisfying, $C(s) \neq \emptyset$ whereas $C(P(s)) = \emptyset$.

For a leaf node s in Γ we pick one of the vertices in $C(s)$ as its representative R_s. For an internal node s, we pick one of its children s' satisfying $R_{s'} \in C(s)$ and set R_s to $R_{s'}$. For every node $s \notin Roots$, we define its neighborhood $N(s)$ as the set of all nodes at depth d_s which are within a distance of 2^{-d_s} from s. To define the neighborhood of a node $s \in Roots$, we introduce some new definitions. For a node s, define

$$region(s) = \{s' \in \Gamma \mid dist(s, s') \text{ lies in } [2^{-d_s+1}, 2^{-d_s+2}] \text{ and } d_s = d_{s'}\},$$

$$Region(s) = \bigcup_{s' = s \text{ or } s' \text{ is an ancestor of } s \text{ in } \Gamma} region(s') .$$

Empty squares can be ignored. The neighborhood of a node $s \in Roots$ is now defined as

$$N(s) = Region(s) \cup \{s' \in \Gamma \mid s' \text{ is at most } 2^{-d_s+1} \text{ from } s \text{ and } d_s = d_{s'}\} .$$

The idea behind creating the regions is to ensure that disks that intersect any disk centered in s, have their centers either close to s or inside a node of $Region(s)$ (Lemma 1). A similar idea has been used in [10] in the construction of spanners of disk graphs. See also Figure 2.

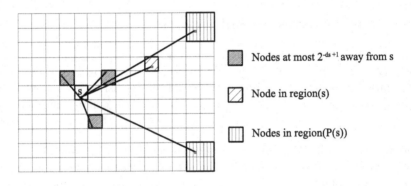

Fig. 2. Neighborhood of a node s belonging to $Roots$

Lemma 1. *Let u be a disk center contained in node $s \in Roots$. Then for every edge (u, v) in G, v is contained in $N(s)$.*

Proof. Trivially if $d(u, v) \leq 2^{-d_s+1}$ then $v \in N(s)$. If $2^{-d_s+1} < d(u, v) \leq 2^{-d_s+2}$ then v is contained in $region(s)$. If $2^{-d_s+2} < d(u, v) \leq 2^{-d_s+3}$ then v is contained in $region(P(s))$. Similarly for every increase in the distance by a factor of 2, we move one position up in Γ to finish the proof. □

The vertices of G' are the representatives of the nodes in Γ and the edge $(R_s, R_{s'})$ is added to G' if $s' \in N(s)$. The forest Γ can be constructed in $O(n \log \rho(\mathcal{D}))$ time. The graph G' can be constructed in $O(n\epsilon^{-2} \log \rho(\mathcal{D}))$ time.

Representative path of an edge: We now define a *representative path* in G' for every edge of G. For a vertex w in G, let $s(w)$ denote the deepest level node in Γ

containing w. Consider an edge (u, v) of the graph G. In the rest of the discussion we assume w.l.o.g. that $r_u \leq r_v$. The representative path $P(u, v)$ starts at $R_{s(u)}$ and ends at $R_{s(v)}$. The following lemma is useful for defining the path.

Lemma 2. *Let u and v be two vertices in G' such that there exists a node $s \in \Gamma$ containing both u and v, with $v \in C(s)$. Then there exists a path from u to v in G'.*

Proof. The proof is by induction over the number of nodes contained in s. The base case is when s contains only one vertex. Now in the subtree of Γ rooted at s, consider the node s_1 at which u and v split, i.e., children of s_1 containing u and v are different. The splitting is guaranteed as u and v are representatives for different nodes at the deepest level in Γ. Let s_2 be the closest descendant of s_1 containing u with $C(s_2) \neq \emptyset$. Let s_3 be the child of s_1 containing v. In G' there exists an edge (R_{s_2}, R_{s_3}). By the inductive hypothesis we know there exists a path between u and R_{s_2} and between v and R_{s_3}. Thus in G' there exists a path $(u, \ldots, R_{s_2}, R_{s_3}, \ldots, v)$. □

Let a be the deepest node in *Roots* containing u. Let b be a node in Γ containing v with $d_a = d_b$. Since u and $R_{s(u)}$ (similarly v and $R_{s(v)}$) are always contained in the same node in Γ, we get that $R_{s(u)}$ is contained in a and $R_{s(v)}$ is contained in b. The representative path $P(u, v)$ for an edge (u, v) is defined using the following case distinction:

Case 1: If the distance between a and b is greater than $2^{-d_a + 1}$, then by Lemma 1, we know that there exists some $c \in Region(a)$ containing v (and thus also $R_{s(v)}$). Since $dist(a, c) \geq 2^{-d_c + 1}$, we also know that $r_v \geq 2^{-d_c}$ and $v \in C(c)$. In G' there is an edge between R_a and R_c. From Lemma 2, we know there exists a path in G' between $R_{s(u)}$ and R_a and between $R_{s(v)}$ and R_c. Define the representative path $P(u, v)$ as $(R_{s(u)}, \ldots, R_a, R_c, \ldots, R_{s(v)})$.

Case 2: Otherwise, consider the deepest nodes f, g in Γ, such that (i) $R_{s(u)}$ is contained in f and $R_{s(v)}$ is contained in g, and (ii) there exists an edge (R_f, R_g) in G'. From Lemma 2, we know there exists a path between $R_{s(u)}$ to R_f and between $R_{s(v)}$ and R_g in G'. Define the representative path $P(u, v)$ as $(R_{s(u)}, \ldots, R_f, R_g, \ldots, R_{s(v)})$.

For every representative path, we also define a pair of nodes in Γ as its *covering nodes*. If $P(u, v)$ is defined using Case 1, then the nodes a and c are the covering nodes. If $P(u, v)$ is defined using Case 2, then f and g are the covering nodes. In both cases, all vertices in $P(u, v)$ are contained in one of the covering nodes with u and v contained in different covering nodes.

Lemma 3. *Let (u, v) be an edge in G of length greater than $2^{-l_{min}}$. Let $P(u, v)$ be its representative path in G' with p and q as the covering nodes. Then $dist(p, q) \geq max\{c_1 2^{-d_p}, c_1 2^{-d_q}\}$, for some constant c_1.*

Proof. We again use the same case distinction as in defining the path $P(u, v)$. Assume p contains u, and q contains v. If $P(u, v)$ is defined using Case 1, then $d_p \geq d_q$. Let r be the ancestor of p which is at the same depth as q. Since q is at least $2^{-d_q + 1}$ away from r, we get that that $dist(p, q) \geq c_1 2^{-d_q}$ (p is a square inside r).

If $P(u, v)$ is defined using Case 2, then $d_p = d_q$. Let p' be the child of p containing u. Let q' be child of q containing v. Since there is no edge between $R_{p'}$ and $R_{q'}$, we conclude $dist(p', q') \geq 2^{-d_q-1}$. This implies that $dist(p, q) \geq c_1 2^{-d_q}$. The existence of nodes p' and q' is guaranteed if $d(u, v) \geq (1 + \epsilon/\sqrt{2})2^{-l_{min}}$. Otherwise, $dist(p, q) = \Omega(2^{l_{min}})$ and again $dist(p, q) \geq c_1 2^{-d_q}$.   □

Separator in G': We now show that a $\sqrt{n}\epsilon^{-1} + \epsilon^{-2} \log \rho(\mathcal{D})$-vertex separator in G' with 7/9-split[1] can be found in $O(n \log n)$. Similar ideas were used in [10] for finding a separator in the spanner of a disk graph. We use recursive partitions of rectangles. Let $\mathcal{D}(l)$ be the set of all vertices (disk centers) which were chosen as representatives at level l of Γ. For a rectangle $\mathcal{R}$, let $\mathcal{X}(\mathcal{R})$ be the sorted list of x-coordinates of vertices of G' which are contained in $\mathcal{R}$ (similarly define $\mathcal{Y}(\mathcal{R})$ for y-coordinates). We say a vertex crosses a given line if any edge incident on it in G' crosses the line.

At every step, the algorithm focuses on one rectangle, which we call the *active rectangle*. An active rectangle $\mathcal{R}$ has at least 2/3 of the vertices of G' inside it and there exists a set of $O(\sqrt{n}\epsilon^{-1} + \epsilon^{-2} \log \rho(\mathcal{D}))$ vertices which when removed ensures that no remaining vertex of G' has an edge that crosses the boundary of $\mathcal{R}$.

A *vertical double line separator* of an active rectangle is a set of at most two vertical line segments that partitions the active rectangle, such that there exists a set of $O(\sqrt{n}\epsilon^{-1} + \epsilon^{-2} \log \rho(\mathcal{D}))$ vertices which when removed ensures that no remaining vertex crosses the vertical line segment (similarly define *horizontal double line separator*). Our algorithm recursively partitions an active rectangle alternatively with a vertical or a horizontal double line separator and stops when none of the new rectangles created contains enough vertices to become active.

Starting from the topmost level in Γ, we do the following step at every level l of Γ. The initial rectangle is the bounding box (and is active).

Constructing double line separators at level l: Let $\mathcal{R}$ be the currently active rectangle. Let $\mathcal{X}(\mathcal{R}) = \{x_1, x_2, \ldots\}$. We maintain sorted doubly linked lists for both the x- and y- coordinates with pointers between elements representing the same point in the two lists. Therefore in going to $\mathcal{R}$ from the previous active rectangle $\mathcal{R}_p$, the lists $\mathcal{X}(\mathcal{R})$ and $\mathcal{Y}(\mathcal{R})$ can be constructed in time proportional to the number of points removed from $\mathcal{R}_p$ to $\mathcal{R}$.

Let x_m denote the median of $\mathcal{X}(\mathcal{R})$. Starting at x_m we scan over the lower half list of $\mathcal{X}$ till we encounter the first x_l, such that $x_l - x_{l-\sqrt{n}\epsilon-1} \geq (1 + \epsilon/\sqrt{2})2^{-l+1}$. We define a vertical line segment (L_1) at $x = x_l$ with its end-points at boundaries of $\mathcal{R}$. Similarly starting from x_m, we scan over the upper half list of $\mathcal{X}(\mathcal{R})$ to find the first x_r such that $x_{r+\sqrt{n}\epsilon-1} - x_r \geq (1 + \epsilon/\sqrt{2})2^{-l+1}$. Again we define a similar vertical line segment (L_2) at $x = x_r$.

L_1 and L_2 divide $\mathcal{R}$ into at most 3 rectangles, of which only the rectangle $(\mathcal{R}_c)$ between L_1 and L_2 could be active. If $\mathcal{R}_c$ is active, we repeat the same procedure with $\mathcal{Y}(\mathcal{R}_c)$ to find horizontal line segments L_3 and L_4. The active rectangle (if any) created is associated with the level $l + 1$.

---

[1] A superior 2/3-split can be achieved by a more careful analysis.

Lemma 4. *The distance between vertical line separators L_1 and L_2 defined at depth l is less than $(1 + \epsilon/\sqrt{2})2^{-l+1}\sqrt{n}\epsilon$.*

Proof. We move in the lower (or upper list) only if there are $\sqrt{n}\epsilon^{-1}$ vertices to the left (or right) within a distance of $(1 + \epsilon/\sqrt{2})2^{-l+1}$. Since there are only n disks, the result follows. □

Remark. The same distance bound holds also for the distance between L_3 and L_4. As a consequence of the above lemma we also get that the lengths of L_1 and L_2 for $l \geq 1$ is at most $(1 + \epsilon/\sqrt{2})2^{-l+2}\sqrt{n}\epsilon$ (consider separation between line separators at level $l - 1$). We use the upper bound of $2^{-l+3}\sqrt{n}\epsilon$ for the lengths of L_1 and L_2 in the proof.

Lemma 5. *Let $\mathcal{R}$ be an active rectangle at depth l. L_1 and L_2 define a vertical double line separator for $\mathcal{R}$.*

Proof. We work with L_1. The case for L_2 is similar. Every edge in G' is attributed to the disk of larger radius, i.e., an edge (u, v) in G' is attributed to u if $r_u \geq r_v$. First consider the edges attributed to vertices in $\bigcup_{l' \geq l} D(l')$. These edges have length at most $(1 + \epsilon/\sqrt{2})2^{-l+1}$. Any such edge crossing L_1 must have both its end points within a distance of $(1 + \epsilon/\sqrt{2})2^{-l+1}$ from L_1. By construction we ensure that this number is $O(\sqrt{n}\epsilon^{-1})$. Therefore there exists a set of $O(\sqrt{n}\epsilon^{-1})$ vertices which when removed ensures that no edge of length less than or equal $(1 + \epsilon/\sqrt{2})2^{-l+1}$ crosses L_1. So if $l = 0$ we are done.

We now consider edges attributed to vertices in $\bigcup_{l' < l} D(l')$. Fix any level $l' < l$. From Lemma 4, and the remark above we know that the length of L_1 is less than $2^{-l+3}\sqrt{n}\epsilon$. Consider four lines, two vertical lines drawn at distance $2^{-l'+2}$ on both sides of L_1 and two horizontal lines drawn above and below L_1 at a distance of $2^{-l'+2}$ from the end points. The rectangular region $\mathcal{R}_{l'}$ formed by these four lines has an area of at most $2^{-l'+3}(2^{-l+3}\sqrt{n}\epsilon + 2^{-l'+3})$. Since each node (square) in level l' has an area of $(\epsilon 2^{-l'})^2$ and at most one vertex from each square can be in $D(l')$, we get that the number of vertices of $D(l')$ present in this area is at most $2^{-l+l'+6}\sqrt{n}\epsilon^{-1} + 128\epsilon^{-2}$.

Each edge attributed to vertices in $D(l')$ and crossing L_1 should have both its end points in $\mathcal{R}_{l'}$. Summing over all $l' < l$ we get that the total number of vertices in $\bigcup_{l' < l} D(l')$ that could have edges crossing L_1 is at most

$$\sum_{0 \leq l' < l} (2^{l'-l+6}\sqrt{n}\epsilon^{-1} + 128\epsilon^{-2}) = O(\sqrt{n}\epsilon^{-1} + \epsilon^{-2}l) .$$

Since there are only $\log \rho(\mathcal{D})$ levels in Γ, we can upper bound l by $\log \rho(\mathcal{D})$. Therefore there exist $O(\sqrt{n}\epsilon^{-1} + \epsilon^{-2}\log \rho(\mathcal{D}))$ vertices such that every edge crossing L_1 is incident on one of them. □

Final Shape of the Separator: We stop the algorithm when no rectangle is still active. Among the rectangles formed by partitioning the last active rectangle, the rectangle $\mathcal{R}_f$ with the largest number of disk centers forms one component of the separator. Since the last active rectangle had at least 2/3 of the vertices of G' and it gets divided into at most 3 rectangles, $\mathcal{R}_f$ contains at least $2/9$ of the vertices of G'. The vertices which are outside of $\mathcal{R}_f$ and are connected to some vertices inside form the vertex separator. The proof of termination follows from Lemma 5, because when the algorithm considers level l_{min}, the horizontal line separator divides the currently active rectangle into exactly two non-active rectangles.

A naive implementation of this algorithm runs in $O(n \log n + n \log \rho(\mathcal{D}))$ time. This can be improved to $O(n \log n)$ using the data structures introduced in [10]. Again we have that,

Theorem 3. *Let G be a disk graph on $\mathcal{D}$ and G' be the cluster graph constructed from G as described above. An $O(\sqrt{n}\epsilon^{-1} + \epsilon^{-2} \log \rho(\mathcal{D}))$-separator decomposition with $7/9$-splits of G' can be found in $O(n \log n)$ time.*

Strong $(1 + \epsilon)$-approximate answer to distance queries: Once we have the separator in G' the preprocessing and query procedure for estimating the distance is the same as with unit disk graphs. The following theorem shows that we get a strongly $(1 + \epsilon)$-approximation.

Theorem 4. *Let G be a disk graph on $\mathcal{D}$ with $m = \Omega(n \log n)$ edges. The graph G can be preprocessed in $O(m\sqrt{n}\epsilon^{-1} + m\epsilon^{-2} \log \rho(\mathcal{D}))$ time, producing a data structure of size $O(n^{3/2}\epsilon^{-1} + n\epsilon^{-2} \log \rho(\mathcal{D}))$, such that subsequent distance queries can be answered approximately, in $O(\sqrt{n}\epsilon^{-1} + \epsilon^{-2} \log \rho(\mathcal{D}))$ time. The outputs produced are strong $(1 + \epsilon)$-approximate distance estimates.*

Proof. If there is an edge between u and v in G, then the actual distance is the estimate. As in Theorem 2, consider a shortest path $P_s = (u_1, u_2 \ldots, u_k)$ between $u_1(= u)$ and $u_k(= v)$ with no edge shortcuts. Since there is always an edge if the distance between two vertices is less than $2^{-l_{min}+1}$, we get $d(u_{j-1}, u_{j+1}) \geq 2^{-l_{min}+1}$ and $max\{d(u_{j-1}, u_j), d(u_j, u_{j+1})\} \geq 2^{-l_{min}}$ for any j between 2 and $k-1$. Let $(s(u_1), s(u_2), \ldots, s(u_k))$ be the sequence of squares at the deepest level in Γ such that $s(u_i)$ contains u_i. Let t be the deepest node in $T_{G'}$ with $R_{s(u_1)}, R_{s(u_k)} \in V(t)$.

If any vertex $R_{s(u_i)}$ from $\{R_{s(u_1)}, \ldots, R_{s(u_k)}\}$ is in the separator $S(t)$, then we do a single source shortest path from $R_{s(u_i)}$ on $H(t)$. Now $R_{s(u_i)}$ is at most $\epsilon 2^{-l_{min}+1/2}$ distance away from u_i and there exists an edge between $R_{s(u_i)}$ and u_i in G. This proves the estimate produced is only $O(\epsilon)$ greater than the shortest path. The strong $(1 + \epsilon)$-approximation follows as there is an edge in P_s which is of length at least $2^{-l_{min}}$.

Otherwise, consider two vertices $R_{s(u_a)}$ and $R_{s(u_b)}$ from $\{R_{s(u_1)}, \ldots, R_{s(u_k)}\}$ which are separated into different components by $S(t)$ and there exists an edge (u_a, u_b) in G (by the assumption about the path P_s we get $|a - b| = 1$). Since there is no edge between $R_{s(u_a)}$ and $R_{s(u_b)}$, we get that $d(u_a, u_b) \geq 2^{-l_{min}}$. Consider the representative path $P(u_a, u_b)$ in G'. There exists at least one vertex (say z) on the path $P(u_a, u_b)$ in the separator $S(t)$. Let p and q be the two covering nodes for $P(u_a, u_b)$. By definition all the vertices in $P(u_a, u_b)$ are contained in either p or q. Assume w.l.o.g. that z and u_a are contained in p.

In G there exists edges (z, R_p) and (R_p, u_a). Since z, R_p and u_a are all contained in p we get $d(z, R_p), d(R_p, u_a) \leq \epsilon 2^{-d_p+1/2}$. On the other hand, $dist(p, q) \geq max\{c_1 2^{-d_p}, c_1 2^{-d_q}\}$ by Lemma 3. This also implies that $d(u_a, u_b) \geq c_2 2^{-d_p}$ for some constant c_2. Putting all together we get $d(z, R_p) \leq c_3 \epsilon d(u_a, u_b)$ and $d(R_p, u_a) \leq c_3 \epsilon d(u_a, u_b)$ for some constant c_3, implying in G there is a path from z to u_a of length at most $2c_3 \epsilon d(u_a, u_b)$. Therefore when we do the single source shortest path computation from z on $H(t)$, the error we make in taking the detour can be bounded by some constant times $\epsilon d(u_a, u_b)$. Implying that our estimate is a strong $(1 + \epsilon)$-approximation.

The running time analysis follows as in Theorem 2. $\qquad\square$

4.1 Extension to Higher Dimensions

In $\mathbb{R}^k$, squares become k-dimensional hypercubes and the lines used in the separator algorithm become $(k-1)$-dimensional hyperplanes. Given a ball graph G in $\mathbb{R}^k$, the corresponding cluster graph G' can be constructed in $O(n\epsilon^{-k} \log \rho(\mathcal{D}))$ time. Using the same algorithm for answering distance queries we have the following.

Theorem 5. *Let G be a k-dimensional ball graph on $\mathcal{D}$ with $m = \Omega(n \log n)$ edges. The graph G can be preprocessed in $O(mn^{1-1/k} \epsilon^{-k+1} + m\epsilon^{-k} \log \rho(\mathcal{D}))$ time, producing a data structure of size $O(n^{2-1/k} \epsilon^{-k+1} + n\epsilon^{-k} \log \rho(\mathcal{D}))$, such that subsequent distance queries can be answered approximately, in $O(n^{1-1/k} \epsilon^{-k+1} + \epsilon^{-k} \log \rho(\mathcal{D}))$ time. The outputs produced are strong $(1+\epsilon)$-approximate distance estimates.*

5 Concluding Remarks

Our results are also applicable in the case where the disk graphs are directed (such models are used in wireless networks to capture radio interferences [14]). All the results presented in this paper can also be extended to cases when intersections are between squares, or regular polygons, or other disk-like objects, as well as their higher dimensional versions. This follows as our algorithms don't use any special properties of balls (or disks). However, the partitioning scheme only works if the objects have almost the same aspect ratio. An open problem is to extend the results to intersection graphs between objects not having this property.

References

1. Mead, C., Conway, L.: Introduction to VLSI System. Addison-Wesley, Reading (1980)
2. Krumke, S.O., Marathe, M.V., Ravi, S.S.: Models and approximation algorithms for channel assignment in radio networks. Wireless Networks **7**(6) (2001) 575–584
3. Li, X.Y.: Algorithmic, geometric and graphs issues in wireless networks. Wireless Communications and Mobile Computing **3**(2) (2003) 119–140
4. Li, X.Y., Wan, P.J., Frieder, O.: Coverage in wireless ad hoc sensor networks. IEEE Transactions on Computers **52**(6) (2003) 753–763

5. Srinivas, A., Modiano, E.: Minimum energy disjoint path routing in wireless ad-hoc networks. In: MOBICOM '03, ACM (2003) 122–133
6. Gao, J., Zhang, L.: Well-separated pair decomposition for the unit-disk graph metric and its applications. SIAM Journal on Computing **35**(1) (2005) 151–169
7. Thorup, M., Zwick, U.: Approximate distance oracles. Journal of ACM **52**(1) (2005) 1–24
8. Zwick, U.: Exact and approximate distances in graphs - A survey. In: ESA '01. Volume 2161., Springer (2001) 33–48
9. Callahan, P.B., Kosaraju, S.R.: A decomposition of multidimensional point sets with applications to K-nearest-neighbors and N-body potential fields. Journal of ACM **42**(1) (1995) 67–90
10. Fürer, M., Kasiviswanathan, S.P.: Spanners for geometric intersection graphs. Available at: http://www.cse.psu.edu/~kasivisw/research.html (2006)
11. Miller, G.L., Teng, S.H., Vavasis, S.A.: A unified geometric approach to graph separators. In: FOCS '01, IEEE (1991) 538–547
12. Arikati, S.R., Chen, D.Z., Chew, L.P., Das, G., Smid, M.H.M., Zaroliagis, C.D.: Planar spanners and approximate shortest path queries among obstacles in the plane. In: ESA '96. Volume 1136., Springer (1996) 514–528
13. Schieber, B., Vishkin, U.: On finding lowest common ancestors: Simplification and parallelization. SIAM Journal on Computing **17**(6) (1988) 1253–1262
14. Balakrishnan, H., Barrett, C.L., Kumar, V.S.A., Marathe, M.V., Thite, S.: The distance-2 matching problem and its relationship to the mac-layer capacity of ad hoc wireless networks. IEEE Journal on Selected Areas in Communications **22** (2004) 1069–1079

Appendix A: Separators in Cluster Graph for the Unit Disk Case

We describe the algorithm for the case when all the disks have the same radius. It can easily be modified for the case where the disk radius differ only by a constant factor.

Let $\mathcal{X} = \{x_1, x_2, \ldots\}$ be the sorted list of x-coordinates of vertices in G'. Similarly define $\mathcal{Y}$ for y-coordinates. Let x_m denote the median of $\mathcal{X}$. Starting at x_m we scan over the lower half list of $\mathcal{X}$ till we encounter the first x_l, such that $x_l - x_{l-\sqrt{n}\epsilon^{-1}} \geq 2 + \sqrt{2}\epsilon$. Define a vertical line (L_1) at $x = x_l$. We add all the vertices to the left of x_l and crossing L_1 into the separator. Similarly starting from x_m, we scan over the upper half list of $\mathcal{X}$ to find the first x_r such that $x_{r+\sqrt{n}\epsilon^{-1}} - x_r \geq 2 + \sqrt{2}\epsilon$. Define a vertical line (L_2) at $x = x_r$. Again we add all the vertices to the right of x_r crossing L_2 into the separator. This entire procedure can be done in linear time if the coordinates are sorted.

Final shape of the separator: Let $\mathcal{P}_b$ be the set of all vertices of G' lying between $x = x_l$ and $x = x_r$. Let $\mathcal{Y}_b$ be the sorted list of y-coordinates of vertices in $\mathcal{P}_b$. Let $\mathcal{P}_l$ be the set of vertices to the left of L_1. Choose the element in $\mathcal{Y}_b$ with rank $|\mathcal{Y}|/2 - |\mathcal{P}_l|$, say y_d. Define the horizontal line segment (L_3) at $y = y_d$ with (x_l, y_d) and (x_r, y_d) as its endpoints. We add all the vertices crossing this horizontal line segment into the separator. After removing the separator, the

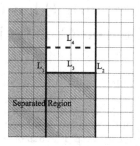

Fig. 3. The line segments L_1, L_2, and L_3 are as defined by the algorithm. The shaded region forms one separated component.

union of vertices in $\mathcal{P}_l$ and the vertices in $\mathcal{P}_b$ whose y-coordinates are less than y_d forms one component. See Figure 3.

Lemma 6. *The number of vertices of G' crossing the horizontal line segment L_3 is $O(\sqrt{n}\epsilon^{-1})$.*

Proof. Since the maximum edge length in G' is $2 + \sqrt{2}\epsilon$, from the construction we know that there are at most $\sqrt{n}\epsilon^{-1}$ vertices to the left of L_1 crossing L_1 and potentially L_3 (similarly from the right of L_2). Consider the vertical strip between L_1 and L_2. By construction we also get that distance between L_1 and L_2 is $O(\sqrt{n}\epsilon)$ and hence the length of L_3 is $O(\sqrt{n}\epsilon)$. We now only consider the area within the strip.

Define another horizontal line segment L_4 at $y = y_d + 2 + \sqrt{2}\epsilon$ with x_l and x_r defining its endpoints (the case where L_4 is defined as $y = y_d - 2 - \sqrt{2}\epsilon$ is symmetric). Consider the rectangular region formed by L_1, L_2, L_3 and L_4. Any edge crossing the line segment L_3 would have an endpoint in this rectangular region. The area of this rectangular region is at most $O(\sqrt{n}\epsilon)$. Since there is only one vertex of G' within each square of *Roots*, there exist at most $O(\sqrt{n}\epsilon^{-1})$ vertices within this rectangular region. Thus the total number of vertices of G' crossing L_3 is $O(\sqrt{n}\epsilon^{-1})$. $\square$

The $1/2$-split is guaranteed by the construction. The running time for finding the separator decomposition is dominated by the time for sorting.

Network Design with Edge-Connectivity and Degree Constraints[*]

Takuro Fukunaga and Hiroshi Nagamochi

Department of Applied Mathematics and Physics,
Graduate School of Informatics, Kyoto University, Japan
{takuro,nag}@amp.i.kyoto-u.ac.jp

Abstract. We consider the following network design problem; Given a vertex set V with a metric cost c on V, an integer $k \geq 1$, and a degree specification b, find a minimum cost k-edge-connected multigraph on V under the constraint that the degree of each vertex $v \in V$ is equal to $b(v)$. This problem generalizes metric TSP. In this paper, we propose that the problem admits a ρ-approximation algorithm if $b(v) \geq 2$, $v \in V$, where $\rho = 2.5$ if k is even, and $\rho = 2.5 + 1.5/k$ if k is odd. We also prove that the digraph version of this problem admits a 2.5-approximation algorithm and discuss some generalization of metric TSP.

1 Introduction

It is a main concern in the field of network design to construct a graph of the least cost which satisfies some connectivity requirement. Actually many results on this topic have been obtained so far. In this paper, we consider a network design problem that asks to find a minimum cost k-edge-connected multigraph on a metric edge cost under degree specification. This provides a natural and flexible framework for treating many network design problems. For example, it generalizes the vehicle routing problem with m vehicles (m-VRP) [3,7], which will be introduced below, and hence contains a well-known metric traveling salesperson problem (TSP), which has already been applied to numerous practical problems [8].

Let $\mathbb{Z}_+$ and $\mathbb{Q}_+$ denote the sets of non-negative integers and non-negative rational numbers, respectively. Let $G = (V, E)$ be a multigraph with a vertex set V and an edge set E, where a multigraph may have some parallel edges but is not allowed to have any loops. For two vertices u and v, an edge joining u and v is denoted by uv. Since we consider multigraphs in this paper, we distinguish two parallel edges $e_1 = uv$ and $e_2 = uv$, which may be simply denoted by uv and uv. For a non-empty vertex set $X \subset V$, $d(X; G)$ (or $d(X)$) denotes the number of edges whose one end vertex is in X and the other is in $V - X$. In particular $d(v; G)$ (or $d(v)$) denotes the degree of vertex v in G. The edge-connectivity $\lambda(u, v; G)$ (or $\lambda(u, v)$) between u and v is the maximum number

---

[*] This research was partially supported by the Scientific Grant-in-Aid from Ministry of Education, Culture, Sports, Science and Technology of Japan.

T. Erlebach and C. Kaklamanis (Eds.): WAOA 2006, LNCS 4368, pp. 188–201, 2006.

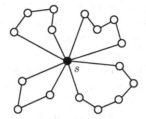

Fig. 1. A solution for 4-VRP

of edge-disjoint paths between them in G. The edge-connectivity $\lambda(G)$ of G is defined as $\min_{u,v \in V} \lambda(u, v; G)$. If $\lambda(G) \geq k$ for some $k \in \mathbb{Z}_+$, then G is called k-*edge-connected*. For a function $r : \binom{V}{2} \to \mathbb{Z}_+$, G is called r-*edge-connected* if $\lambda(u, v; G) \geq r(u, v)$ for every $u, v \in V$. Edge cost $c : \binom{V}{2} \to \mathbb{Q}_+$ is called *metric* if it obeys the triangle inequality, i.e., $c(uv) + c(vw) \geq c(uw)$ for every $u, v, w \in V$.

For a degree specification $b : V \to \mathbb{Z}_+$, a multigraph G with $d(v; G) = b(v)$ for all $v \in V$ is called a *perfect b-matching*. In this paper, we focus on the following network design problem.

k-edge-connected multigraph with degree specification (k-ECMDS):
A vertex set V, a metric edge cost $c : \binom{V}{2} \to \mathbb{Q}_+$, a degree specification $b : V \to \mathbb{Z}_+$, and a positive integer k are given. We are asked to find a minimum cost perfect b-matching $G = (V, E)$ of edge-connectivity k. $\qquad\qquad\square$

In this paper, we suppose that $b(v) \geq 2$ for all $v \in V$ unless stated otherwise, and propose approximation algorithms to k-ECMDS in this case.

Problem k-ECMDS is a generalization of m-VRP, which asks to find a minimum cost set of m cycles, each containing a designated initial city s, such that each of the other cities is covered by exactly one cycle (see Fig. 1). Observe that this problem is 2-ECMDS where $b(s) = 2m$ for the initial city $s \in V$ and $b(v) = 2$ for every $v \in V - s$. If $m = 1$, then m-VRP is exactly TSP. Since TSP is known to be NP-hard [11] even if a given cost is metric (metric TSP), k-ECMDS is also NP-hard. If a given cost is not metric, TSP cannot be approximated unless P = NP [11]. For m-VRP, there is a 2-approximation algorithm based on the primal-dual method [7].

It is well studied to find a minimum cost multigraph either with k-edge-connectivity or with degree specification. It is known that finding a minimum cost k-edge-connected graph is NP-hard since it is equivalent to metric TSP when $k = 2$ and a given edge cost is metric. On the other hand, it is known that a minimum cost perfect b-matching can be constructed in polynomial time (for example, see [10]). As a prior result on problems equipped with both edge-connectivity requirements and degree constraints, Frank [1] showed that it is polynomially solvable to find a minimum cost r-edge-connected multigraph G with $\ell(v) \leq d(v; G) \leq u(v)$, $v \in V$ for degree lower and upper bounds $\ell, u : V \to \mathbb{Z}_+$ and a metric edge cost c such that $c(uv)$ is defined by $w(u) + w(v)$ for some weight $w : V \to \mathbb{Q}_+$ (in particular, $c(uv) = 1$ for every $uv \in \binom{V}{2}$).

Recently Fukunaga and Nagamochi [4] presented approximation algorithms for a network design problem with a general metric edge cost and some degree bounds; For example, they presented a $(2 + 1/\lfloor \min_{u,v \in V} r(u,v)/2 \rfloor)$-approximation algorithm for constructing a minimum cost r-edge-connected multigraph that meets a local-edge-connectivity requirement r with $r(u,v) \geq 2$, $u, v \in V$ under a uniform degree upper bound. Afterwards Fukunaga and Nagamochi [5] gave a 3-approximation algorithm for the case where $r(u,v) \in \{1,2\}$ for every $u, v \in V$ and $\ell(v) = u(v)$ for each $v \in V$. In this paper, we extend the 3-approximation result [5] to k-ECMDS. Concretely, we prove that k-ECMDS is ρ-approximable if $b(v) \geq 2$, $v \in V$, where $\rho = 2.5$ if k is even and $\rho = 2.5 + 1.5/k$ if k is odd. To design our algorithms for k-ECMDS, we take a similar approach with famous 2- and 1.5-approximation algorithms for metric TSP.

Furthermore, we also generalize k-ECMDS to a network design problem in digraphs. We denote an arc (i.e., a directed edge) from a vertex u to another vertex v by uv. Two arcs from u to v are called *parallel*. Let $D = (V, A)$ be a multi-digraph, where a multi-digraph may have some parallel arcs but is not allowed to have any loops. For an ordered pair of vertices u and v, $\lambda(u, v; D)$ (or $\lambda(u, v)$) denotes the arc-connectivity from u to v, i.e., the maximum number of arc-disjoint paths from u to v in D. The arc-connectivity $\lambda(D)$ of D is defined as $\min_{u,v \in V} \lambda(u, v; D)$. If $\lambda(D) \geq k$ for some $k \in \mathbb{Z}_+$, D is called k-*arc-connected*. Moreover, $d^-(v; D)$ (or $d^-(v)$) and $d^+(v; D)$ (or $d^+(v)$) denote in- and out-degree of vertex v in digraph D, respectively. Arc cost $c : V \times V \to \mathbb{Q}_+$ is called *symmetric* if $c(uv) = c(vu)$ for every $u, v \in V$, and *metric* if it obeys the triangle inequality, i.e., $c(uv) + c(vz) \geq c(uz)$ for every $u, v, z \in V$.

We call a multi-digraph D with $d^-(v; D) = b^-(v)$ and $d^+(v; D) = b^+(v)$ for all $v \in V$ *perfect* (b^-, b^+)-*matching* for in- and out-degree specifications $b^-, b^+ : V \to \mathbb{Z}_+$. A minimum cost perfect (b^-, b^+)-matching can be found by computing a minimum cost perfect b-matching in a bipartite graph. The digraph version of the problem is described as follows.

k-arc-connected multi-digraph with degree specification (k-ACMDS):
A vertex set V, a symmetric metric arc cost $c : V \times V \to \mathbb{Q}_+$, in- and out-degree specifications $b^-, b^+ : V \to \mathbb{Z}_+$, and a positive integer k are given. We are asked to find a minimum cost perfect (b^-, b^+)-matching $D = (V, A)$ of arc-connectivity k.   □

This paper is organized as follows. Section 2 presents an algorithm for k-ECMDS. Section 3 provides a 2.5-approximation algorithm for k-ACMDS problem.

2   Algorithm for k-ECMDS

This section describes an approximation algorithm for k-ECMDS. Before describing the algorithm, we consider how to check the feasibility of a given instance.

For some degree specification b, there is no perfect b-matching. The following theorem shows provides a necessary and sufficient condition for a degree specification to admit a perfect b-matching. Note that $b(v)$ can be 1 in this theorem.

Theorem 1. *Let V be a vertex set with $|V| \geq 2$ and $b : V \to \mathbb{Z}_+$ be a degree specification. Then there exists a perfect b-matching if and only if $\sum_{v \in V} b(v)$ is even and $b(v) \leq \sum_{u \in V - v} b(u)$ for each $v \in V$.*

Proof. Omitted due to the space limitation. □

Theorem 1 does not mention the edge-connectivity. For existence of connected perfect b-matchings, we additionally need the condition that $\sum_{v \in V} b(v) \geq 2(|V| - 1)$ [5]. This is always satisfied if $b(v) \geq 2$, $v \in V$, which we assume for 1-ECMDS. For $k \geq 2$, the conditions in Theorem 1 and $b(v) \geq k$, $v \in V$ are sufficient for the existence of k-edge-connected perfect b-matchings as our algorithm will construct such b-matchings under the conditions.

Now we describe our algorithm to k-ECMDS. Let (V, b, c, k) be an instance of k-ECMDS. The conditions appeared in Theorem 1 and $b(v) \geq k$ for all $v \in V$ can be verified in polynomial time, where they are apparently necessary for an instance to have k-edge-connected perfect b-matchings. Hence our algorithm checks them, and if some of them are violated, it outputs message "INFEASIBLE". In the following, we suppose the existence of perfect b-matchings with $b(v) \geq k$ for all $v \in V$. If $2 \leq |V| \leq 3$, then every perfect b-matching is k-edge-connected because any non-empty vertex set $X \subset V$ is $\{v\}$ or $V - \{v\}$ for some $v \in V$, and then $d(X) = d(v) \geq k$. Hence we can assume without loss of generality that $|V| \geq 4$.

For an edge set F on V, we denote graph (V, F) by G_F. Let M be a minimum cost edge set such that G_M is a perfect b-matching. In addition, let H be an edge set of a Hamiltonian cycle spanning V constructed by the 1.5-approximation algorithm for TSP due to Christofides [11].

Initialization: After testing the feasibility of a given instance, our algorithm first prepares M and $k' = \lceil k/2 \rceil$ copies $H_1, \ldots, H_{k'}$ of H. Let E denote the union $M \cup H_1 \cup \cdots \cup H_{k'}$ of them. Notice that G_E is $2k'$-edge-connected by the existence of edge-disjoint k' Hamiltonian cycles. We call a vertex v in a handling graph G an *excess vertex* if $d(v; G) > b(v)$ (otherwise a *non-excess vertex*). In G_E, all vertices are excess vertices since $d(v; G_E) = b(v) + 2k'$. In the following steps, the algorithm reduces the degree of excess vertices until no excess vertex exists while generating no loops and keeping k-edge-connectivity (Notice that $k < 2k'$ if k is odd). This is achieved by two phases, Phase 1 and Phase 2, as follows.

Phase 1: In this phase, we modify only edges in M while keeping edges in $H_1, \ldots, H_{k'}$ unchanged. We define the following two operations on an excess vertex $v \in V$.

Operation 1: If v has two incident edges xv and yv in M with $x \neq y$, replace xv and yv by new edge xy.

Operation 2: If v has two parallel edges uv in M with $d(u) > b(u)$, remove those edges.

Phase 1 repeats Operations 1 and 2 until none of them is executable. For avoiding ambiguity, we let M' denote M after executing Phase 1, and M denote the original set in what follows. Moreover, let $E' = M' \cup H_1 \cup \cdots \cup H_{k'}$. Note that $d(v) - b(v)$ is always a non-negative even integer throughout (and after) these operations because $d(v; G_E) - b(v) = 2k'$ and each operation decreases the degree of a vertex by 2. If no excess vertex remains in $G_{E'}$, then we are done. We consider the case in which there remain some excess vertices, and show some properties on M' before describing Phase 2.

Lemma 1. *Every excess vertex in $G_{E'}$ has at least one incident edge in M' and its neighbors in $G_{M'}$ are unique.*

Proof. Since $d(v; G_{E'}) - b(v)$ is a positive even integer for an excess vertex v in $G_{E'}$, it holds $d(v; G_{M'}) = d(v; G_{E'}) - d(v; G_{H_1 \cup \cdots \cup H_{k'}}) \geq (b(v) + 2) - 2k' > 0$, Hence v has at least one incident edges in M'. If neighbors of v in $G_{M'}$ are not unique, Operation 1 can be applied to v. □

For an excess vertex v in $G_{E'}$, let $n(v)$ denote the unique neighbor of v in $G_{M'}$. If $n(v)$ is also an excess vertex in $G_{E'}$, we call the pair $\{v, n(v)\}$ by a *strict pair*.

Lemma 2. *Let $\{v, n(v)\}$ be a strict pair. Then $d(v; G_{M'}) = d(n(v); G_{M'}) = 1$, k is odd, and $b(v) = b(n(v)) = k$.*

Proof. By Lemma 1, $d(v; G_{M'}) = d(n(v); G_{M'})$. If $d(v; G_{M'}) = d(n(v); G_{M'}) > 1$, Operation 2 can be applied to v and $n(v)$, a contradiction. Hence $d(v; G_{M'}) = d(n(v); G_{M'}) = 1$ holds. Let $u \in \{v, n(v)\}$. Then it holds that $d(u; G_{E'}) = d(u; G_{H_1 \cup \cdots \cup H_{k'}}) + d(u; G_{M'}) = 2k' + 1 = 2\lceil k/2 \rceil + 1$. Since $d(u; G_{E'}) - b(u)$ is even, $b(u)$ must be odd. This fact and $d(u, G_{E'}) > b(u) \geq k$ indicates that $b(u) = k$ and k is odd. □

By definition, the existence of excess vertices which are in no strict pairs indicate that of some non-excess vertices. Upon completion of Phase 1, let N denote the set of non-excess vertices in $G_{E'}$, and S denote the set of strict pairs in $G_{E'}$. If $N = \emptyset$, all excess vertices are in some strict pairs. By Lemma 2, k is an odd integer in this case, and furthermore $k \geq 3$ by the assumption that $b(v) \geq 2$, $v \in V$ if $k = 1$. From this fact and $|V| \geq 4$, $N = \emptyset$ implies that at least two strict pairs exist (i.e., $|S| \geq 2$).

Phase 2: Now we describe Phase 2. First, we deal with a special case in which V consists of only two strict pairs.

Lemma 3. *If V consists of two strict pairs after Phase 1, we can transform $G_{E'}$ into a k-edge-connected perfect b-matching without increasing the cost.*

Proof. Let $V = \{u, v, w, z\}$ and $H = \{uv, vw, wz, zu\}$. Now $E' = M' \cup H_1 \cup \cdots \cup H_{k'}$ $(k \geq 2)$. Then either $M' = \{uv, wz\}$ (or $\{vw, zu\}$) or $M' = \{uw, vz\}$

holds. In both cases, we replace $M' \cup H_1 \cup H_2$ by $E'' = \{uv, vw, wz, zu, uw, vz\}$ (see Fig. 2). Then, we can see that $d(v; G_{E''}) = 3$ for all $v \in V$ and $G_{E''}$ is 3-edge-connected. Since $d(v; G_{H_i}) = 2$ for $v \in V, i = 3, \ldots, k'$ and G_{H_i} is 2-edge-connected for $i = 3, \ldots, k'$, it holds that $d(v; G_{E'' \cup H_3 \cup \cdots \cup H_{k'}}) = 3 + 2(k' - 2) = k = b(v)$ for $v \in V$ and the edge-connectivity of $G_{E'' \cup H_3 \cup \cdots \cup H_{k'}}$ is $3 + 2(k'-2) = k$ (The existence of strict pair implies that k is odd by Lemma 2.).

Hence it suffices to show that $c(E'') \leq c(M') + c(H_1) + c(H_2)$. If $M' = \{uw, vz\}$ (or $\{vw, zu\}$), then it is obvious since $E'' = M' \cup H_1 \subseteq M' \cup H_1 \cup H_2$. Let us consider the other case, i.e., $M' = \{uv, wz\}$. From $M' \cup H_1 \cup H_2$, remove $\{uv, uv\}$, replace $\{wz, zu\}$ by $\{wu\}$, and replace $\{vw, wz\}$ by $\{vz\}$. Then the edge set becomes E'' without increasing edge cost, as required.   □

In the following, we assume that $|S| \geq 3$ when $N = \emptyset$. In this case, Phase 2 modifies only edges in H_i, $i = 1, \ldots, k'$ while keeping the edges in M' unchanged. Let $V(H_i)$ denote the set of vertices spanned by H_i. We define *detaching v from cycle H_i* to be an operation that replaces the pair $\{uv, vw\} \subseteq H_i$ of edges incident to v by a new edge uw. Note that this decreases $d(v)$ by 2, but H_i remains a cycle on $V(H_i) := V(H_i) - \{v\}$. For each excess vertex v in $G_{E'}$, Phase 2 reduces $d(v)$ to $b(v)$ by detaching v from $(d(v; G_{E'}) - b(v))/2$ cycles in $H_1, \ldots, H_{k'}$. We notice that $(d(v; G_{E'}) - b(v))/2 \leq k'$ by $d(v; G_{E'}) - b(v) \leq d(v; G_E) - b(v) = 2k'$. One important point is to keep $|V(H_i)| \geq 2$ for each $i = 1, \ldots, k'$ during Phase 2. In other words, we always select H_i with $|V(H_i)| \geq 3$ to detach an excess vertex. This is necessary because, if we detach a vertex from H_i with $V(H_i) = 2$, then H_i becomes a loop. In addition, we detach the two excess vertices u and v in a strict pair from different cycles in $H_1, \ldots, H_{k'}$, respectively. This is in order to maintain the k-edge-connectivity of $G_{E'}$ as will be explained below.

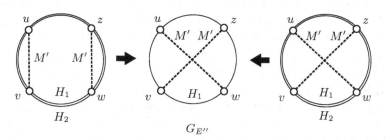

Fig. 2. Operations when V consists of two strict pairs

Lemma 4. *It is possible to decrease the degree of each excess vertex v in $G_{E'}$ to $b(v)$ by detaching from some cycles in $H_1, \ldots, H_{k'}$ so that $|V(H_i)|$ remains at least 2 for $i = 1, \ldots, k'$ and the two excess vertices in each strict pair are detached from H_i and H_j with $i \neq j$, respectively.*

Proof. First, let us consider the case of $S \neq \emptyset$. Recall $k \geq 3$ and $k' = \lceil k/2 \rceil \geq 2$ in this case. For each strict pair $\{u, v\} \in S$, we detach u and v from different cycles in $H_1, \ldots, H_{k'}$. On the other hand, we detach excess vertex z from arbitrary $(d(z; G_{E'}) - b(z))/2$ cycles. After this, each of $H_1, \ldots, H_{k'}$ is incident to at least one vertex of any strict pair in S in addition to all non-excess vertices in N. By the relation between $|S|$ and $|N|$ we explained in the above, it holds that $|V(H_i)| \geq |S| + |N| \geq 2$ for each $i = 1, \ldots, k'$, as required.

Next, let us consider the case of $S = \emptyset$. As explained in the above, $|N| \geq 1$ holds for this case. If $|N| \geq 2$, the claim is obvious since each of $H_1, \ldots, H_{k'}$ is always incident to all vertices in N. Hence suppose that $|N| = 1$, and let x be the unique non-excess vertex in N. Then all edges in M' are incident to x, since otherwise $S = \emptyset$ implies that Operation 1 or 2 would be applicable to some vertex in $V - x$. In other words, $b(x) = d(x; G_{E'}) = |M'| + 2k'$ holds before Phase 2. Moreover $\sum_{v \in V - x} b(v) \geq b(x)$ also holds by the assumption that perfect b-matchings exist. Now assume that we have converted some excess vertices in $G_{E'}$ into non-excess vertices by detaching them from some of $H_1, \ldots, H_{k'}$ while keeping $|V(H_i)| \geq 2$, $i = 1, \ldots, k'$, and yet an excess vertex $y \in V - x$ remains. Hence $\sum_{v \in V} d(v) > \sum_{v \in V} b(v)$. Then there remains a cycle H_i with $|V(H_i)| > 2$ because

$$2 \sum_{1 \leq i \leq k'} |V(H_i)| = \sum_{v \in V} d(v; G_{H_1 \cup \cdots \cup H_{k'}}) = \sum_{v \in V} d(v) - 2|M'|$$
$$> \sum_{v \in V - \{x\}} b(v) + b(x) - 2|M'| \geq 2(b(x) - |M'|) \geq 4k'. \qquad \square$$

In the following, we let H_i' denote H_i after Phase 2, and H_i denote the original Hamiltonian cycle for $i = 1, \ldots, k'$. Moreover let $E'' = M' \cup H_1' \cup \cdots \cup H_{k'}'$. The algorithm outputs $G_{E''}$. The entire algorithm is described as follows.

Algorithm UNDIRECT(k)

Input: A vertex set V, a degree specification $b : V \to \mathbb{Z}_+$, a metric edge cost $c : V \to \mathbb{Q}_+$, and a positive integer k

Output: A k-edge-connected perfect b-matching or "INFEASIBLE"

1: **if** $\sum_{v \in V} b(v)$ is odd, $\exists v : b(v) > \sum_{u \in V - v} b(u)$ or $k > b(v)$ **then**
2:     Output "INFEASIBLE" and halt
3: **end if**;
4: Compute a minimum cost perfect b-matching G_M;
5: **if** $|V| \leq 3$ **then**
6:     Output G_M and halt
7: **end if**;
8: Compute a Hamiltonian cycle G_H on V by Christofides' algorithm;
9: $k' := \lceil k/2 \rceil$; Let $H_1, \ldots, H_{k'}$ be k' copies of H;

   # Phase 1
10: $M' := M$;
11: **while** Operation 1 or 2 is applicable to a vertex $v \in V$
    with $d(v; G_{M' \cup H_1 \cup \cdots \cup H_{k'}}) > b(v)$ **do**
12:     **if** $\exists \{xv, vy\} \subseteq M'$ such that $x \neq y$ **then**
13:         $M' := (M' - \{xv, vy\}) \cup \{xy\}$ # Operation 1
14:     **else**

15:     **if** $\exists\{xv, vx\} \subseteq M'$ such that $d(x; G_{M' \cup H_1 \cup \cdots \cup H_{k'}}) > b(x)$ **then**
16:         $M' := M' - \{xv, vx\}$ # Operation 2
17:     **end if**
18:     **end if**
19: **end while**;

    # Phase 2
20: **if** V consists of two strict pairs **then**
21:     Rename vertices so that $H = \{uv, vw, wz, zu\}$;
22:     $H'_2 := \emptyset$; $M' := \{uw, vz\}$;
23:     Output $G_{M' \cup H'_1 \cup \cdots \cup H'_{k'}}$ and halt
24: **end if**;
25: $H'_i := H_i$ for each $i = 1, \ldots, k'$;
26: **while** $\exists v \in V$ with $d(v; G_{M' \cup H'_1 \cup \cdots \cup H'_{k'}}) > b(v)$ **do**
27:     **if** v and $n(v)$ forms a strict pair **then**
28:         Detach v from H'_i and $n(v)$ from H'_j, where $i \neq j$
29:     **else**
30:         Detach v from H'_i with $V(H'_i) > 2$
31:     **end if**
32: **end while**;
33: $E'' := M' \cup H'_1 \cup \cdots \cup H'_{k'}$;
34: Output $G_{E''}$

Lemma 5. $G_{E''}$ *is a k-edge-connected perfect b-matching.*

Proof. We have already seen the case in which V consists of two strict pairs. Hence we suppose the other case in the following. Moreover we have already observed that $d(v; G_{E''}) = b(v)$ holds for each $v \in V$. Furthermore $G_{E''}$ is loopless since G_E is loopless and no operations in the algorithm generate loops. Hence we prove the k-edge-connectivity of $G_{E''}$ below.

Let $u, v \in V$. (i) First suppose that u and v are in some (possibly different) strict pairs in $G_{E'}$. Moreover, let $u \notin V(H'_i)$ and $v \notin V(H'_j)$ (hence $u \in V(H'_{i'})$ for $i' \neq i$ and $v \in V(H'_{j'})$ for $j' \neq j$). For each $\ell \in \{1, \ldots, k'\} - \{i, j\}$, $\lambda(u, v; G_{H'_\ell}) = 2$ holds because $u, v \in V(H'_\ell)$. If $i = j$, $\lambda(u, v; G_{H'_i \cup M'}) = 1$ holds because $d(u; G_{M'}) = d(v; G_{M'}) = 1$ and $n(u), n(v) \in V(H'_i)$. Then it holds that $\lambda(u, v; G_{E''}) = 2(k' - 1) + 1 = k$ in this case (Recall that the existence of strict pairs implies that k is odd by Lemma 2). Hence we let $i \neq j$, and show that $\lambda(u, v; G_{H'_i \cup H'_j \cup M'}) \geq 3$ from now on, from which $\lambda(u, v; G_{E''}) \geq 2(k' - 2) + 3 = k$ can be derived.

Let N and S denote the sets of non-excess vertices and strict pairs in $G_{E'}$ after Phase 1, respectively. Suppose that $V(H'_i) \cap V(H'_j) = \emptyset$. In this case, it can be seen that $N = \emptyset$, and hence $|S| \geq 3$ by the assumption about the relation between N and S. Since at least one vertex of each strict pair is spanned by each cycle in $H'_1, \ldots, H'_{k'}$, we can see that M' contains at least three vertex-disjoint edges that join vertices in $V(H'_i)$ and in $V(H'_j)$, two of which are u and v. This indicates that $\lambda(u, v; G_{H'_i \cup H'_j \cup M'}) \geq 3$ holds (see the graph of Fig. 3 (b)).

Let us consider the case of $V(H_i') \cap V(H_j') \neq \emptyset$ in the next. By the existence of u and v, $|S| \geq 1$ holds. If u and v forms a strict pair (i.e., $uv \in M'$), $\lambda(u,v;G_{M'}) = 1$ holds. Since $V(H_i') \cap V(H_j') \neq \emptyset$ implies $\lambda(G_{H_i' \cup H_j'}) \geq 2$, we see that $\lambda(u,v; G_{H_i' \cup H_j' \cup M'}) \geq 3$ in this case. Thus let u and v belong to different strict pairs (i.e., $|S| \geq 2$). Then there exists two vertex-disjoint edges in M' joins vertices in $V(H_i')$ and in $V(H_j')$ (see Fig. 3 (a)). If we split each vertex $w \in V(H_i') \cap V(H_j')$ into two vertices w' and w'' so that H_i' and H_j' are vertex-disjoint cycles, and add new edges $w'w''$ joining those two split vertices to M', then we can reduce this case to the case of $V(H_i') \cap V(H_j') = \emptyset$, in which $\lambda(u,v; G_{H_i' \cup H_j' \cup M'}) \geq 3$ has already been observed in the above (see Fig. 3). Accordingly, we have $\lambda(u,v; G_{H_i' \cup H_j' \cup M'}) \geq 3$ if u and v are in some strict pairs, as required.

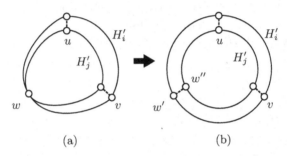

Fig. 3. Reduction to the case of $V(H_i') \cap V(H_j') = \emptyset$

(ii) In the next, let u and v be not in any strict pairs. For $z \in \{u,v\}$, let $n'(z)$ denote z itself if $z \in N$, and $n(z)$ otherwise. Notice that $n'(z) \in N$ for any $z \in \{u,v\}$, i.e., it is spanned by $H_1', \ldots, H_{k'}'$. If $z \in \{u,v\}$ is not spanned by $p > 0$ cycles in $H_1', \ldots, H_{k'}'$ (and hence z is an excess vertex in $G_{E''}$), then z has at least $k - 2(k'-p)$ incident edges in M' because $d(z; G_{M'}) = b(z) - d(z; G_{H_1' \cup \cdots \cup H_{k'}'}) \geq k - 2(k'-p)$. Hence $\lambda(z, n'(z); G_{E''}) \geq 2(k'-p) + k - 2(k'-p) = k$ holds for each $z \in \{u,v\}$, where we define $\lambda(z,z; G_{E''}) = +\infty$. Moreover it is obvious that $\lambda(n'(u), n'(v); G_{E''}) \geq 2k'$. Therefore, it holds that

$$\lambda(u,v; G_{E''}) \geq \min\{\lambda(u, n'(u); G_{E''}), \lambda(n'(u), n'(v); G_{E''}), \lambda(n'(v), v; G_{E''})\} \geq k.$$

(iii) Finally, let us consider the remaining case, i.e., u is in a strict pair and v is a vertex which is not in any strict pair. Let us define $n'(v)$ as in the above. Then $\lambda(v, n'(v); G_{E''}) \geq k$ holds. Without loss of generality, let u be detached from H_1', and spanned by $H_2', \ldots, H_{k'}'$. Since $un(u) \in M'$ and $n(u), n'(v) \in V(H_1')$, it holds that $\lambda(u, n(u); G_{M' \cup H_1'}) = 1$, and $\lambda(n(u), n'(v); G_{M' \cup H_1'}) \geq 2$. Then,

$$\lambda(u, n'(v); G_{E''}) \geq \min\{\lambda(u, n(u); G_{M' \cup H_1'}), \lambda(n(u), n'(v); G_{M' \cup H_1'})\}$$
$$+ \lambda(u, n'(v); G_{H_2' \cup \cdots \cup H_{k'}'}) \geq 1 + 2(k'-1) = 2k' - 1 = k.$$

Therefore,

$$\lambda(u, v; G_{E''}) \geq \min\{\lambda(u, n'(v); G_{E''}), \lambda(v, n'(v); G_{E''})\} \geq k,$$

holds, as required.   □

Let us consider the cost of the graph $G_{E''}$. The following theorem on the Christofides' algorithm gives us an upper bound on $c(H)$. Here, we let $\delta(U)$ denote the set of edges whose one end vertex is in U and the other is in $V - U$ for nonempty $U \subset V$.

Theorem 2 ([6,12]). *Let*

$$\begin{aligned}
OPT_{TSP} = \min &\ \textstyle\sum_{e \in E} c(e)x(e) \\
subject\ to &\ \textstyle\sum_{e \in \delta(U)} x(e) \geq 2\ for\ each\ nonempty\ U \subset V, \\
&\ x(e) \geq 0 \qquad\quad for\ each\ e \in E.
\end{aligned}$$

Christofides' algorithm for TSP outputs a solution of cost at most $1.5 OPT_{TSP}$.
    □

Lemma 6. $c(E'')$ *is at most $1 + 3\lceil k/2 \rceil/k$ times the optimal cost of k-ECMDS.*

Proof. No operation in Phases 1 and 2 increases the cost of the graph since the edge cost is metric. Hence it suffices to show that $c(M \cup H_1 \cup \cdots \cup H_{k'})$ is at most $(1 + 3\lceil k/2 \rceil/k) \cdot c(G)$, where G denotes an optimal solution of k-ECMDS. Since G is a perfect b-matching, $c(M) \leq c(G)$ obviously holds. Thus it suffices to show that $c(H_i) \leq 3c(G)/k$ for $1 \leq i \leq k'$, from which the claim follows.

Let $x_G : \binom{V}{2} \to \mathbb{Z}_+$ be the function such that $x_G(uv)$ denotes the number of edges joining u and v in G. Since G is k-edge-connected, $\sum_{e \in \delta(U)} x_G(e) \geq k$ holds for every nonempty $U \subset V$. Hence $2x_G/k$ is feasible for the linear programming in Theorem 2, which means that $OPT_{TSP} \leq 2c(G)/k$. By Theorem 2, $c(H_i) \leq 1.5 OPT_{TSP}$. Therefore we have $c(H_i) \leq 3c(G)/k$, as required.   □

Lemmas 5 and 6 establish the next.

Theorem 3. *Algorithm UNDIRECT(k) is a ρ-approximation algorithm for k-ECMDS, where $\rho = 2.5$ if k is even and $\rho = 2.5 + 1.5/k$ if k is odd.*   □

Algorithm UNDIRECT(k) always outputs a solution for $k \geq 2$ as long as there exists a perfect b-matching and $b(v) \geq k$ for all $v \in V$. This fact and Theorem 1 imply the following corollary.

Corollary 1. *For $k \geq 2$, there exists a k-edge-connected perfect b-matching if and only if $\sum_{v \in V} b(v)$ is even and $k \leq b(v) \leq \sum_{u \in V - v} b(u)$ for all $v \in V$.*   □

We close this section with a few remarks. The operations in Phases 1 and 2 are equivalent to a graph transformation called *splitting*, followed by removing generated loops if any. There are many results on the conditions for splitting to maintain the edge-connectivity [2,9]. However, the splittings in these results may generate loops. Hence algorithm UNDIRECT(k) needs to specify a sequence of splitting so that removing loops does not make the degrees lower than the degree specification.

One may consider that a perfect $(b - 2k')$-matching is more appropriate than a perfect b-matching as a building block of our algorithm, since there is no excess vertex for the union of a perfect $(b - 2k')$-matching and k' Hamiltonian cycles. However, there is a degree specification b that admits a perfect b-matching, and no perfect $(b - 2k')$-matching. Furthermore, even if there exits a perfect $(b - 2k')$-matching, the minimum cost of the perfect $(b - 2k')$-matching may not be a lower bound on the optimal cost of k-ECMDS. Therefore we do not use a perfect $(b - 2k')$-matching in general case. When $b(v) = \ell$ for all $v \in V$ with some integer $\ell \geq k$, we can show that a perfect $(b - 2k')$-matching exist and its cost can be estimated although we do not present the detail due to the space limitation. By exploiting this fact, the approximation factor for k-ECMDS can be improved to $\frac{\ell-k}{\ell} + 1.5$ in this case.

3   Algorithm for k-ACMDS

This section shows that k-ACMDS is 2.5-approximable. The algorithm for k-ACMDS can be designed analogously with that for k-ECMDS. Before describing the algorithm, we consider the feasibility of k-ACMDS.

A problem to find a minimum cost perfect (b^-, b^+)-matching in a digraph can be reduced to find a minimum cost perfect b-matching in an undirected bipartite graph. From this reduction and Frobenius' classic theorem on the relation-ship between vertex covers and matchings in an undirected bipartite graph (see [10] for example), we can immediately derive a condition for a digraph to have a perfect (b^-, b^+)-matching.

Theorem 4. *Let V be a vertex set, and $b^-, b^+ : V \to \mathbb{Z}_+$ be in- and out- degree specifications, respectively. There exists a perfect (b^-, b^+)-matching if and only if $\sum_{v \in V} b^-(v) = \sum_{v \in V} b^+(v)$, $b^-(v) \leq \sum_{u \in V-v} b^+(u)$ for each $v \in V$, and $b^+(v) \leq \sum_{u \in V-v} b^-(u)$ for each $v \in V$.*

Proof. Omitted due to the space limitation.   □

We are ready to explain the algorithm for k-ACMDS. In the following, we assume that $b^-(v), b^+(v) \geq k$ for each $v \in V$ and a perfect (b^-, b^+)-matching exists.

Let M be a minimum cost perfect (b^-, b^+)-matching and H be a directed Hamiltonian cycle constructed by Christofides' algorithm for the edge cost obtained from c by ignoring the direction of arcs (Recall that c is symmetric). Moreover let $H_1, \ldots, H_k$ be k copies of H, $A = M \cup H_1 \cup \cdots \cup H_k$, and D_F denote the digraph (V, F) for an arc set F. A vertex $v \in V$ is called an *excess vertex* if $d^-(v) > b^-(v)$ or $d^+(v) > b^+(v)$ (otherwise v is called a *non-excess vertex*). Notice that $d^-(v; D_A) - b^-(v) = d^+(v; D_A) - b^+(v)$. This condition is maintained throughout the algorithm, i.e., $d^-(v) > b^-(v)$ is equivalent to $d^+(v) > b^+(v)$. Our algorithm for k-ACMDS decreases the degree of excess vertices as k-ECMDS. One difference between algorithms for k-ECMDS and for

k-ACMDS is the definition of Operations 1 and 2. These will be executed for a pair of arcs entering and leaving the same vertex as follows.

Operation 1: If an excess vertex v has two incident arcs xv and vy in M with $x \neq y$, replace xv and vy by new edge $xy \in M$.

Operation 2: If an excess vertex v has two arcs uv and vu in M with $d^-(u) > b^-(u)$ (and $d^+(v) > b^+(v)$), remove these arcs.

Phase 1 of our algorithm modifies edges in M by repeating Operations 1 and 2 until none of them is executable. We let M' denote M after Phase 1, and M denote the original set in the following. Moreover let $A' = M' \cup H_1 \cup \cdots \cup H_k$, and N denote the set of non-excess vertices in $D_{A'}$. Note that the number of arcs in M' entering (resp., leaving) each excess vertices v in $D_{A'}$ has $d^-(v; D_{A'}) - k \geq d^-(v; D_{A'}) - b^-(v)$ (resp., $d^-(v; D_{A'}) - b^-(v) > d^+(v; D_{A'}) - b^+(v)$) arcs. The other end vertex of them is unique and in N (i.e., a non-excess vertex in $D_{A'}$) since otherwise Operation 1 or 2 can be applied to v. This situation is simpler than after Phase 2 of UNDIRECT(k) since no correspondence of strict pairs exists. Notice that $N \neq \emptyset$ always holds here.

Phase 2 of our algorithm for k-ACMDS modifies edges in $H_1, \ldots, H_k$ so as to decrease the degrees of all excess vertices as in UNDIRECT(k). We repeat *detaching* each excess vertex from some of $H_1, \ldots, H_k$, where detaching a vertex v from H_i is defined as an operation that replaces the pair $\{uv, vw\} \subseteq H_i$ of arcs entering and leaving v by new arc uw. We can prove that it is possible to detach excess vertices from Hamiltonian cycles while keeping $V(H_i) \geq 2$ for $1 \leq i \leq k$ as in UNDIRECT(k).

Lemma 7. *It is possible to decrease the degree of each excess vertex v to $b(v)$ by detaching v from some cycles in $H_1, \ldots, H_k$ so that $|V(H_i)|$ remains at least two for all $i = 1, \ldots, k$.*

Proof. Recall that $N \neq \emptyset$. If $|N| \geq 2$, the claim is obvious since each of $H_1, \cdots, H_k$ is incident to all vertices in N. Hence suppose that $|N| = 1$, and let x be the unique vertex in N. Then all arcs in M' are incident to x since otherwise Operation 1 or 2 would be applicable to some vertex in $V - x$. In other words, it hold $|M'| = d^-(x; D_{M'}) + d^+(v; D_{M'}) = b^-(x) + b^+(x) - 2k$. Recall that $\sum_{v \in V-x} b^+(v) \geq b^-(x)$ and $\sum_{v \in V-x} b^-(v) \geq b^+(x)$ hold by the assumption that perfect (b^-, b^+)-matchings exist. Now assume that we have converted some excess vertices in $D_{A'}$ into non-excess vertices by detaching them from some of $H_1, \ldots, H_k$ while keeping $|V(H_i)| \geq 2$, $i = 1, \ldots, k$, and yet an excess vertex $y \in V - x$ remains. Then there remains a cycles H_i with $|V(H_i)| > 2$ because

$$\sum_{1 \leq i \leq k} |V(H_i)| = \sum_{v \in V} d^-(v; D_{H_1 \cup \cdots \cup H_k}) = \sum_{v \in V} d^-(v; D_{E'}) - |M'|$$

$$> \sum_{v \in V - \{x\}} b^-(v) + d^-(x; D_{E'}) - |M'| \geq b^+(x) + b^-(x) - |M'| \geq 2k.$$

Hence we can detach y from such H_i, implying the claim also for $|N| = 1$.   □

In the following, we let H_i' denote H_i after Phase 2, and H_i denote the original Hamiltonian cycle for $i = 1, \ldots, k$ in order to avoid the ambiguity. Moreover let $A'' = M' \cup H_1' \cup \cdots \cup H_k'$. Our algorithm outputs $D_{A''}$ as a solution.

Algorithm DIRECT(k)

Input: A vertex set V, in- and out-degree specification $b^-, b^+ : V \rightarrow \mathbb{Z}_+$, a symmetric metric arc cost $c : V \times V \rightarrow \mathbb{Q}_+$, and a positive integer k

Output: A k-arc-connected perfect (b^-, b^+)-matching or "INFEASIBLE"

1: **if** $\sum_{v \in V} b^-(v) \neq \sum_{v \in V} b^+(v)$, $\exists v : b^-(v) > \sum_{u \in V-v} b^+(u)$, $\exists v : b^+(v) > \sum_{u \in V-v} b^-(u)$, $\exists v : k > b^-(v)$, or $\exists v : k > b^+(v)$ **then**
2:    Output "INFEASIBLE" and halt
3: **end if**;
4: Compute a minimum cost perfect (b^-, b^+)-matching D_M;
5: Compute a Hamiltonian cycle D_H on V by Christofides' algorithm; Let $H_1, \ldots, H_k$ be k copies of H;

Phase 1

6: $M' := M$;
7: **while** Operation 1 or 2 is applicable to a vertex $v \in V$
   with $d^-(v; D_{M' \cup H_1 \cup \cdots \cup H_k}) > b^-(v)$ **do**
8:    **if** $\exists \{xv, vy\} \subseteq M'$ such that $x \neq y$ **then**
9:       $M' := (M' - \{xv, vy\}) \cup \{xy\}$ # Operation 1
10:    **else if** $\exists \{xv, vx\} \subseteq M'$ such that $d^-(x; D_{M' \cup H_1 \cup \cdots \cup H_k}) > b^-(x)$ **then**
11:       $M' := M' - \{xv, vx\}$ # Operation 2
12:    **end if**
13: **end while**;

Phase 2

14: $H_i' := H_i$ for each $i = 1, \ldots, k$;
15: **while** $\exists v \in V$ with $d^-(v; D_{M' \cup H_1' \cup \cdots \cup H_k'}) > b^-(v)$ **do**
16:    Detach v from H_i' with $V(H_i') > 2$
17: **end while**;
18: $A'' := M' \cup H_1' \cup \cdots \cup H_k'$;
19: Output $D_{A''}$

Let OPT denote the optimal cost of k-ACMDS. We can show that $D_{A''}$ is k-arc-connected, $c(M) \leq$ OPT and $c(H_i) \leq 1.5$OPT$/k$ for $1 \leq i \leq k$, similarly for UNDIRECT(k) although we leave the proof to the readers. As a conclusion, we have the following theorem.

Theorem 5. *Algorithm DIRECT(k) is a 2.5-approximation algorithm for k-ACMDS.* $\square$

Algorithm DIRECT(k) always outputs a solution when there exists a perfect (b^-, b^+)-matching and $b^-(v) \geq k$, $b^+(v) \geq k$ for all $v \in V$. This fact and Theorem 4 implies the following corollary.

Corollary 2. *For $k \geq 1$, there exists a k-arc-connected perfect (b^-, b^+)-matching if and only if $\sum_{v \in V} b^-(v) = \sum_{v \in V} b^+(v)$, $k \leq b^-(v) \leq \sum_{u \in V-v} b^+(u)$ for each $v \in V$, and $k \leq b^+(v) \leq \sum_{u \in V-v} b^-(u)$ for each $v \in V$.*   □

References

1. A. Frank, Augmenting graphs to meet edge-connectivity requirements, SIAM Journal on Discrete Mathematics 5 (1992) 25–53.
2. A. Frank, On a theorem of Mader, Discrete Mathematics 191 (1992) 49–57.
3. G. N. Frederickson, M. S. Hecht, C. E. Kim, Approximation algorithms for some routing problems, SIAM Journal of Computing 7 (1978) 178–193.
4. T. Fukunaga, H. Nagamochi, Approximating minimum cost multigraphs of specified edge-connectivity under degree bounds, Proceedings of the 9th Japan-Korea Joint Workshop on Algorithm and Computation (2006) 25-32.
5. T. Fukunaga, H. Nagamochi, Approximating a generalization of metric TSP, IEICE Transactions on Information and Systems, to appear.
6. M. X. Goemans, D. J. Bertsimas, Survivable networks, linear programming relaxations and the parsimonious property, Mathematical Programming 60 (1993) 145–166.
7. M. X. Goemans, D. P. Williamson, The primal-dual method for approximation algorithms and its application to network design problems, PWS, 1997, Ch. 4, pp. 144–191.
8. E. L. Lawler, J. K. Lenstra, A. H. G. Rinnooy Kan, D. B. Shmoys (Eds.), The Traveling Salesman Problem: A Guided Tour of Combinatorial Optimization, John Wiley & Sons, 1985.
9. W. Mader, A reduction method for edge-connectivity in graphs, Annals of Discrete Mathematics 3 (1978) 145–164.
10. A. Schrijver, Combinatorial Optimization: Polyhedra and Efficiency, Springer, 2003.
11. V. Vazirani, Approximation Algorithm, Springer, 2001.
12. L. A. Wolsey, Heuristic analysis, linear programming and branch and bound, Mathematical Programming Study 13 (1980) 121–134.

Approximating Maximum Cut with Limited Unbalance

Giulia Galbiati[1] and Francesco Maffioli[2]

[1] Dipartimento di Informatica e Sistemistica, University of Pavia (Italy)
giulia.galbiati@unipv.it
[2] Dipartimento di Elettronica e Informazione, Politecnico di Milano (Italy)
maffioli@elet.polimi.it

Abstract. We present polynomial time randomized approximation algorithms with non trivial performance guarantees for the problem of partitioning the vertices of a weighted graph into two sets of sizes that differ at most by a given threshold B, so as to maximize the weight of the crossing edges. For B equal to 0 this problem is known as Max Bisection, whereas for B equal to the number n of nodes it is the Maximum Cut problem. The approximation results are obtained by extending the methodology used by Y. Ye for Max Bisection and by combining this technique with another one that uses the algorithm of Goemans and Williamson for the Maximum Cut problem. When B is equal to zero the approximation ratio achieved coincides with the one obtained by Y. Ye; otherwise it is always above this value and tends to the value obtained by Goemans and Williamson as B approaches the number n of nodes.

Keywords: randomized algorithm, approximation algorithm, semidefinite programming.

1 Introduction

Problems addressing optimum cuts are often considered in theoretical computer science and in combinatorial opimization; recently unbalanced graph cuts have received attention [8]. Here we address the following problem: given an undirected graph $G = (V, E)$, with vertex set V of cardinality n and edge set E, where each edge (i, j) has a non-negative weight w_{ij}, and given a constant B, $0 \leq B < n$, find a cut $(S, V \setminus S)$ of G of maximum weight such that the difference between the cardinalities of the two shores of the cut is not greater than B. We refer to this problem with the name Maximum Cut with Limited Unbalance (MaxCUT-LU for short). When B is equal to zero it is known as the Max Bisection problem and the algorithm in [14] gives an approximation ratio equal to 0.699. When B is equal to $n - 1$ it is the well-known Maximum Cut problem and the famous algorithm of [7] gives an approximation ratio equal to 0.87856. It is also a classic result of Trevisan et al. [12] and Håstad [9] that the Maximum Cut problem cannot be approximated by a deterministic algorithm having an approximation ratio strictly exceeding 16/17, unless P=NP. For a

T. Erlebach and C. Kaklamanis (Eds.): WAOA 2006, LNCS 4368, pp. 202–213, 2006.
© Springer-Verlag Berlin Heidelberg 2006

specific class of graphs an approximation ratio strictly exceeding 16/17 has been recently obtained in [10].

In [11] several applications of Maximum Cut are reported in different fields such as network planning, circuit design, scheduling, cryptanalysis, logic, psychology. For most of them the generalization to MaxCUT-LU makes sense. For instance, in circuit design, the problem of dividing the vertex set of the graph underlined by the circuit into two parts of equal cardinalities is of interest, and relaxing the equal cardinality constraint to that of limited unbalance can allow to get better results as far as approximating the optimum weight of the cut obtained, without affecting the suitability of the partition from the point of view of the circuit designer. MaxCUT-LU might also make sense for balancing signed graphs when special constraints arise, with applications e.g. in psychology [2]. The strict relation between Maximum Cut and Maximum 2-Satisfiability problems is well known [7]. Here the extension to MaxCUT-LU can allow to consider constraints on the number of variables set to "true".

In this paper we present polynomial time randomized approximation algorithms with non-trivial performance guarantees for MaxCUT-LU. Our results are obtained by extending to this problem the methodology used in [14] for Max Bisection and by combining this technique with another one that uses the algorithm of [7]. When B is equal to zero the approximation ratio achieved coincides with the one of [14], which is equal to 0.699; otherwise it is always greater than this value, and tends to the 0.87856 value of the algorithm of [7] when B approaches the number n of nodes. Our main results are summarized in Theorem 3, Proposition 1 and Theorem 4. Both Theorem 3 and Theorem 4 give compact and precise expressions for the approximation ratios achieved, that lend themselves very well to machine computations. Sample examples have in fact been computed; in Table 1 at the end of this work we report the approximation ratios r obtained for some values of η, where $\eta = B/n$ is our unbalance parameter, and of another parameter θ used in the algorithm, as described in Subsection 3.1.

A problem related to MaxCUT-LU and addressed in [1] and [5] is the problem of finding a cut of maximum weight such that the cardinalitiy of one shore of the cut is equal to a given integer k. The approximation ratios that the authors achieve with sophisticated techniques are naturally weaker than ours: for instance in [5], for the case $k = B = n/3$, they achieve a 0.58 ratio against our 0.797.

Other problems concerning optimal cuts with side constraints have recently received attention, as described in the introductory section of [8].

The formulation of MaxCUT-LU is introduced in the next section. The algorithms used to solve the problem are presented in the subsequent sections, the first devoted the the case when the parameter η is small, the other when the unbalance parameter is large. The last section concludes the work, summarizes the results, and presents some of them in Table 1.

2   The Formulation

MaxCUT-LU can be formulated by assigning to each vertex i a binary variable $x_i \in \{-1, 1\}$, with vertices on the same shore of the cut receiving the same value, and by setting $w_{ij} = 0$ if $(i, j) \notin E$, as:

$$w^* := \max \left\{ \frac{1}{4} \sum_{i,j} w_{ij}(1 - x_i x_j) : \sum_{i,j} x_i x_j \leq B^2; x_i \in \{-1, 1\}, i = 1, \ldots, n \right\}.$$
(1)

The semidefinite relaxation of this binary quadratic program can be formulated as follows:

$$w^{SDP} := \max \left\{ \frac{1}{4} \sum_{i,j} w_{ij}(1 - X_{ij}) : \sum_{i,j} X_{ij} \leq B^2; X_{ii} = 1, i = 1, .., n; \mathbf{X} \in M_n \right\}$$
(2)

where M_n is the set of real, symmetric, positive semidefinite matrices of order n. It is easy to see that any solution $\mathbf{x}$ of (1) yields a solution $\mathbf{X}$ of (2) with $X_{ij} = x_i x_j$. Hence obviously $w^* \leq w^{SDP}$.

It is known (see e.g. [3]) that such SDP program can be solved to any degree of accuracy in polynomial time. From an almost optimal solution, one can then derive a solution of the integer program using appropriate rounding techniques.

Rounding techniques applied to the solution of SDP relaxations of Combinatorial Optimization problems in order to get integral solutions of guaranteed degree of approximation have been pioneered by Goemans and Williamson [7] for the MAX CUT and MAX SAT problems. Frieze and Jerrum [6] have developed such techniques further, addressing the MAX BISECTION and the MAX-k CUT problems. Yinyu Ye [14] has improved the approximation ratio for MAX BISECTION using a more sophisticated rounding technique. With respect to MAX CUT, problem MaxCUT-LU presents, as MAX BISECTION, the extra difficulty of having to deal with two objectives: the weight of the cut and the size of its shores.

3   The Algorithm for Small Unbalance

We now present our first algorithm, suitable for solving problem MaxCUT-LU when B is small. In the algorithm, I indicates the identity matrix, the parameters θ and k are fixed in an appropriate way, as specified in Subsection 3.1 entirely devoted to this aspect. The algorithm uses the following technique, introduced in [14], which refines the one in [7]: from a solution $\widetilde{X}$ of the SDP relaxation first it constructs a new matrix X as a convex combination of $\widetilde{X}$ and the identity matrix I; then to matrix X, which is positive definite, it applies the Cholesky decomposition to obtain vectors $(\mathbf{v_1}, ..., \mathbf{v_n})$ on the unit n-dimensional sphere S_n. The algorithm then uses the so called random hyperplane technique, i.e.

it repeatedly generates a uniformly distributed vector $\mathbf{r}$ on the unit sphere, computes vector $\mathbf{u} = (\mathbf{r} \cdot \mathbf{v_1}, ..., \mathbf{r} \cdot \mathbf{v_n})$ and then rounds $\mathbf{u}$ to a vector $\widehat{\mathbf{x}}$ with $\widehat{x}_i \in \{-1, 1\}$, and $\widehat{x}_i = -1$ iff $u_i \geq 0$, $i = 1, ..., n$. Each vector $\widehat{\mathbf{x}}$ hence identifies a cut $(S, V \backslash S)$ of G, where $S = \{i : \widehat{x}_i = 1\}$ or $S = \{i : \widehat{x}_i = -1\}$; in our algorithm we always choose wlog S to be the set of vertices with the larger cardinality.

In the analysis of our algorithm, for the sake of clarity, we assume that $\widetilde{X}$ is an optimum solution of the SDP relaxation and that the vectors of the Cholesky decomposition exactly satisfy the equalities $(\mathbf{v_i} \cdot \mathbf{v_j}) = X_{ij}$. It can be shown that the inaccuracies resulting from using an almost optimal solution $\widetilde{X}$ and an almost exact Cholesky decomposition can be absorbed into the approximation factor presented in Theorem 3. This ensures that the algorithm runs in polynomial time.

Finally function $rebalance(S)$, when invoked by the algorithm, moves the nodes which least contribute to the weight of the cut from set S to the other set of the cut, so as to reduce the number of nodes in S to $(n + B)/2$.

Throughout this paper, $w(S)$ denotes the weight of the cut $(S, V \backslash S)$, N is set equal to $\frac{n^2}{4} - \frac{B^2}{4}$.

Algorithm 1

> 1 - *Solve the SDP problem (2) and let* $\widetilde{X}_{ij}$ *be the solution matrix;*
> 2 - *fix a value* θ *with* $0 \leq \theta < 1$ *and a positive integer* k;
> 3 - *let* $X = \theta \widetilde{X} + (1 - \theta)I$;
> 4 - *apply Cholesky decomposition to* X *to obtain vectors* $(\mathbf{v_1}, ..., \mathbf{v_n})$;
> 5 - $S_R = \phi$;
> 6 - *repeat for* k *times the following {*
>   6.1 - *generate a uniformly distributed vector* $\mathbf{r}$ *on the unit sphere;*
>   6.2 - *compute* $\mathbf{u} = (\mathbf{r} \cdot \mathbf{v_1}, ..., \mathbf{r} \cdot \mathbf{v_n})$;
>   6.3 - *round* $\mathbf{u}$ *to vector* $\widehat{\mathbf{x}} \in \{-1, 1\}^n$ *identifyng a cut* $(S, V \backslash S)$;
>   6.4 - *if* $|S| \leq (n + B)/2$ /* *the cut is feasible for MaxCUT-LU* */
>   *let* $\widetilde{S} = S$ *else let* $\widetilde{S} = rebalance(S)$;
>   6.5 - *if* $w(\widetilde{S}) > w(S_R)$ /* *a better cut for MaxCUT-LU is found* */
>   *let* $S_R = \widetilde{S}$;
> *}*
> 7 - *return* S_R.

In order to analyze the quality of the solution S_R returned by the algorithm, we define:

$$\alpha(\theta) := \min_{-1 \leq y < 1} \frac{1 - \frac{2}{\pi} \arcsin(\theta y)}{1 - y} \tag{3}$$

and

$$\beta(\theta, \eta) := (1 - \frac{1}{n}) \frac{1}{1 - \eta^2} b(\theta) + c(\theta) \tag{4}$$

with

$$b(\theta) = 1 - \frac{2}{\pi} \arcsin(\theta) \text{ and } c(\theta) = \min_{-1 \le y < 1} \frac{2}{\pi} \frac{\arcsin(\theta) - \arcsin(\theta y)}{1 - y}. \quad (5)$$

Notice that the definition of $\alpha(\theta)$ is as in [14] whereas that of $\beta(\theta, \eta)$ is different.

Lemma 1. *If functions $\alpha(\theta)$ and $\beta(\theta, \eta)$ are defined as in (3) and (4), then for the cut $(S, V \backslash S)$ generated by Algorithm 1 at line 6.3, we have that $Ex[w(S)] \ge \alpha(\theta)w^*$ and $Ex[|S|(n - |S|)] \ge \beta(\theta, \eta)N$, where $w(S)$ is the random variable associated to the cut.*

Proof. (sketched) In [7], [6] it is proved that the probability that vertices i and j are separated in the cut identified by S is equal to $1 - \frac{2}{\pi} \arcsin(X_{ij})$. It follows easily that $Ex[\hat{x}_i \hat{x}_j] = \frac{2}{\pi} \arcsin(X_{ij})$ and hence that

$$Ex[w(S)] = \frac{1}{4} \sum_{i,j} w_{ij} (1 - \frac{2}{\pi} \arcsin(X_{ij})). \quad (6)$$

Since $\arcsin(X_{ii}) = \pi/2$, for each $i = 1, ..., n$, and $X_{ij} = \theta \tilde{X}_{ij}$ when $i \ne j$, we conclude from (3) that the value in (6) is:

$$\ge \frac{1}{4} \sum_{i,j} w_{ij} \alpha(\theta)(1 - \tilde{X}_{ij}) = \alpha(\theta)w^{SDP} \ge \alpha(\theta)w^*.$$

We also can derive that:

$$Ex[|S|(n - |S|] = \frac{1}{4} \sum_{i,j} (1 - \frac{2}{\pi} \arcsin(X_{ij}))$$

$$= \frac{1}{4} \sum_{i \ne j} (1 - \frac{2}{\pi} \arcsin(\theta) + \frac{2}{\pi} \arcsin(\theta) - \frac{2}{\pi} \arcsin(\theta \tilde{X}_{ij}))$$

$$\ge \frac{1}{4} \sum_{i \ne j} (b(\theta) + c(\theta)(1 - \tilde{X}_{ij})) \quad (7)$$

Now, noticing that $\sum_{i \ne j} \tilde{X}_{ij} \le B^2 - n$, from (7) we derive that:

$$Ex[|S|(n - |S|] \ge \frac{1}{4}[(n^2 - n)b(\theta) + (n^2 - n)c(\theta) + c(\theta)(n - B^2)]$$

$$= [(1 - \frac{1}{n})\frac{1}{1 - \eta^2}b(\theta) + c(\theta)]\frac{n^2 - B^2}{4} = \beta(\theta, \eta)N.$$

$\square$

Lemma 2. *For every cut $(S, V \backslash S)$ generated by Algorithm 1 at line 6.3 we have that $\frac{w(S)}{w^*} \le 2$ and $\frac{|S|(n - |S|)}{N} \le \frac{1}{1 - \eta^2}$.*

Proof. Let $S = \{i_1, ..., i_s\}$. If $s \leq (n + B)/2$ then by definition $w(S) \leq w^*$. Otherwise denote by $\delta^c(i)$ the contribution of vertex i to the weight of the cut $(S, V \backslash S)$, i.e. $\delta^c(i) = \sum_{j \notin S} w_{ij}$. Obviously $w(S) = \sum_{i \in S} \delta^c(i)$. Suppose wlog that $\delta^c(i_1) \leq ... \leq \delta^c(i_s)$. Since $s > (n + B)/2$ we may remove from S the first $s - (n + B)/2$ vertices and we let S' be the reduced set. The weight of the cut has decreased by at most $\frac{w(S)}{s}(s - \frac{n+B}{2})$. By definition $w(S') \leq w^*$ but $w(S') \geq w(S) - \frac{w(S)}{s}(s - \frac{n+B}{2})$ and this implies $\frac{w(S)}{w^*} \leq 2s/(n + B) \leq 2$. The second inequality of the lemma follows immediately from the fact that $\frac{|S|(n-|S|)}{N} \leq \frac{n^2}{4}\frac{4}{n^2-B^2}$. $\qquad\square$

Let us now fix a value $\gamma > 0$ and study the random variable

$$Z = \frac{w(S)}{w^*} + \gamma \frac{|S|(n - |S|)}{N}. \tag{8}$$

The two preceding lemmas imply that $Z \leq 2 + \frac{\gamma}{1-\eta^2}$ and that $Ex[Z] \geq \alpha(\theta) + \gamma\beta(\theta, \eta)$. Hence for small values of η (say $\eta < \sqrt{2/3} \leq 0.8165$) variable Z is bounded above, so that for any $\epsilon > 0$ and for constant k sufficiently large, Algorithm 1 generates a set S for which:

$$Z \geq [\alpha(\theta) + \gamma\beta(\theta, \eta)](1 - \epsilon). \tag{9}$$

Theorem 1. *For any $\gamma > 0$, if random variable Z satisfies (9), then for the corresponding set $\widetilde{S}$ computed by Algorithm 1 at line 6.4, we have:*

$$w(\widetilde{S}) \geq \min(g_1, g_2)w^* \tag{10}$$

with:

$$g_1 = 2(\sqrt{\gamma[\alpha(\theta) + \gamma\beta(\theta, \eta)]\frac{(1 + \eta)(1 - \epsilon)}{1 - \eta}} - \frac{\gamma}{1 - \eta}) \tag{11}$$

$$g_2 = [\alpha(\theta) + \gamma\beta(\theta, \eta)](1 - \epsilon) - \frac{\gamma}{1 - \eta^2}. \tag{12}$$

Proof. When the algorithm finds a set S satisfying (9) we let $\delta = |S|/n$ and $\lambda = w(S)/w^*$. From (8) and (9) it follows that:

$$\lambda \geq [\alpha(\theta) + \gamma\beta(\theta, \eta)](1 - \epsilon) - 4\gamma\delta(1 - \delta)\frac{1}{1 - \eta^2} \tag{13}$$

There are two possibilities for $\widetilde{S}$: either $\widetilde{S} = rebalance(S)$ or $\widetilde{S} = S$. In the first case it is easy to see that $w(\widetilde{S}) \geq \frac{n+B}{2}\frac{w(S)}{|S|} = \frac{1+\eta}{2\delta}\lambda w^*$, whereas in the second case we obviously have $w(\widetilde{S}) = \lambda w^*$. Hence $w(\widetilde{S}) \geq \min(\frac{1+\eta}{2\delta}\lambda, \lambda)w^*$ and, using inequality (13) for λ, we get:

$$w(\widetilde{S}) \geq \min(f_1, f_2)w^* \tag{14}$$

with

$$f_1 = [\alpha(\theta) + \gamma\beta(\theta, \eta)]\frac{(1 + \eta)(1 - \epsilon)}{2\delta} - 2\gamma\frac{1 - \delta}{1 - \eta} \tag{15}$$

$$f_2 = [\alpha(\theta) + \gamma\beta(\theta,\eta)](1-\epsilon) - 4\gamma\delta(1-\delta)\frac{1}{1-\eta^2}. \tag{16}$$

In order to simplify (15) and (16) and to remove the dependence on δ we study functions f_1 and f_2 for $\delta \geq 0$. Simple calculations show that function f_1 has a minimum at $\delta_1 = \sqrt{\frac{[\alpha(\theta)+\gamma\beta(\theta,\eta)](1-\eta^2)(1-\epsilon)}{4\gamma}}$, where it assumes the value $2(\sqrt{\gamma[\alpha(\theta) + \gamma\beta(\theta,\eta)]\frac{(1+\eta)(1-\epsilon)}{1-\eta}} - \frac{\gamma}{1-\eta})$ which is, by definition, the value of function g_1. Instead function f_2 has a minimum at $\delta_2 = 1/2$, where it takes on the value $[\alpha(\theta) + \gamma(\beta(\theta,\eta)](1-\epsilon) - \frac{\gamma}{1-\eta^2}$ which again is, by definition, the value of function g_2. $\square$

Our aim now is to find the value of γ that maximizes $\min(g_1, g_2)$.

It can be seen that function g_1 is concave, is equal to zero for $\gamma = 0$ and for $\gamma_R = \frac{\alpha(\theta)(1-\eta^2)(1-\epsilon)}{1-\beta(\theta,\eta)(1-\eta^2)(1-\epsilon)}$ and has a maximum at

$$\gamma_M = \frac{\alpha(\theta)}{2\beta(\theta,\eta)}(\frac{1}{\sqrt{1 - \beta(\theta,\eta)(1-\eta^2)(1-\epsilon)}} - 1).$$

Of course $\gamma_M \leq \gamma_R$.

The graph of function g_2 on the other hand is a line. For $\gamma = 0$ g_2 has value $\alpha(\theta)(1-\epsilon)$ and then decreases until it vanishes, quite surprisingly, again in γ_R.

We have the following result.

Theorem 2. *For each η, $0 \leq \eta < 1$, we have that $g_2 \leq g_1$ iff $\gamma_L \leq \gamma \leq \gamma_R$, where we let $\gamma_L = \frac{\alpha(\theta)(1-\eta^2)(1-\epsilon)}{(1+2\eta)^2 - \beta(\theta,\eta)(1-\eta^2)(1-\epsilon)}$ and $\gamma_R = \frac{\alpha(\theta)(1-\eta^2)(1-\epsilon)}{1-\beta(\theta,\eta)(1-\eta^2)(1-\epsilon)}$. Moreover $\gamma_L = \gamma_R$ iff $\eta = 0$.*

Proof. For simplicity of notation we let $\xi = [\alpha(\theta) + \gamma\beta(\theta,\eta)]$. Then by definition we have that $g_2 \leq g_1$ iff:

$$\xi(1-\eta^2)(1-\epsilon) - \gamma \leq 2(1-\eta^2)(\sqrt{\gamma\xi\frac{(1+\eta)(1-\epsilon)}{1-\eta}} - \frac{\gamma}{1-\eta})$$

and hence iff:

$$\xi(1-\eta^2)(1-\epsilon) - \gamma + 2\gamma(1+\eta) \leq 2\sqrt{\xi\gamma(1-\eta^2)(1-\epsilon)(1+\eta)^2}. \tag{17}$$

Now if we let $x^2 = \xi\gamma(1-\eta^2)(1-\epsilon)$ then inequality (17) becomes:

$$x^2 - 2x\gamma(1+\eta) + \gamma^2(1+2\eta) \leq 0 \tag{18}$$

which has solutions for $\gamma \leq x \leq \gamma(1+2\eta)$. It can easily be seen that $\gamma \leq \sqrt{\xi\gamma(1-\eta^2)(1-\epsilon)}$ iff $\gamma \leq \frac{\alpha(\theta)(1-\eta^2)(1-\epsilon)}{1-\beta(\theta,\eta)(1-\eta^2)(1-\epsilon)}$ and that $\sqrt{\xi\gamma(1-\eta^2)(1-\epsilon)} \leq \gamma(1+2\eta)$ iff $\frac{\alpha(\theta)(1-\eta^2)(1-\epsilon)}{(1+2\eta)^2 - \beta(\theta,\eta)(1-\eta^2)(1-\epsilon)} \leq \gamma$. $\square$

Now from Theorem 1 and Theorem 2 we obtain the following.

Corollary 1. *The value of γ that maximizes $min(g_1, g_2)$ is γ_M if $\gamma_M \leq \gamma_L$ otherwise it is γ_L. Moreover $max_{\gamma>0} \, min(g_1, g_2)$ is equal to the value assumed by function g_1 in γ_M, if $\gamma_M \leq \gamma_L$, or to the value assumed in γ_L, if $\gamma_M > \gamma_L$.*

As a consequence of this corollary we can obtain a more precise evaluation of (10). Since it can be shown that $\eta^2 \leq \frac{1-\beta(\theta,\eta)(1-\epsilon)}{4-\beta(\theta,\eta)(1-\epsilon)}$ iff $\gamma_M \leq \gamma_L$, the following lemma follows from Theorem 1 and Corollary 1; long but straightforward computations show that ρ_1(resp. ρ_2) is equal to the value assumed by function g_1 in γ_M (resp. γ_L).

Lemma 3. *If the random variable Z satisfies (9) for $\gamma \in \{\gamma_M, \gamma_L\}$, then for the corresponding set $\widetilde{S}$ computed by Algorithm 1 at line 6.4 , we have:*

$$w(\widetilde{S}) \geq \rho_1 w^* \quad if \quad \eta^2 \leq \frac{1-\beta(\theta,\eta)(1-\epsilon)}{4-\beta(\theta,\eta)(1-\epsilon)} \quad and \quad \gamma = \gamma_M \qquad (19)$$

$$w(\widetilde{S}) \geq \rho_2 w^* \quad if \quad \eta^2 \geq \frac{1-\beta(\theta,\eta)(1-\epsilon)}{4-\beta(\theta,\eta)(1-\epsilon)} \quad and \quad \gamma = \gamma_L \qquad (20)$$

with:

$$\rho_1 = \frac{\alpha(\theta)}{\beta(\theta,\eta)(1-\eta)}(1 - \sqrt{1-\beta(\theta,\eta)(1-\eta^2)(1-\epsilon)}) \qquad (21)$$

$$\rho_2 = \frac{4\eta\alpha(\theta)(1+\eta)(1-\epsilon)}{(1+2\eta)^2 - \beta(\theta,\eta)(1-\eta^2)(1-\epsilon)}. \qquad (22)$$

Notice that functions ρ_1 and ρ_2, for each $\epsilon > 0$, depend on η and also on the value fixed in the algorithm for θ. For $\eta = 0$ and $\epsilon = 0$, ρ_1 coincides with the bound given in [14]; function ρ_2 has instead no counterpart in [14].

3.1   The Appropriate Choice of θ and k

In this subsection we discuss the choices that Algorithm 1 makes at line 2. Let us first consider the choice of θ. Since ρ_1 (ρ_2), for fixed $\epsilon > 0$, is a function of η and θ, then, for any given η, it is possible to compute, and to use in the algorithm, the value of θ that maximizes ρ_1 (ρ_2). The ratios reported in the first three groups of lines of Table 1 have been computed with this strategy, for n sufficiently large ($n \geq 10^4$) and $\epsilon = 0$. The figures below show the behavior of the two functions; function ρ_1 has been plotted for $\theta \in [0.8..1]$ and $\eta \in [0..0.2]$, function ρ_2 for $\theta \in [0.8..1]$ and $\eta \in [0.2..0.8]$. It is evident that the value of θ that maximizes ρ_1 (ρ_2) is a value in $[0.88..1)$.

For what concerns the choice of k we make the following considerations.

If we let $x = [\alpha(\theta) + \gamma\beta(\theta,\eta)](1 - \epsilon)$ and $p = Pr\{Z < x\}$ then we have, for bounded values of η and hence of Z, that $Ex[Z] \leq px + (1 - p)\max(Z)$. This inequality, together with the fact that from Lemma 1 and 2 it follows that $Z \leq 2 + \frac{\gamma}{1-\eta^2}$ and $Ex[Z] \geq \alpha(\theta) + \gamma\beta(\theta,\eta)$, implies that:

$$p \leq \frac{2 + \frac{\gamma}{1-\eta^2} - (\alpha(\theta) + \gamma\beta(\theta,\eta))}{2 + \frac{\gamma}{1-\eta^2} - (\alpha(\theta) + \gamma\beta(\theta,\eta))(1-\epsilon)}. \qquad (23)$$

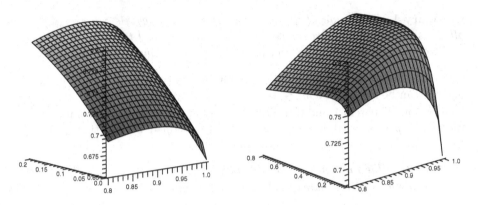

Fig. 1. On the left: function ρ_1 for $\theta \in [0.8..1]$ and $\eta \in [0..0.2]$. On the rigth: function ρ_2 for $\theta \in [0.8..1]$ and $\eta \in [0.2..0.8]$.

It can be verified that $(\frac{2+\frac{\gamma}{1-\eta^2}-(\alpha(\theta)+\gamma\beta(\theta,\eta))}{2+\frac{\gamma}{1-\eta^2}-(\alpha(\theta)+\gamma\beta(\theta,\eta))(1-\epsilon)})^k \leq \epsilon$ if we choose $k = \frac{1}{\epsilon}\log\frac{1}{\epsilon}$ for small value of η and $\gamma = \gamma_M$, or $k = \frac{1}{\epsilon}\log\frac{1}{\epsilon^2}$ for large value of η and $\gamma = \gamma_L$.

In Algorithm 1 we therefore fix the values of θ and k according to these considerations. The overall performance of the algorithm then may finally be specified by the following theorem.

Theorem 3. *Let ϵ be a small positive constant. Then Algorithm 1 returns a solution S_R having $Ex[w(S_R)] \geq \rho w^*$ with:*

$$\rho = \rho_1(1-\epsilon) \quad \text{if} \quad \eta^2 \leq \frac{1-\beta(\theta,\eta)(1-\epsilon)}{4-\beta(\theta)(1-\epsilon)} \tag{24}$$

$$\rho = \rho_2(1-\epsilon) \quad \text{if} \quad \eta^2 \geq \frac{1-\beta(\theta,\eta)(1-\epsilon)}{4-\beta(\theta)(1-\epsilon)}, \tag{25}$$

where ρ_1 and ρ_2 are defined in (21) and (22).

Proof. Let $\gamma_1 = \gamma_M$, $\gamma_2 = \gamma_L$. For each $i \in \{1,2\}$, let $x_i = [\alpha(\theta)+\gamma_i\beta(\theta,\eta)](1-\epsilon)$, let Z_i be the random variable defined in (8) with $\gamma = \gamma_i$, and let Z_{M_i} be the random variable assuming the maximum value for Z_i in the loop of Algorithm 1, with $(\widetilde{S}_i, V\setminus\widetilde{S}_i))$ being the corresponding cut. It is straightforward that $Ex[w(S_R)] \geq Ex[w(\widetilde{S}_i)]$ and that $Ex[w(\widetilde{S}_i)] \geq \rho_i w^* \Pr\{w(\widetilde{S}_i) \geq \rho_i w^*\}$, $i \in \{1,2\}$. Now from Lemma 3 it follows that $Pr\{w(\widetilde{S}_1) \geq \rho_1 w^*\} \geq Pr\{Z_1 \geq x_1\}$ when $\eta^2 \leq \frac{1-\beta(\theta,\eta)(1-\epsilon)}{4-\beta(\theta)(1-\epsilon)}$, and also that $Pr\{w(\widetilde{S}_2) \geq \rho_2 w^*\} \geq Pr\{Z_2 \geq x_2\}$ when $\eta^2 \geq \frac{1-\beta(\theta,\eta)(1-\epsilon)}{4-\beta(\theta)(1-\epsilon)}$. From the considerations made on the appropriate choice of k, we derive that $Pr\{Z_{M_i} \geq x_i\} \geq 1-\epsilon$, for each $i \in \{1,2\}$, and the conclusion follows. $\square$

Hence the following proposition can be stated and easily proved.

Proposition 1. *For each of the values of η reported in the first three groups of lines of Table 1, the value of ρ guaranteed by Theorem 3 is larger than the value of r reported in the table, for n sufficiently large.*

Proof. The values reported in the table have been computed, by truncation at the third decimal, for each η, using the value of θ that maximizes ρ_1 (ρ_2) for n sufficiently large ($n \geq 10^4$) and $\epsilon = 0$. Since $\rho_i(1 - \epsilon)$ tends, for $\epsilon \to 0$, to a value greater or equal to the one reported in the table, the result follows.   □

4   The Algorithm for Large Unbalance

In this section we use the following very simple algorithm, that uses function *rebalance(S)*, introduced in Section 3. Here with w^M we indicate the weight of a maximum cut.

Algorithm 2
- *use the algorithm in [7] to obtain a cut $(S, V \backslash S)$ having*
  $w(S) \geq 0.87856 \, w^M$;
- *denote by S the set of vertices with larger cardinality;*
- *if $|S| \leq (n + B)/2$ /* the cut is feasible for MaxCUT-LU */*
       let $\widetilde{S} = S$ else let $\widetilde{S} = rebalance(S)$;
- *return $\widetilde{S}$.*

Theorem 4. *Algorithm 2 returns a set $\widetilde{S}$ having:*

$$w(\widetilde{S}) \geq 0.87856\frac{1 + \eta}{2}w^*. \qquad (26)$$

Proof. If $\widetilde{S} = S$ the result follows easily since $w^M \geq w^*$ and $\eta \leq 1$. Otherwise, as in the proof of Lemma 2, the removal from S of the $|S| - (n + B)/2$ vertices that contribute less to the weight $w(S)$ of the cut reduces the weight by at most $\frac{w(S)}{|S|}(|s| - \frac{n+B}{2})$. Hence $w(\widetilde{S}) \geq w(S) - \frac{w(S)}{|S|}(|S| - \frac{n+B}{2}) = w(S)\frac{n+B}{2|S|}$. Since $w(S) \geq 0.87856 \, w^M$ the result follows.   □

5   Conclusions

We have presented two polynomial time randomized approximation algorithms giving non-trivial performance guarantees for the MaxCUT-LU problem. The approximation ratios have been obtained by extending to this problem the methodology used in [14] for Max Bisection and by combining this technique with another one that uses the algorithm of [7]. Depending on the value of η, (24) or (25) or (26) give our best approximation result. These approximation ratios are expressed by compact and precise expressions that lend themselves very well to machine computations. In Table 1 we report the ratios r obtained

Table 1.

η	0.0000	0.0500	0.1000	0.1050	0.1065
θ	0.888	0.890	0.894	0.895	0.895
r	0.699	0.731	0.759	0.761	0.762

η	0.1065	0.2000	0.3333	0.4000	0.4500
θ	0.893	0.941	0.966	0.972	0.975
r	0.762	0.788	0.797	0.798	0.798

η	0.4930	0.5000	0.6000	0.7000	0.8000
θ	0.977	0.977	0.980	0.982	0.984
r	0.797	0.797	0.795	0.793	0.790

η	0.8000	0.8500	0.9000	0.9500	0.9999
r	0.790	0.8126	0.834	0.856	0.878

for some values of the parameter η. The values in the first group are given by (24), those in the last group are given by (26), the others by (25). Moreover the values reported in the first three groups have been computed, for n sufficiently large ($n \geq 10^4$) and $\epsilon = 0$, with truncation at the third decimal, using the values of θ, also reported in the table, that maximize the ratios.

For smaller n, e.g. $n = 10^3$, some of the approximation ratios in the table decrease only by 10^{-3}. Note that the breaking point between Algorithms 1 and 2 occurs for $\eta < \sqrt{2/3}$, as we have assumed.

References

1. Ageev, A.A., Sviridenko, M.I.: Approximation algorithms for Maximum Coverage and Max Cut with Given Sizes of Parts. In: Cornuéjols, G., Burkard, R.E., Woeginger, G.J.(eds.): Integer Programming and Combinatorial Optimization. Lecture Notes in Computer Science, Vol. 1610. Springer-Verlag, Berlin Heidelberg New York (1999) 17–30
2. Akiyama, J., Avis, D., Chvatal, V., Era, H.: Balancing signed graphs. Discrete Applied Mathematics **3** (1981) 227–233
3. Alizadeh, F.: Interior point methods in semidefinite programming with applications to combinatorial optimization. SIAM J. Optimization **5** (1995) 13–51
4. Arora, S., Lund, C., Motwani, R., Sudan, M., Szegedy, M.: Proof verification and the hardness of approximation problems. Journal of the ACM **45** (1998) 501–555
5. Feige, U., Langberg, M.: Approximation algorithms for maximization problems in graph partitioning. J. of Algorithms **41** (2001) 1074–211
6. Frieze, A., Jerrum, M.: Improved approximation algorithms for MAX k-CUT and MAX BISECTION. Algorithmica **18** (1997) 67–81
7. Goemans, M.X., Williamson, D.P.: Improved approximation algorithms for maximum cut and satisfiability problems using semidefinite programming. Journal of ACM **42** (1995) 1115–1145
8. Hayrapetyan, A., Kempe, D., Pal, M., Svitkina, Z.: Unbalanced Graph Cuts. In: Brodal, G.S., Leonardi, S.(eds.): Algorithms - ESA 2005 Lecture Notes in Computer Science, Vol. 3669. Springer-Verlag, Berlin Heidelberg New York (2005) 191–202

9. Håstad, J.: Some optimal inapproximability results. Journal of the ACM **48** (2001) 798-869
10. Kaporis, A.C., Kirousis, L.M., Stavropoulos, E.C.: Approximating Almost All Instances of MAX-CUT Within a Ratio Above the Håstad Threshold. In: Azar, Y., Erlebach, T. (eds.): Algorithms-ESA 2006. Lecture Notes in Computer Science, Vol. 4168. Springer-Verlag, Berlin Heidelberg New York (2006) 432–443
11. Poljak, S., Tuza, Z.: Maximum cuts and large bipartite subgraphs. In: Cook, W., Lovasz, L., Seymour, P.(Eds.): Combinatorial Optimization. AMS - DIMACS Series in Discrete Mathematics and Theoretical Computer Science, Vol. 20. American Mathematical Society, Providence, RI (1995) 181–244
12. Trevisan, L., Sorkin, G., Sudan, M., Williamson, D.: Gadgets, approximation, and linear programming. SIAM Journal on Computing **29(6)** (2000) 2074-2097
13. Vazirani, V.V.: Approximation Algorithms. Springer-Verlag, (2001), chapter 26
14. Ye, Y.: A .699-approximation algorithm for Max-Bisection. Math. Programming Ser. A (90) (2001) 101–111

Worst Case Analysis of Max-Regret, Greedy and Other Heuristics for Multidimensional Assignment and Traveling Salesman Problems

Gregory Gutin[1,2], Boris Goldengorin[3,4], and Jing Huang[5,6]

[1] Department of Computer Science, Royal Holloway University of London,
Egham, Surrey TW20 OEX, UK
gutin@cs.rhul.ac.uk
[2] Department of Computer Science, University of Haifa, Israel
[3] Department of Econometrics and Operations Research, University of Groningen,
P.O. Box 800, 9700 AV Groningen, The Netherlands
B.Goldengorin@rug.nl
[4] Department of Applied Mathematics, Khmelnitsky National University, Ukraine
[5] Department of Mathematics and Statistics, P.O. Box 3045,
University of Victoria, Canada V8W 3P4
jing@math.uvic.ca
[6] School of Mathematics and Computer Science,
Nanjing Normal University, Nanjing, China

Abstract. Optimization heuristics are often compared with each other to determine which one performs best by means of worst-case performance ratio reflecting the quality of returned solution in the worst case. The domination number is a complement parameter indicating the quality of the heuristic in hand by determining how many feasible solutions are dominated by the heuristic solution. We prove that the Max-Regret heuristic introduced by Balas and Saltzman finds the unique worst possible solution for some instances of the s-dimensional ($s \geq 3$) assignment and asymmetric traveling salesman problems of each possible size. We show that the Triple Interchange heuristic (for $s = 3$) also introduced by Balas and Saltzman and two new heuristics (Part and Recursive Opt Matching) have factorial domination numbers for the s-dimensional ($s \geq 3$) assignment problem.

1 Introduction

The Multidimensional Assignment Problem (abbreviated s-AP in the case of s dimensions) is a well-known optimization problem with a host of applications (see, e.g., [3,8] for 'classic' applications and [5,24] for recent applications in solving systems of polynomial equations and centralized multisensor multitarget tracking). In fact, several applications described in [5,8,24] naturally require the use of s-AP for values of s larger than 3. The Asymmetric Traveling Salesman Problem (ATSP) has a large variety of applications, see, e.g., [14] for a recent one in geophysical seismic acquisitions. However, most of ATSP research was

T. Erlebach and C. Kaklamanis (Eds.): WAOA 2006, LNCS 4368, pp. 214–225, 2006.
© Springer-Verlag Berlin Heidelberg 2006

concentrated on its symmetric special case, see, e.g., [13] and more research of the general case heuristics is required [20].

Both well-known GREEDY algorithm and so-far-less-investigated MAX-REGRET algorithm [3,9,24] are fast construction heuristics that build a solution element by element without an attempt to improve it. We perform worst case analysis of MAX-REGRET for s-AP and ATSP by means of domination analysis.

While computational experiments in [3] show that MAX-REGRET significantly outperforms GREEDY for s-AP ($s \geq 3$), more extensive experiments in [24] indicate that neither of the two heuristics dominates the other. This conclusion is confirmed in our paper. Moreover, we prove that GREEDY and MAX-REGRET find the unique worst assignments for some instances of s-AP ($s \geq 3$) of every possible size. We introduce and discuss heuristics that perform much better in the worst case than GREEDY and MAX-REGRET. Such heuristics can be more reliable alternatives to both GREEDY and MAX-REGRET especially when we deal with previously uninvestigated families of instances of s-AP.

Experimental results in [9] indicate that a version of MAX-REGRET, MAX-REGRET-FC (called R-R-GREEDY in [9]), clearly outperforms GREEDY for ATSP. Nevertheless, we prove that, like GREEDY, both MAX-REGRET and MAX-REGRET-FC find the unique worst tour for some instances of ATSP of each possible size. This, in particular, settles the problem of finding good bounds for the domination number of MAX-REGRET-FC stated in [9].

The paper is organized as follows. We provide basic notions on domination analysis and GREEDY in Section 2. In Section 3, we describe MAX-REGRET for s-AP and prove that, for each $n \geq 1$ and $s \geq 3$, there is an s-AP instance of size n^s for which MAX-REGRET constructs the unique worst assignment. For 2-AP we only prove that there are instances for which MAX-REGRET finds an assignment which is worse than at least $n! - 2^{n-1}$ assignments. We conjecture that, in fact, the domination number of MAX-REGRET for 2-AP is exactly 2^{n-1}. Section 4 is devoted to three s-AP heuristics which *always* find assignments that are not worse that $((n-1)!)^{s-1}$ assignments. Two of the heuristics are new and might well be of interest in practice. In Section 5 we describe MAX-REGRET and its version MAX-REGRET-FC for ATSP and prove that, for each $n \geq 2$, there is an ATSP instance on n vertices for which both heuristics find the unique worst tour. Conclusions appear in Section 6.

2   Domination Analysis and Greedy

Research on combinatorial optimization (CO) heuristics has produced a large variety of heuristics especially for well-known CO problems and, thus, it is important to develop ways of selecting the best ones among them. In most of the literature, heuristics are compared by means of computational experiments and, while experimental analysis is of definite importance, it cannot cover all possible families of instances of the CO problem at hand and, in particular, it usually does not cover the hardest instances. Worst case analysis is normally performed by approximation analysis [2], where upper or lower bounds for the worst case performance ratio are of interest. Introduced in [11], domination analysis provides an alternative and a complement to approximation analysis. In domination analysis, we are interested in the domination number or domination ratio of the heuristic solution. We define these parameters below.

Pros and cons of domination analysis are discussed in [15] and, in our view, it is advantageous to have bounds for both performance ratio and domination ratio of a heuristic whenever it is possible. Roughly speaking this would enable us to see a 2D picture rather than a 1D picture.

Let $\mathcal{P}$ be a minimization CO problem, let $\mathcal{I}$ be an instance of $\mathcal{P}$, let $S(\mathcal{I})$ denote the set of feasible solutions of $\mathcal{I}$, and let H be a heuristic for $\mathcal{P}$. The size of $\mathcal{I}$ is denoted by $|\mathcal{I}|$ and the solution obtained by H for $\mathcal{I}$ is denoted by $H(\mathcal{I})$. When considering the weight of a solution y we write $w(y)$.

The *domination number* of a heuristic H is

$$\text{domn}(H, n) = \min_{\mathcal{I} \in \mathcal{P}: \ |\mathcal{I}|=n} \text{domn}(H, \mathcal{I}),$$

where $\text{domn}(H, \mathcal{I}) = |\{y \in S(\mathcal{I}) : w(H(\mathcal{I})) \leq w(y)\}|$. In other words, the domination number $\text{domn}(H, n)$ is the maximum integer such that the solution $H(\mathcal{I})$ obtained by H for *any* instance $\mathcal{I}$ of $\mathcal{P}$ of size n is not worse than at least $\text{domn}(H, n)$ feasible solutions of $\mathcal{I}$ (including $H(\mathcal{I})$). The *domination ratio* of H is

$$\text{domr}(H, n) = \min_{\mathcal{I} \in \mathcal{P}: \ |\mathcal{I}|=n} \frac{\text{domn}(H, \mathcal{I})}{|S(\mathcal{I})|}.$$

In many cases, domination analysis is very useful. For example, the greedy algorithm has domination number 1 for many CO problems, see, e.g., [4,17,23]. In other words, the greedy algorithm, in the worst case, produces the unique worst possible solution. This is reflected in computational experiments with the greedy algorithm for the asymmetric traveling salesman problem (ATSP), see, e.g., [20], where it was concluded that the greedy algorithm 'might be said to self-destruct.' The fact that the greedy algorithm is of domination number 1 for s-AP ($s \geq 3$) as well (see Theorem 1) implies that the algorithm should be used with great care for s-AP. In [6], domination analysis is used to establish which of the two heuristics for the generalized ATSP is a better choice (both heuristics exhibited similar behavior in computational experiments, but the domination number of one of them turned out to be significantly larger than that of the other heuristic). Bounds for domination numbers/ratios were obtained for many CO heuristics, see, e.g., [1,15,18,21,22].

Many CO problems can be formulated as follows. We are given a pair $(E, \mathcal{F})$, where E is a finite set and $\mathcal{F}$ is a family of subsets of E, and a weight function w that assigns a real weight $w(e)$ to every element of E. A maximal (with respect to inclusion) set $B \in \mathcal{F}$ is called a *base*. The weight $w(S)$ of $S \in \mathcal{F}$ is defined as the sum of the weights of the elements of S. The objective is to find a base $B \in \mathcal{F}$ of minimum weight.

The well-known GREEDY algorithm proceeds as follows. It starts from the empty set X. In every iteration GREEDY adds a minimum weight element e to the current set X provided $e \notin X$ and $X \cup \{e\}$ is a subset of a set in $\mathcal{F}$. The algorithm stops when a base has been constructed.

Unfortunately, both computational experiments and domination analysis point out that GREEDY is often a poor choice for heuristic even if it is only used to generate initial solutions that will be improved by more sophisticated heuristics. Thus, other heuristics are of definite interest. MAX-REGRET algorithm studied in [3,24] for 3-AP seems to be a promising and quite universal heuristic. Variations of MAX-REGRET were introduced and investigated in [9] for ATSP. Our analysis for both s-AP ($s \geq 3$) and ATSP indicates that MAX-REGRET is of similar quality in the worst case as GREEDY, namely, the domination number of MAX-REGRET for both problems equals 1. Recently, Bendall and Margot [7] studied an extension of GREEDY, which is of domination number 1 for many CO problems as well.

3 Greedy, s-AP-Max-Regret and s-AP-Max-Regret-FC

For a fixed $s \geq 2$, the s-AP is stated as follows. Let $X_1 = X_2 = \cdots = X_s = \{1, 2, \ldots, n\}$. We will consider only vectors that belong to the Cartesian product $X = X_1 \times X_2 \times \cdots \times X_s$. Each vector e is assigned a weight $w(e)$. For a vector e, e_j denotes its jth coordinate, i.e., $e_j \in X_j$. A *partial assignment* is a collection $e^1, e^2, \ldots, e^t$ of $t \leq n$ vectors such that $e_j^i \neq e_j^k$ for each $i \neq k$ and $j \in \{1, 2, \ldots, s\}$. An *assignment* is a partial assignment with n vectors. The *weight* of a partial assignment $A = \{e^1, e^2, \ldots, e^t\}$ is $w(A) = \sum_{i=1}^{t} w(e^i)$. The objective is to find an assignment of minimum weight.

We will start from GREEDY for s-AP. Using Theorem 2.1 in [17] one can prove that, for each $s \geq 2$, $n \geq 2$, there exists an instance of s-AP for which GREEDY will find the unique worst possible assignment. We will give a short direct proof of this result, which is also of interest later in this section.

A vector h is *backward* if $\min\{h_i : 2 \leq i \leq s\} < h_1$; a vector h is *horizontal* if $h_1 = h_2 = \cdots = h_s$. A vector is *forward* if it is not horizontal or backward.

Lemma 1. *Let F be an assignment of s-AP ($s \geq 2$). Either all vectors of F are horizontal or F contains a backward vector.*

Proof. Let $F = \{f^1, f^2, \ldots, f^n\}$, where $f_1^i = i$ for each $1 \leq i \leq n$. Assume that not every vector of F is horizontal. We show that F has a backward vector. Suppose it is not true. Then F has a forward vector f^i. Thus, there is a subscript j such that $f_j^i > i$. By the pigeonhole principle, there exists a superscript $k > i$ such that $f_j^k < k = f_1^k$, i.e., f^k is backward; a contradiction. □

Theorem 1. *For each $s \geq 2$, $n \geq 2$, there exists an instance of s-AP for which GREEDY will find the unique worst possible assignment.*

Proof. Let $M > n$ and let $E = \{e^1, e^2, \ldots, e^n\}$, where $e^i = (i, i, \ldots, i)$ for every $1 \leq i \leq n$. We define the required instance $\mathcal{I}$ as follows: $w(e^i) = iM$ for each $1 \leq i \leq n$ and, for each $f \notin E$, $w(f) = \min\{f_i : 1 \leq i \leq s\} \cdot M + 1$.

Observe that GREEDY will construct E. Let $F = \{f^1, f^2, \ldots, f^n\}$ be any other assignment, where $f_1^i = i$ for each $1 \le i \le n$. By Lemma 1, F has a backward vector f^k and

$$w(f^k) \le (k-1)M + 1 \tag{1}$$

By the definition of the weights and (1),

$$
\begin{aligned}
w(F) = \sum_{i=1}^{n} w(f^i) &= \sum_{i \ne k} w(f^i) + w(f^k) \\
&\le \sum_{i \ne k} (iM + 1) + (k-1)M + 1 \\
&= \sum_{i=1}^{n} iM + n - M \\
&< \sum_{i=1}^{n} iM = w(E) \qquad \qquad \square
\end{aligned}
$$

The heuristic s-AP-MAX-REGRET described next was first introduced in [3] for 3-AP and its modifications for s-AP were considered in [5]. The authors of [12] gave a general approach that extends MAX-REGRET heuristics.

s-AP-MAX-REGRET proceeds as follows. Set $W_j = A = \emptyset$ for each $j = 1, 2, \ldots, s$. While $|X_1| \ne |W_1|$ do the following: For each $i \in \{1, 2, \ldots, s\}$ and $a \in X_i \setminus W_i$, find two lightest vectors $e^{i,a}$ and $f^{i,a}$ ($w(e^{i,a}) \le w(f^{i,a})$) in the set

$$H = \{h \in X : \ h_i = a, h_j \in X_j \setminus W_j, j \in \{1, 2, 3, \ldots, s\} \setminus \{i\}\}$$

and compute the difference (called *regret*) $\Delta_{i,a} = w(f^{i,a}) - w(e^{i,a})$. Compute the *max-regret*

$$\Delta_{i_0, a_0} = \max\{\Delta_{i,a} : \ i \in \{1, 2, \ldots, s\}, \ a \in X_i \setminus W_i\}.$$

Add e^{i_0, a_0} to A and each $e_j^{i_0, a_0}$ to W_j, $j = 1, 2, \ldots, s$.

A modification of s-AP-MAX-REGRET that computes the regrets only for the first coordinates, i.e., only $\Delta_{1,a}$'s will be denoted s-AP-MAX-REGRET-FC[1].

Remark 1. In s-AP-MAX-REGRET and s-AP-MAX-REGRET-FC, when $|H| = 1$ we set $\Delta_{i,a} = 0$. Since we perform the worst case analysis, when breaking ties, we will follow the choice leading to the worst solution among possible options.

Theorem 2. *The domination number of both s-AP-MAX-REGRET and s-AP-MAX-REGRET-FC equals 1 for each $s \ge 3$.*

Proof. Consider the instance $\mathcal{I}$ described in the proof of Theorem 1. Observe that $\Delta_{i,1} = (M+1) - M = 1$ for each i and $\Delta_{i,a} = (M+1) - (M+1) = 0$ for

---

[1] FC abbreviates First Coordinate.

each $a > 1$. Thus, both s-AP-MAX-REGRET and s-AP-MAX-REGRET-FC will choose e^1 first. Similarly, we can see that both heuristics will uniquely choose $e^2, \ldots, e^n$ one by one. In Theorem 1, we showed that $E = \{e^1, e^2, \ldots, e^n\}$ is unique worst possible for $\mathcal{I}$. $\qquad\qquad\square$

Notice that the proof of Theorem 2 cannot be extended to 2-AP-MAX-REGRET or 2-AP-MAX-REGRET-FC. Moreover, it was proved in [9] that 2-AP-MAX-REGRET-FC is of domination number 2^{n-1}. We believe that 2^{n-1} is also the domination number for 2-AP-MAX-REGRET, but we are unable to prove it. In support of this conjecture we prove the following:

Theorem 3. *The domination number of* 2-AP-MAX-REGRET *is at most* 2^{n-1}.

Proof. Choose n positive numbers $d_1 > d_2 > \cdots > d_n$ arbitrarily and consider the following instance of 2-AP: $w(i,i) = -d_i$ for each $i = 1, 2, \ldots, n$, $w(i,j) = 0$ for each $1 \leq i < j \leq n$ and $w(i,j) = -\sum_{k=j}^{i} d_k$ for each $1 \leq j < i \leq n$.

Initially 2-AP-MAX-REGRET computes the regrets as follows: $\Delta_{1,k} = d_1$ and $\Delta_{2,k} = d_n$ for each $k = 1, 2, \ldots, n$. We may assume that 2-AP-MAX-REGRET chooses $(1,1)$ (see Remark 1). Similarly, we can see that 2-AP-MAX-REGRET chooses $(2,2), (3,3), \ldots, (n,n)$ one by one. Thus, the weight of the assignment $M = \{(1,1), (2,2), \ldots, (n,n)\}$ built by 2-AP-MAX-REGRET equals $-\sum_{i=1}^{n} d_i$.

For an integer $p \geq 1$, let $\text{Op}(i,p)$ denote an operation that replaces in M the vectors $\{(i,i), (i+1,i+1), \ldots, (i+p,i+p)\}$ by the vectors $\{(i,i+1), (i+1,i+2), \ldots, (i+p-1,i+p), (i+p,i)\}$. The operation $\text{Op}(i,0)$ does nothing. Consider the following procedure. It starts from $i := 1$. It chooses an arbitrary integer p with $0 \leq p \leq n - i$, performs $\text{Op}(i,p)$, sets $i := i + p + 1$ and continues this loop while $i < n$.

Notice that $\text{Op}(i,p)$ preserves the weight of the assignment and, thus, every assignment obtained by the procedure is of weight $w(M)$. Let $f(n)$ be the number of all possible assignments that can be obtained by the procedure. Clearly, $f(1) = 1$ and set $f(0) = 1$. To compute $f(n)$ observe that after using $\text{Op}(1,p)$ we will have $f(n - p - 1)$ possible assignments. Thus, for each $n \geq 2$ we have $f(n) = f(n-1) + f(n-2) + \ldots + f(0)$. This implies that $f(n) = 2^{n-1}$ for $n \geq 1$.

To show that any assignment that cannot be constructed by the procedure is of weight smaller than $w(M)$, build a complete digraph DK_n with vertices $\{1, 2, \ldots, n\}$ and with a loop on every vertex. For arbitrary $1 \leq i, j \leq n$, the arc (i,j) of DK_n corresponds to the vector (i,j) and we set the weight of arc (i,j) equal $w(i,j)$. We call an arc (i,j) with $i < j$ *forward* and with $i \geq j$ *backward*. Notice that the weight of every forward arc is 0.

An assignment corresponds to a *cycle factor* of DK_n, which is a collection of disjoint cycles (some of them may be loops) that cover all vertices of DK_n. In particular, the weight of an assignment equals the weight of the corresponding cycle factor in DK_n. Notice that the weight of every forward arc is 0 and, thus, the weight of a cycle factor equals the sum of the weights of its backward arcs. We call a pair $(i,j), (i',j')$ of backward arcs *intersecting* if the intervals $[j,i]$ and $[j',i']$ of real line intersect (one of these intervals may be just a point). Observe that if a cycle factor does not have intersecting backward arcs, then its weight equals $-\sum_{i=1}^{n} d_i = w(M)$ and every such cycle factor corresponds to an assignment that can be obtained by the procedure above. Thus, there are exactly $f(n) = 2^{n-1}$ cycle factors without intersecting backward arcs.

Now suppose that a cycle factor F has an intersecting pair $(i, j), (i', j')$ of backward arcs. Thus, there is an integer k such that $k \in [j, i] \cap [j', i']$. By the definition of a cycle factor, $k < n$. Observe that the above arguments imply that $w(F) \leq -\sum_{i=1}^{n} d_i - d_k < w(M)$.

So, there are only 2^{n-1} assignments of weight not smaller than $w(M)$.   □

4 s-AP Heuristics of Large Domination Number

For ATSP, there are several heuristics with domination number at least $(n-2)!$ (see, e.g., [18]). In this section, we will demonstrate that s-AP admits a number of heuristics of domination number at least $((n-1)!)^{s-1}$. We introduce two such new heuristics PART and RECURSIVE OPT MATCHING, which might be of interest in practice. The key lemma is the following result similar to the corresponding result in [16].

The average weight of an assignment (denoted by $\bar{w}$) is the total weight of all assignments divided by the number of assignments. The average weight of a vector in X is $w(X)/n^s$. Thus, by linearity of expectation, the average weight of an assignment equals $\bar{w} = w(X)/n^{s-1}$.

Lemma 2. *Let H be a heuristic that for each instance of s-AP constructs an assignment of weight at most the average weight of an assignment. Then the domination number of H is at least $((n-1)!)^{s-1}$.*

Proof. Consider an instance $\mathcal{I}$ of s-AP. Let C denote the set of all vectors of $\mathcal{I}$ with the first coordinate equal 1. Consider $\mathcal{P} = \{A_f : f^1 \in C\}$, where $A_f = \{f^1, f^2, \ldots, f^n\}$ is an assignment with $f_j^i = f_j^1 + i - 1$ (modulo n), $j = 1, 2, \ldots, s$. Observe that each vector is in exactly one A_f and, thus, $\mathcal{P}$ is a partition of $X = X_1 \times X_2 \times \cdots \times X_s$ into assignments. Since $\sum_{f \in C} w(A_f) = w(X)$, $|C| = n^{s-1}$ and $\bar{w} = w(X)/n^{s-1}$, the heaviest assignment A_h in $\mathcal{P}$ is of weight at least $\bar{w}$.

Let $S(X_i)$ be the set of all permutations on X_i ($2 \leq i \leq s$) and let $\pi_2 \in S(X_2), \pi_3 \in S(X_3), \ldots, \pi_s \in S(X_s)$. To obtain $\mathcal{P}(\pi_2, \pi_3, \ldots, \pi_s)$ from $\mathcal{P}$, replace f_j^i with $\pi_j(f_j^i)$ for each $j \geq 2$ and $i = 1, 2, \ldots, n$. Thus, we obtain a family

$$\mathcal{F} = \{\mathcal{P}(\pi_2, \pi_3, \ldots, \pi_s) : \pi_2 \in S(X_2), \pi_3 \in S(X_3), \ldots, \pi_s \in S(X_s)\}$$

of partitions of X into assignments. The family consists of $(n!)^{s-1}$ partitions. We may choose the heaviest assignment in each partition and, thus, obtain a family $\mathcal{A}$ of assignments of weight at least $\bar{w}$.

However, we can have several occurrences of the same assignment in $\mathcal{A}$. We claim that no assignment $G = \{g^1, g^2, \ldots, g^n\}$ (with $g_1^i = i$ for $i = 1, 2, \ldots, n$) can be in more than n^{s-1} partitions of $\mathcal{F}$. We may assume that $G \in \mathcal{P}$. Let G be also in some $\mathcal{P}(\pi_2, \pi_3, \ldots, \pi_s)$. By definition, there is an assignment $\{d^1, d^2, \ldots, d^n\}$ in $\mathcal{P}$ with $d_1^i = i$ for $i = 1, 2, \ldots, n$ such that $g_j^i = \pi_j(d_j^i)$ for each $j = 2, 3, \ldots, s$ and $i = 1, 2, \ldots, n$. These relations uniquely define the permutations $\pi_2, \pi_3, \ldots, \pi_s$. Thus, $\{g^1, g^2, \ldots, g^n\}$ can be repeated in $\mathcal{F}$ at most $|\mathcal{P}| = n^{s-1}$ times.

So, each assignment in $\mathcal{A}$ is of weight at least $\bar{w}$, no assignment in $\mathcal{A}$ can be repeated more than n^{s-1} times, and $\mathcal{A}$ has $(n!)^{s-1}$ assignments with repetitions. Therefore, we can find (in $\mathcal{A}$) $((n-1)!)^{s-1}$ distinct assignments of weight at least $\bar{w}$. Since $w(H(\mathcal{I})) \leq \bar{w}$ and $\mathcal{I}$ is arbitrary, we conclude that H is of domination number at least $(n!)^{s-1}$.   □

Consider a new heuristic PART that finds a partition $\mathcal{P}$ of X into assignments and computes an assignment in $\mathcal{P}$ of minimum weight. The proof above shows that PART is of domination number at least $((n-1)!)^{s-1}$. This heuristic is fast (of time complexity $O(n^s)$) and might be of interest at least for producing initial assignments for local improvement heuristics such as the TRIPLE INTERCHANGE introduced in [3] for 3-AP. Before studying TRIPLE INTERCHANGE we consider another new heuristic RECURSIVE OPT MATCHING for s-AP.

RECURSIVE OPT MATCHING proceeds as follows. Compute a new weight $\bar{w}(i, j) = w(X_{ij})/n^{s-2}$, where X_{ij} is the set of all vectors with last two coordinates equal i and j, respectively. Solving the 2-AP with the new weights to optimality, find an optimal assignment $\{(i, \pi_s(i)) : i = 1, 2, \ldots, n\}$, where π_s is a permutation on X_s. While $s \neq 1$, introduce $(s-1)$-AP with weights given as follows: $w'(f^i) = w(f^i, \pi_s(i))$ for each vector $f^i \in X'$, where $X' = X_1 \times X_2 \times \cdots \times X_{s-1}$, with last coordinate equal i and apply RECURSIVE OPT MATCHING recursively. As a result we have obtained permutations $\pi_s, \pi_{s-1}, \ldots, \pi_2$. The output is the assignment $\{(i, \pi_2(i), \pi_3(\pi_2(i)), \ldots, \pi_s(\pi_{s-1}(\ldots(\pi_2(i)))\ldots)) : i = 1, 2, \ldots, n\}$.

Theorem 4. *For each $s \geq 2$,* RECURSIVE OPT MATCHING *is of domination number at least* $((n-1)!)^{s-1}$.

Proof. By Lemma 2, it suffices to show that the assignment obtained by RECURSIVE OPT MATCHING is of weight at most $\bar{w} = w(X)/n^{s-1}$, the average weight of an assignment. Our proof is by induction on $s \geq 2$. Clearly the assertion holds for $s = 2$ and consider $s \geq 3$. Observe that

$$\frac{w(X)}{n^{s-1}} = \bar{w} = \frac{1}{n} \sum_{i=1}^{n} \sum_{j=1}^{n} \bar{w}(i, j) \geq \sum_{i=1}^{n} \bar{w}(i, \pi_s(i)) = \frac{w'(X')}{n^{s-2}}.$$

Let $A = \{(g^1, \pi_s(1)), \ldots, (g^n, \pi_s(n))\}$ be an assignment obtained by RECURSIVE OPT MATCHING, where $g^i \in X'$ such that $g^i_{s-1} = i$ for every $i = 1, \ldots, n$. Let $A' = \{g^1, \ldots, g^n\}$. Then by induction hypothesis, $\bar{w}' = w'(X')/n^{s-2} \geq w'(A') = w(A)$ and we are done.   □

It is straightforward to see that for any fixed $s \geq 3$, RECURSIVE OPT MATCHING is of running time merely $O(n^s)$.

Consider 3-AP. TRIPLE INTERCHANGE is a local search heuristic that at every step tries to improve an assignment $D = \{d^1, d^2, \ldots, d^n\}$ by looking at a triple of vectors d^i, d^j, d^k. It compares $w(d^i) + w(d^j) + w(d^k)$ with the weight of each of 35 triples $(h^i, h^j, h^k) \neq (d^i, d^j, d^k)$ such that $h_1^i = d_1^i, h_1^j = d_1^j, h_1^k = d_1^k$, $\{h_2^i, h_2^j, h_2^k\} = \{d_2^i, d_2^j, d_2^k\}$ and $\{h_3^i, h_3^j, h_3^k\} = \{d_3^i, d_3^j, d_3^k\}$. If TRIPLE INTERCHANGE finds a triple h^i, h^j, h^k lighter than d^i, d^j, d^k, it replaces d^i, d^j, d^k with h^i, h^j, h^k in D. The heuristic stops when no triple in the current assignment D can be replaced by a lighter one.

The following theorem does not depend on the initial assignment in TRIPLE INTERCHANGE.

Theorem 5. *Consider 3-AP. The domination number of* TRIPLE INTERCHANGE *is at least* $((n-1)!)^2$.

Proof. Assume that $E = \{e^1, e^2, \ldots, e^n\}$, where $e^i = (i, i, i)$, is an assignment that cannot be improved using TRIPLE INTERCHANGE. The set of all vectors $X = Y \cup Z \cup E$, where Y is the set of vectors with exactly two equal coordinates and Z is the set of vectors with all coordinates being different. Clearly, $w(X) = w(Y) + w(Z) + w(E)$. We will prove that $w(Y) \geq 3(n-1)w(E)$ and $w(Z) \geq (n-1)(n-2)w(E)$, which imply that $w(E) \leq \bar{w} = w(X)/n^2$ and the result of the theorem follows from Lemma 2.

Observe that $|Y| = 3n(n-1)$ (there are 3 ways to choose which coordinate is different from the other two, n ways to choose value from $\{1, 2, \ldots, n\}$ for this coordinate and $n - 1$ ways to choose value for the two coordinates). The set Y can be partitioned into $|Y|/2$ pairs of the form f^i, f^j such that f^i has one coordinate equal i and two coordinates equal j and f^j has one coordinate equal j and two coordinates equal i. For each such pair f^i, f^j, we have $w(f^i) + w(f^j) \geq w(e^i) + w(e^j)$ as otherwise we could improve e^i, e^j, e^k by f^i, f^j, e^k ($k \neq i, j$). Summing up all the inequalities we obtain $w(Y) \geq 3(n-1)w(E)$.

Observe that $|Z| = n(n-1)(n-2)$. For a vector $f = (i, j, k)$, let $f^+ = (k, i, j)$ and $f^- = (j, k, i)$. Let $F = \{(i, j, k) \in X : i < j < k\}$ and $G = \{(i, j, k) \in X : j < i < k\}$. Then $Z = \{\{f, f^+, f^-\} : f \in F \cup G\}$ is a partition of Z into $|Z|/3$ triples. Observe that for a triple $h = (i, j, k), h^+, h^-$, we have $w(h) + w(h^+) + w(h^-) \geq w(e^i) + w(e^j) + w(e^k)$. This implies $w(Z) \geq (n-1)(n-2)w(E)$. $\square$

The PAIR INTERCHANGE heuristic also described in [3] is similar to TRIPLE INTERCHANGE, but tries to improve pairs of vectors in the current assignment. PAIR INTERCHANGE does not always produce an assignment whose weight is at most the average weight of an assignment. To see that consider an instance of 3-AP with the following weights: $w(i, i, i) = 0$ for each $i = 1, 2, \ldots, n$, $w(i, j, k) = 1$ for each triple i, j, k in which exactly two members equal, and $w(i, j, k) = -n^3$ for each triple i, j, k in which all members of different. The assignment $E = \{(1, 1, 1), (2, 2, 2), \ldots, (n, n, n)\}$ cannot be improved by PAIR INTERCHANGE, but $w(E) = 0$ and the average weight of an assignment is negative for each $n \geq 3$.

5   ATSP–Max-Regret and ATSP-Max-Regret-FC

A variation of MAX-REGRET for ATSP, ATSP-MAX-REGRET-FC, was first introduced in [9] under a different name, R-R-GREEDY. The authors of [9] found an exponential upper bound on the domination number of ATSP-MAX-REGRET-FC and stated a problem to obtain a nontrivial lower bound for the domination number. Extensive computational experiments in [9] demonstrated a clear superiority of ATSP-MAX-REGRET-FC over GREEDY and several other construction heuristics in [10]. Therefore, the result of Theorem 6 is somewhat unexpected.

Let K_n^* be a complete digraph with vertices $V = \{1, 2, \ldots, n\}$. The weight of an arc (i, j) is denoted by w_{ij}. The *ATSP* is the problem of finding a tour (i.e., a Hamilton cycle) of K_n^* of total minimum weight. Let Q be a collection of disjoint paths in K_n^*. An arc $a = (i, j)$ is a *feasible addition* to Q if $Q + a$ is either a collection of disjoint paths or a tour in K_n^*. Consider ATSP-MAX-REGRET-FC and ATSP-MAX-REGRET.

ATSP-MAX-REGRET-FC proceeds as follows. Set $W = T = \emptyset$. While $V \neq W$ do the following: For each $i \in V \setminus W$, compute two lightest arcs (i, j) and (i, k) that are feasible additions to T, and compute the difference $\Delta_i = |w_{ij} - w_{ik}|$. For $i \in V - W$ with maximum Δ_i choose the lightest arc (i, j), which is a feasible addition to T and add (i, j) to M and i to W.

ATSP-MAX-REGRET proceeds as follows. Set $W^+ = W^- = T = \emptyset$. While $V \neq W^+$ do the following: For each $i \in V \setminus W^+$, compute two lightest arcs (i, j) and (i, k) that are feasible additions to T, and compute the difference $\Delta_i^+ = |w_{ij} - w_{ik}|$; for each $i \in V \setminus W^-$, compute two lightest arcs (j, i) and (k, i) that are feasible additions to T, and compute the difference $\Delta_i^- = |w_{ji} - w_{ki}|$. Compute $i' \in V \setminus W^+$ with maximum $\Delta_{i'}^+$ and $i'' \in V \setminus W^-$ with maximum $\Delta_{i''}^-$. If $\Delta_{i'}^+ \geq \Delta_{i''}^-$ choose the lightest arc (i', j'), which is a feasible addition to T and add (i', j') to M, i' to W^+ and j' to W^-. Otherwise, choose the lightest arc (j'', i''), which is a feasible addition to T and add (j'', i'') to M, i'' to W^- and j'' to W^+.

Remark 2. In ATSP-MAX-REGRET-FC, if $|V \setminus W| = 1$ we set $\Delta_i = 0$. A similar remark applies to ATSP-MAX-REGRET.

Theorem 6. *The domination number of both* ATSP-MAX-REGRET-FC *and* ATSP-MAX-REGRET *equals 1 for each* $n \geq 2$.

Proof. Since the proofs for both heuristics use the same family of instances and are similar, we restrict ourselves only to ATSP-MAX-REGRET-FC.

Consider an instance of ATSP on the complete digraph with vertex set $\{1, 2, \ldots, n\}$, $n \geq 2$. Let the weights be as follows: $w_{ik} = \min\{0, i - k\}$ for each $1 \leq i \neq k \leq n$, $i \neq n$, and $w_{nk} = -k$ for each $1 \leq k \leq n - 1$. We will slightly modify the weights: $w_{ij}' = w_{ij}$ unless $j = i + 1$ modulo n. We set $w_{i,i+1}' = -1 - \frac{1}{n+1}$ for $1 \leq i \leq n - 1$ and $w_{n,1}' = -1 - \frac{1}{n+1}$. ATSP-MAX-REGRET-FC will use the weight function w'.

ATSP-MAX-REGRET-FC constructs the tour $T_{MR} = (1, 2, 3, \ldots, n, 1)$ by first choosing the arc $(n - 1, n)$, then the arc $(n - 2, n - 1)$, etc. The last two arcs are $(1, 2)$ and $(n, 1)$ (they must be included in the tour). Indeed, initially $\Delta_{n-1} = \frac{n+2}{n+1} > \Delta_i$ for each $i \neq n - 1$. Once $(n - 1, n)$ is added to T_{MR}, $\Delta_{n-2} = \frac{n+2}{n+1}$ becomes maximal, etc.

Let T', T'' be a pair of tours. Since $\sum_{(i,j) \in K_n^*} |w_{ij} - w'_{ij}| < 1$, $w(T') < w(T'')$ implies $w'(T') < w'(T'')$. Thus, to prove that $w'(T) < w'(T_{MR})$ for each tour $T \neq T_{MR}$, it suffices to show that $w(T) < w(T_{MR})$.

Observe that $w(T_{MR}) = -n$. Let $T = (i_1, i_2, \ldots, i_n, i_1)$ be an arbitrary tour, where $i_1 = 1$. Suppose that $i_s = n$. Observe that the weight of the path $P = (i_1, i_2, \ldots, i_s)$ equals $\sum_{k=1}^{s-1} \min\{0, i_k - i_{k+1}\}$. Thus, $w(P) \leq 1 - n$ and $w(P) = 1 - n$ if and only if $i_1 < i_2 < \cdots < i_s$. Since $i_s = n$, the weight of the arc (i_s, i_{s+1}) equals $-i_{s+1}$. Thus, $w(T) \leq 1 - n - i_{s+1}$ and $w(T) \geq w(T_{MR})$ if and only if $i_{s+1} = 1$ and $i_1 < i_2 < \cdots < i_s$. We conclude that $w(T) \geq w(T_{MR})$ if and only if $T = T_{MR}$. $\qquad \square$

6 Conclusions

We have carried out worst-case analysis of MAX-REGRET for the Multidimensional Assignment Problem (s-AP, $s \geq 3$) and Asymmetric Traveling Salesman Problem (ATSP). We proved that MAX-REGRET for both problems may find unique worst possible solution. Thus, like GREEDY, MAX-REGRET should be used with great care and, possibly, avoided all together when instances of previously unstudied families are to be solved. In such a case heuristics of factorial domination number that have a proven excellent computational record (such as Helsgaun's version of Lin-Kernighan heuristic for the Symmetric TSP, see [19,22]) appear to be a much better choice.

Acknowledgements. We are thankful to all anonymous reviewers for useful comments and suggestions. Most of this paper was written when the first author was visiting Department of Mathematics and Statistics, University of Victoria, Canada. He would like to thank the department for its hospitality. His research was supported in part by the IST Programme of the European Community, under the PASCAL Network of Excellence, IST-2002-506778. The second author acknowledges support by the DFG project SI 657/5.

References

1. N. Alon, G. Gutin and M. Krivelevich, Algorithms with large domination ratio. *J. Algorithms* **50** (2004) 118–131.
2. G. Ausiello, P. Crescenzi, G. Gambosi, V. Kann, A. Marchetti-Spaccamela and M. Protasi, *Complexity and Approximation*, Springer, Berlin, 1999.
3. E. Balas and M.J. Saltzman, An algorithm for the three-index assignment problem. *Operations Research* **39** (1991) 150–161.
4. J. Bang-Jensen, G. Gutin and A. Yeo, When the greedy algorithm fails. *Discrete Optimization* **1** (2004) 121–127.
5. H. Bekker, E.P. Braad and B. Goldengorin, Using bipartite and multidimentional matchings to select roots of a system of polynomial equations. In Proc. ICCSA'05, Lecture Notes in Computer Science 3483 (2005) 397–406.
6. D. Ben-Arieh, G. Gutin, M. Penn, A. Yeo and A. Zverovitch, Transformations of generalized ATSP into ATSP. *Operations Research Letters* **31** (2003) 357–365.

7. G. Bendall and F. Margot, Greedy Type Resistance of Combinatorial Problems. To appear in *Discrete Optimization.*

8. R. Burkard and E. Cela, Linear assignment problems and extensions. In *Handbook of Combinatorial Optimization*, Z. Du and P. Pardalos (Eds). Kluwer Academic Publishers: Dordrecht, 1999; 75–149.

9. D. Ghosh, B. Goldengorin, G. Gutin and G. Jäger, Tolerance based greedy algorithms for the traveling salesman problem. To appear in *Communic. in DQM.*

10. F. Glover, G. Gutin, A. Yeo and A. Zverovich, Construction heuristics for the asymmetric TSP. *European Journal of Operational Research* **129** (2001) 555–568.

11. F. Glover and A. Punnen, The traveling salesman problem: New solvable cases and linkages with the development of approximation algorithms, *J. Oper. Res. Soc.* **48** (1997) 502–510.

12. B. Goldengorin, G. Jäger and P. Molitor. Some Basics on Tolerances. Proc. AAIM'06, S.-W. Cheng and C.K. Poon (Eds.), Lecture Notes in Computer Science 4041 (2006) 194-206.

13. G. Gutin and A. Punnen, eds., *The Traveling Salesman Problem and its Variations*, Kluwer, Dordrecht, 2002.

14. G. Gutin, H. Jakubowicz, S. Ronnen and A. Zverovitch, Seismic vessel problem. *Communic. in DQM* **8** (2005) 13–20.

15. G. Gutin and A. Yeo, Domination Analysis of Combinatorial Optimization Algorithms and Problems. *Graph Theory, Combinatorics and Algorithms: Interdisciplinary Applications* (M. Golumbic and I. Hartman, eds.), Springer-Verlag, 2005.

16. G. Gutin and A. Yeo, Polynomial approximation algorithms for the TSP and QAP with a factorial domination number. *Discrete Appl. Math.* **119** (2002) 107–116.

17. G. Gutin and A. Yeo, Anti-matroids. *Operations Research Letters* **30** (2002) 97–99.

18. G. Gutin, A. Yeo and A. Zverovitch, Exponential Neighborhoods and Domination Analysis for the TSP. In *The Traveling Salesman Problem and its Variations* (G. Gutin and A. Punnen, eds.), Kluwer, Dordrecht, 2002.

19. K. Helsgaun, An effective implementation of the Lin-Kernighan traveling salesman heuristic. *Europ. J. Oper. Res.* **126** (2000) 106–130.

20. D.S. Johnson, G. Gutin, L. McGeoch, A. Yeo, X. Zhang, and A. Zverovitch, Experimental Analysis of Heuristics for ATSP. In *The Traveling Salesman Problem and its Variations* (G. Gutin and A. Punnen, eds.), Kluwer, Dordrecht, 2002.

21. A.E. Koller and S.D. Noble, Domination analysis of greedy heuristics for the frequency assignment problem, *Discrete Math.* **275** (2004) 331-338.

22. A.P. Punnen, F. Margot and S.N. Kabadi, TSP heuristics: domination analysis and complexity. *Algorithmica* **35** (2003) 111–127.

23. A.P. Punnen and S. Kabadi, Domination analysis of some heuristics for the traveling salesman problem. *Discrete Appl. Math.* **119** (2002) 117–128.

24. A.J. Robertson, A set of greedy randomized adaptive local search procedure implementations for the multidimensional assignment problem. *Computational Optimization and Applications* **19** (2001) 145–164.

Improved Online Hypercube Packing

Xin Han[1,*], Deshi Ye[2,**], and Yong Zhou[3]

[1] School of Informatics, Kyoto University, Kyoto 606-8501, Japan
hanxin@kuis.kyoto-u.ac.jp
[2] College of Computer Science, Zhejiang University, Hangzhou, 310027, China
yedeshi@zju.edu.cn
[3] Graduate School of Science, Hokkaido University, Sapporo, Japan
zhou@castor.sci.hokudai.ac.jp

Abstract. In this paper, we study online multidimensional bin packing problem when all items are hypercubes. Based on the techniques in one dimensional bin packing algorithm Super Harmonic by Seiden, we give a framework for online hypercube packing problem and obtain new upper bounds of asymptotic competitive ratios. For square packing, we get an upper bound of 2.1439, which is better than 2.24437. For cube packing, we also give a new upper bound 2.6852 which is better than 2.9421 by Epstein and van Stee.

1 Introduction

The classical one-dimensional Bin Packing is one of the oldest and most well-studied problems in computer science [2], [5]. In the early 1970's it was one of the first combinatorial optimization problems for which the idea of worst-case performance guarantees was investigated. It was also in this domain that the idea of proving lower bounds on the performance of online algorithm was first developed. In this paper, we consider a generalization of the classical bin packing problem: hypercube packing problem.

Problem Definition. Let $d \geq 1$ be an integer. We receive a sequence δ of items $p_1, p_2, ..., p_n$. Each item p is a d-dimensional hypercube and has a fixed size, which is $s(p) \times \cdots \times s(p)$, i.e., $s(p)$ is the size of p in any dimension. We have an infinite number of bins, each of which is a d-dimensional unit hypercube. Each item must be assigned to a position $(x_1(p), ..., x_d(p))$ of some bin, where $0 \leq x_i(p)$ and $x_i(p) + s(p) \leq 1$ for $1 \leq i \leq d$. Further, the positions must be assigned in such a way that no two items in the same bin overlap. Note that for $d = 1$ the problem reduces to the classic bin packing problem. In this paper, we study the *online* version of this problem, i.e., each item must be assigned in turn, without knowledge of the next items.

Asymptotic competitive ratio. To evaluate an online algorithm for bin packing, we use the standard measure *Asymptotic competitive ratio* which is defined as follows.

* Research supported in part by KAKENHI (16092101, 16092215, 16300002).
** Research supported in part by NSFC (10601048).

Given an input list L and an online algorithm A, Let $OPT(L)$ and $A(L)$ be the cost (number of bins used) by an optimal (offline) algorithm and the cost by online algorithm A for packing list L, respectively. The *asymptotic competitive ratio* R_A^∞ of algorithm A is defined by

$$R_A^\infty = \lim_{k\to\infty} \sup \max_L \{A(L)/OPT(L) | OPT(L) = k\}.$$

Previous results. On the classic online bin packing, Johnson et al. [9] showed that the First Fit algorithm has the competitive ratio 1.7. Yao [17] gave an upper bound of 5/3. Lee and Lee [11] showed the Harmonic algorithm has the competitive ratio 1.69103 and improved it to 1.63597. Ramanan et al. [13] improved the upper bound to 1.61217. Currently, the best known upper bound is 1.58889 by Seiden [14]. On the lower bounds, Yao [17] showed no online algorithm has performance ratio less that 1.5. Brown [1] and Liang [10] independently improved this lower bound to 1.53635. The lower bound currently stands at 1.54014, due to van Vliet [16].

On online hypercube packing, Coppersmith and Raghavan [3] showed an upper bound of $43/16 = 2.6875$ for online square packing and an upper bound 6.25 for online cube packing. The upper bound for square packing was improved to $395/162 < 2.43828$ by Seiden and van Stee [15]. For online cube packing, Miyazawa and Wakabayashi [12] showed an upper bound of 3.954. Epstein and van Stee [6] gave an upper bound of 2.2697 for square packing and an upper bound of 2.9421 for online cube packing. By using a computer program, the upper bound for square packing was improved to 2.24437 by Epstein and van Stee [8]. They [8] also gave lower bounds of 1.6406 and 1.6680 for square packing and cube packing, respectively.

Our contributions. When the Harmonic algorithm [11] is extended into the online hypercube packing problem, the items of sizes $1/2 + \epsilon, 1/3 + \epsilon, 1/4 + \epsilon, \ldots$ are still the crucial items related to the asymptotic competitive ratio, where $\epsilon > 0$ is sufficiently small. Using the techniques in one dimensional bin packing, Epstein and van Stee [8] combined the items of size in $(1/2, 1 - \Delta]$ with the items of size in $(1/3, \Delta]$ and improved the Harmonic algorithm for hypercube packing, where Δ is a specified number in $(1/3, 0.385)$. In this paper, we do not only consider the combinatorial packing for the items in $(1/2, 1 - \Delta]$ and $(1/3, \Delta]$, but also other crucial items. Based on the techniques in one dimensional bin packing algorithm Super Harmonic by Seiden [14], we classify all the items into 17 groups and give a framework for online hypercube packing. To analyse our algorithm, we give a weighting system consisting of four weighting functions. By the weighting functions, we show that for square packing, the asymptotic competitive ratio of our algorithm is at most 2.1439 which is better than 2.24437[8], for cube packing, the ratio is at most 2.6852, which is also better than 2.9421[8].

Definition: If an item p of size (side length) $s(p) \leq 1/M$, where M is a fixed integer, then call p *small*, otherwise *large*.

2   Online Packing Small Items

The following algorithm for packing small items is from [4], [7]. The key ideas are below:

1. Classify all *small* squares into M groups. In detail, for an item p of size $s(p)$, we classify it into group i such that $2^k \cdot s(p) \in (1/(i+1), 1/i]$, where $i \in \{M, ..., 2M-1\}$ and k is an integer.
2. Exclusively pack items of the same group into bins, i.e., each bin is used to pack items belonged to the same group. During packing, one bin may be partitioned into sub-bins.

Definition: An item is defined to be of type i if it belongs to group i. A sub-bin which received an item is said to be *used*. A sub-bin which is not used and not cut into smaller sub-bins is called *empty*. A bin is called *active* if it can still receive items, otherwise *closed*.

Given an item p of type i, where $2^k \times s(p) \in (1/(i+1), 1/i]$, *algorithm* AssignSmall(i) works as follows.

1. If there is an empty sub-bin of size $1/(2^k i)$, then the item is simply packed there.
2. Else, in the current bin, if there is no empty sub-bin of size $1/(2^j i)$ for $j < k$, then close the bin and open a new bin and partition it into sub-bins of size $1/i$. If $k = 0$ then pack the item in one of sub-bins of size $1/i$. Else goes to next step.
3. Take an empty sub-bin of size $1/(2^j i)$ for a maximum $j < k$. Partition it into 2^d identical sub-bins. If the resulting sub-bins are larger than $1/(2^k i)$, then take *one* of them and partition it in the same way. This is done until sub-bins of size $1/(2^k i)$ are reached. Then the item is packed into one such sub-bin.

The following results are from [7].

Lemma 1. *In the above algorithm,*

i) at any time, there are at most M active bins.
ii) in each closed bin of type $i \geq M$, the occupied volume is at least $(i^d - 1)/(i + 1)^d \geq (M^d - 1)/(M+1)^d$.

So, roughly speaking, a small item with size x takes at most $\frac{(M+1)^d}{(M^d-1)} \times x^d$ bin.

3   Algorithm $\mathcal{A}$ for Online Hypercube Packing

The key points in our online algorithm are

1. divide all items into *small* and *large* groups.
2. pack small items by algorithm AssignSmall, pack large items by an extended Super Harmonic algorithm.

Before giving our algorithm, we first give some definitions and descriptions about the algorithm, which are similar with the ones in [14], but some definitions are different from the ones in [14].

Classification of large items: Given an integer $M \geq 11$, let $t_1 = 1 > t_2 > \cdots > t_{N+1} = 1/M > t_{N+2} = 0$, where N is a fixed integer. We define the interval I_j to be $(t_{j+1}, t_j]$ for $j = 1, ..., N+1$ and say a large item p of size $s(p)$ has type i if $s(p) \in I_i$.

Definition: An item of size s has type $\tau(s)$, where

$$\tau(s) = j \quad \Leftrightarrow \quad s \in I_j.$$

Parameters in algorithm $\mathcal{A}$: An instance of the algorithm is described by the following parameters: integers N and K; real numbers $1 = t_1 > t_2 > \cdots > t_N > t_{N+1} = 1/M$, $\alpha_1, ..., \alpha_N \in [0, 1]$ and $0 = \Delta_0 < \Delta_1 < \cdots < \Delta_K < 1/2$, and a function $\phi : \{1, ..., N\} \mapsto \{0, ..., K\}$.

Next, we give the operation of our algorithm, essentially, which is quite similar with the Super Harmonic algorithm [14]. Each *large* item of type j is assigned a color, *red* or *blue*. The algorithm uses two sets of counters, $e_1, ..., e_N$ and $s_1, ..., s_N$, all of which are initially zero. s_i keeps track of the total number of type i items. e_i is the number of type i items which get colored red. For $1 \leq i \leq N$, the invariant $e_i = \lfloor \alpha_i s_i \rfloor$ is maintained, i.e. the percentage of type i items colored red is approximately α_i.

We first introduce some parameters used in Super Harmonic algorithm, then give the corresponding ones for d-dimensional packing. In one dimensional packing, a bin can be placed at most $\beta_i = \lfloor 1/t_i \rfloor$ items with size t_i. After packing β_i type i items, there is $\delta_i = 1 - t_i \beta_i$ space left. The rest space can be used for red items.

However, we sometimes use less than δ_i in a bin in order to simplify the algorithm and its analysis, i.e., we use $\mathcal{D} = \{\Delta_1, ..., \Delta_K\}$ instead of the set of δ_i, for all i. $\Delta_{\phi(i)}$ is the amount of space used to hold red items in a bin which holds blue items of type i. We therefore require that ϕ satisfy $\Delta_{\phi(i)} \leq \delta_i$. $\phi(i) = 0$ indicates that no red items are accepted. To ensure that every red item potentially can be packed, we require that $\alpha_i = 0$ for all i such that $t_i > \Delta_K$, that is, there are no red items of type i. Define $\gamma_i = 0$ if $t_i > \delta_K$ and $\gamma_i = \max\{1, \lfloor \Delta_1/t_i \rfloor\}$, otherwise. This is the number of red item of type i placed in a bin.

In d-dimensional packing, we place β_i^d blue items of type i into a bin and introduce a new parameter θ_i instead of γ_i. Let

$$\theta_i = \beta_i^d - (\beta_i - \gamma_i)^d.$$

This is the number of red items of type i that the algorithm places together in a bin. In details, if $t_i > \Delta_K$, then $\theta_i = 0$, i.e., we do not pack type i items as red items. So, in this case, we require $\alpha_i = 0$. Else if $t_i \leq \Delta_1$, then $\theta_i = \beta_i^d - (\beta_i - \lfloor \Delta_1/t_i \rfloor)^d$. If $\Delta_1 < t_i \leq \Delta_K$, we set $\theta_i = \beta_i^d - (\beta_i - 1)^d$.

Here, we illustrate the structure of a bin for $d = 2$.

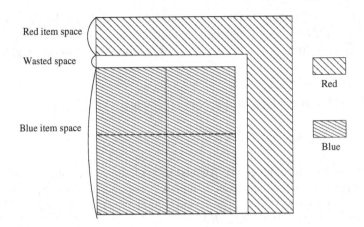

Fig. 1. If the bin is a (i, j) or (i, ?) bin, the amount of area for blue items is $(t_i\beta_i)^2$. The amount of area left is $1 - (t_i\beta_i)^2$. The amount of this area actually used for red items is $1 - (1 - \Delta_{\phi(i)})^2$, where $\Delta_{\phi(i)} \leq \delta_i = 1 - t_i\beta_i$.

Naming bins: Bins are named as follows:

$$\{i|\phi_i = 0, 1 \leq i \leq N,\}$$
$$\{(i, ?)|\phi_i \neq 0, 1 \leq i \leq N,\}$$
$$\{(?, j)|\alpha_j \neq 0, 1 \leq j \leq N,\}$$
$$\{(i, j)|\phi_i \neq 0, \alpha_j \neq 0, \gamma_j t_j \leq \Delta_{\phi(i)}, 1 \leq i, j \leq N\}.$$

We call these groups *monochromatic, indeterminate blue, indeterminate red* and *bichromatic*, respectively. And we call the monochromatic and bichromatic groups *final* groups.

The monochromatic group i contains bins that hold only blue items of type i. There is only one open bin in each of these groups; this bin has fewer than β_i^d items. The closed bins all contain β_i^d items.

The bichromatic group (i, j) contains bins that contain blue items of type i along with red items of type j. A closed bin in this group contains β_i^d type i items and θ_j type j items. There are at most three open bins.

The indeterminate blue group $(i, ?)$ contains bins that hold only blue items of type i. These bins are all open, but only one has fewer than β_i^d items.

The indeterminate red group $(?, j)$ contains bins that hold only red items of type j. These bins are all open, but only one has fewer than θ_j items.

Essentially, the algorithm tries to minimize the number of indeterminate bins, while maintaining all the aforementioned invariants. That is, we try to place red and blue items together whenever possible; when this is not possible we place them in indeterminate bins in hope that they can later be so combined.

Algorithm $\mathcal{A}$: A formal description of algorithm $\mathcal{A}$ is given as blow:

Initialize $e_i \leftarrow 0$ and $s_i \leftarrow 0$ for $1 \leq i \leq M + 1$.

For a small item p, call algorithm AssignSmall.

For a large item p:

    $i \leftarrow \tau(p), \quad s_i \leftarrow s_i + 1.$

    If $e_i < \lfloor \alpha_i s_i \rfloor$:

        $e_i \leftarrow e_i + 1.$

        Color p red.

        If there is an open bin in group $(?, i)$ with fewer than θ_i type i items, then pack p in this bin.

        If there is an open bin in group (j, i) with fewer than θ_i type i items, then pack p in this bin.

        Else if there is some bin in group $(j, ?)$ such that $\Delta_{\phi(j)} \geq \gamma_i t_i$ then place p in it and change the group of this bin to (j, i).

        Otherwise, open a new group $(?, i)$ bin and place p in it.

    Else:

        Color p blue.

        If $\phi(i) = 0$:

            If there is an open bin in group i with fewer than β_i^d items, then place p in it.

            Otherwise, open a new group i bin and pack p there.

        Else:

            If, for any j, there is an open bin (i, j) with fewer than β_i^d items, then place p in this bin.

            Else, if there is some bin in group $(i, ?)$ with fewer than β_i^d items, then place p in this bin.

            Else, if there is some bin in group $(?, j)$ such that $\Delta_{\phi(i)} \geq \gamma_j t_j$ then pack p in it and change the group of this bin to (i, j).

            Otherwise, open a new group $(i, ?)$ bin and pack p there.

4   The Analyses for Square and Cube Packing

In this section, we fix the parameters in the framework given in the last section for square packing and cube packing respectively. Then we analyse the competitive ratios by a corresponding weighting system consisting of four weighting functions.

4.1   An Instance of Algorithm $\mathcal{A}$

Let $M = 11$, i.e., a *small* item has its side length as most $1/11$. And the parameters in $\mathcal{A}$ are given in the following tables. First we classify all the items into 17 groups by fixing the values of t_i, where $1 \leq i \leq 18$. Then we calculate the number of blue type i in a bin, β_i^d. Finally, we define the set $\mathcal{D} = \{\Delta_1, ..., \Delta_K\}$ and the function $\phi(i)$, which are related to how many red items θ_i^d can be accepted in a bin, where $K = 4$. Note that α_i which is the percentage of type i items colored red will be given later. For square packing, we use a set of α_i. While for cube packing, we use another set of α_i.

i	$(t_{i+1}, t_i]$	β_i	δ_i	$\phi(i)$	γ_i
1	$(0.7, 1]$	1	0	0	0
2	$(0.65, 0.7]$	1	0.3	2	0
3	$(0.60, 0.65]$	1	0.35	3	0
4	$(0.5, 0.60]$	1	0.4	4	0
5	$(0.4, 0.5]$	2	0	0	0
6	$(0.35, 0.4]$	2	0.2	1	1
7	$(1/3, 0.35]$	2	0.3	2	1
8	$(0.30, 1/3]$	3	0	0	0
9	$(1/4, 0.30]$	3	0.1	0	1
10	$(1/5, 1/4]$	4	0	0	1
11	$(1/6, 1/5]$	5	0	0	1
12	$(1/7, 1/6]$	6	0	0	1
13	$(1/8, 1/7]$	7	0	0	1
14	$(1/9, 1/8]$	8	0	0	1
15	$(0.1, 1/9]$	9	0	0	1
16	$(1/11, 0.1]$	10	0	0	2
17	$(0, 1/11]$	*	*	*	*

$j = \phi(i)$	Δ_j	Red items accepted
1	0.20	11..16
2	0.30	9..16
3	0.35	7, 9..16
4	0.40	6..7, 9..16

Observation: By the above tables, in any dimension of a $(4, ?)$ bin, the distance between the type 4 item and the opposite edge (face) of the bin is at least $\Delta_4 = 0.4$, since we pack a type 4 item in a corner of a bin. So, all red items with size at most 0.4 can be packed in $(4, ?)$ bins. In the same ways, all red items with size at most 0.35 can be packed in $(4, ?)$ and $(3, ?)$ bins, all red items with size at most 0.30 can be packed in $(4, ?)$, $(3, ?)$, $(7, ?)$ and $(2, ?)$ bins, all red items with size at most 0.2 can be packed in $(4, ?)$, $(3, ?)$, $(7, ?)$, $(2, ?)$, $(6, ?)$ bins.

Next we define the weight function $W(p)$ for a given item p with size x. Roughly speaking, a weight of an item is the maximal portion of a bin that it can occupy. Given a small item p with size x, by the approach in [7], it occupies a $\frac{x^d (11+1)^d}{11^d - 1}$ bin. So, we define

$$W(p) = \frac{x^d (11+1)^d}{11^d - 1}.$$

Given a large item p, we consider four cases to define its weight. Let R_i and B_i be the number of bins containing blue items of type i and red items of type i, respectively. Let E be the number of indeterminate red group bins, i.e., some bins like $(?, i)$. If $E > 0$ then there are some $(?, j)$ bins. Let

$$e = \min\{j | (?, j)\},$$

which is the type of the smallest red item in an indeterminate red group bin. Let $\mathcal{A}(L)$ be the number of bins used by $\mathcal{A}$.

Case 1: $E = 0$, i.e., no indeterminate red bins. Then every red item is packed with one or more blue items. Therefore

$$\mathcal{A}(L) \leq \mathcal{A}(L_s) + \sum_i B_i,$$

where $\mathcal{A}(L_s)$ is the number of bins for small items. Since there are a constant number of active bins and every closed blue bin (i) or $(i, *)$ contains $\frac{1}{\beta_i^d}$ items, we define the weighting function as below:

$$W_{1,1}(p) = \frac{1 - \alpha_i}{\beta_i^d} \quad \text{if } x \in I_i, \text{ for } i = 1..16.$$

Case 2: $E > 0$ and $e = 6$. Then there are some bins $(?, 6)$ and no other bins $(?, j)$ bins, where $j > 6$. Since a type 4 item can be packed into a bin $(?, 6)$, it is impossible to have bins $(4, ?)$. If we count all $(4, j)$ bins as red bins, then

$$\mathcal{A}(L) \leq \mathcal{A}(L_s) + \sum_{i=1..3,5,8} B_i + \sum_{i=6,7,9..16} (R_i + B_i).$$

Else we count all $(4, j)$ bins as blue bins then

$$\mathcal{A}(L) \leq \mathcal{A}(L_s) + \sum_{i=1..16} B_i + R_6.$$

Since there are a constant number of active bins and every closed blue bin (i) or $(i, *)$ contains $\frac{1}{\beta_i^d}$ items, every closed red bin (j, i) or $(?, i)$ contains $\frac{1}{\theta_i}$ items, we define the weighting functions for two subcases as below:

$$W_{2,1}(p) = \begin{cases} \frac{1-\alpha_i}{\beta_i^d} & \text{if } x \in I_i, \text{ for } i = 1, 2, 3, 5, 8. \\ 0 & \text{if } x \in I_4. \\ \frac{1-\alpha_i}{\beta_i^d} + \frac{\alpha_i}{\theta_i} & \text{if } x \in I_i, \text{ for } i = 6, 7, 9..16. \end{cases}$$

and

$$W_{2,2}(p) = \begin{cases} \frac{1-\alpha_i}{\beta_i^d} & \text{if } x \in I_i, \text{ for } i = 1..5, 7..16. \\ \frac{1-\alpha_i}{\beta_i^d} + \frac{\alpha_i}{\theta_i} & \text{if } x \in I_i, \text{ for } i = 6. \end{cases}$$

Case 3: $E > 0$ and $e = 7$. Then there are some bins $(?, 7)$ and no other bins $(?, j)$, where $j > 7$. Since a type 4 or a type 3 item can be packed into a bin $(?, 7)$, it is impossible to have bins $(4, ?)$ and $(3, ?)$. If we count all $(4, j)$ and $(3, j)$ bins as red bins, then

$$\mathcal{A}(L) \leq \mathcal{A}(L_s) + \sum_{i=1,2,5,8} B_i + \sum_{i=6,7,9..16} (R_i + B_i).$$

Else we count all $(4, j)$ and $(3, j)$ bins as blue bins then

$$A(L) \leq A(L_s) + \sum_{i=1..16} B_i + R_6 + R_7.$$

We define the weighting functions for two subcases as below:

$$W_{3,1}(p) = \begin{cases} \frac{1-\alpha_i}{\beta_i^d} & \text{if } x \in I_i, \text{ for } i = 1, 2, 5, 8. \\ 0 & \text{if } x \in I_3, I_4. \\ \frac{1-\alpha_i}{\beta_i^d} + \frac{\alpha_i}{\theta_i} & \text{if } x \in I_i, \text{ for } i = 6, 7, 9..16. \end{cases}$$

and

$$W_{3,2}(p) = \begin{cases} \frac{1-\alpha_i}{\beta_i^d} & \text{if } x \in I_i, \text{ for } i = 1..5, 8..16. \\ \frac{1-\alpha_i}{\beta_i^d} + \frac{\alpha_i}{\theta_i} & \text{if } x \in I_i, \text{ for } i = 6, 7. \end{cases}$$

Case 4: $E > 0$ and $e \geq 9$. Then there are some bins $(?, 9)$. Since a type 2,3,4,7 item can be packed into a bin $(?, 9)$, it is impossible to have bins $(2, ?)$, $(3, ?)$, $(4, ?)$, $(7, ?)$. If we count these bins $(2, j)$, $(3, j)$, $(4, j)$, $(7, j)$ as red bins, then

$$A(L) \leq A(L_s) + \sum_{i=1,5,8} B_i + \sum_{i=6,9..16} (R_i + B_i) + R_7.$$

We define the weighting function as below:

$$W_{4,1}(p) = \begin{cases} \frac{1}{\beta_i^d} & \text{if } x \in I_i, \text{ for } i = 1, 5, 8. \\ 0 & \text{if } x \in I_2, I_3, I_4 \\ \frac{1-\alpha_i}{\beta_i^d} + \frac{\alpha_i}{\theta_i} & \text{if } x \in I_i, \text{ for } i = 6, 9..16 \\ \frac{\alpha_i}{\theta_i} & \text{if } x \in I_i, \text{ for } i = 7. \end{cases}$$

Definition: A set of items X is a feasible set if all items in it can be packed into a bin. And,

$$W_{i,j}(X) = \sum_{p \in X} W_{i,j}(p).$$

Over all feasible sets X, let

$$W_i(X) = \min\{W_{i,j}(X)\}, \ j = 1 \text{ or } 2,$$

and define

$$\mathcal{P}(W) = \max\{W_i(X)\} \text{ for all } i.$$

We defined four sets of weighting functions for all items. This is a weighting system, which is a special case of general weighting system defined in [14]. So, the following lemma follows directly from [14].

Lemma 2. *The asymptotic performance ratio of A is upper bounded by $\mathcal{P}(W)$.*

4.2   Upper Bounds for Square and Cube Packing

In this subsection, we fix the parameters α_i for square packing and cube packing respectively, and get the upper bounds of the asymptotic competitive ratios.

Definition: Let $m_i \geq 0$ be the number of type i items in a feasible set X. Given an item p with size x, define an efficient function $E_{i,j}(p)$ as $W_{i,j}(p)/x^d$.

Theorem 1. *The asymptotic performance ratio of $\mathcal{A}$ for square packing is at most 2.1439.*

Proof. For square packing, we set parameters α_i according to the following table.

i	$1-4$	5	6	7	8	9	10	11	12	13	14	15	16
α_i	0		0	0.12	0.2	0	0.2546	0.2096	0.15	0.1	0.1	0.1	0.1	0.05
θ_i	0		0	3	3	0	5	7	9	11	13	15	17	36
β_i^2	1		4	4	4	9	9	16	25	36	49	64	81	100

Based on the values in the following two tables, we calculate the upper bound of $\mathcal{P}(W) = \max\{W_i(X)\}$.

i	$(t_{i+1}, t_i]$	$W_{1,1}(p)$	$E_{1,1}(p)$	$W_{2,1}(p)$	$E_{2,1}(p)$	$W_{2,2}(p)$	$E_{2,2}(p)$
1	$(0.7, 1]$	1	2.05	1	2.05	1	2.05
2	$(0.65, 0.7]$	1	2.37	1	2.37	1	2.37
3	$(0.6, 0.65]$	1	2.7778	1	2.7778	1	2.7778
4	$(0.5, 0.6]$	1	4	0	0	1	4
5	$(0.4, 0.5]$	$1/4$	1.5625	$1/4$	1.5625	$1/4$	1.5625
6	$(0.35, 0.4]$	0.22	1.8	0.26	2.123	0.26	2.123
7	$(1/3, 0.35]$	0.2	1.8	0.8/3	2.4	0.2	1.8
8	$(0.3, 1/3]$	$1/9$	1.235	$1/9$	1.235	$1/9$	1.235
9	$(1/4, 0.3]$	0.0829	1.327	0.1338	2.141	0.0829	1.327
10..17	$(0, 1/4]$	$1.235x^2$	1.235	$1.99x^2$	1.99	$1.235x^2$	1.235

Case 1: $W_1(X) \leq 2.1439$.

If $m_2 + m_3 + m_4 = 0$, i.e., no type $2, 3, 4$ items in X, then

$$W_1(X) = \sum_{p \in X} E_{1,1}(p)s(p)^2 \leq 2.05 \sum_{p \in X} s(p)^2 \leq 2.05.$$

Else $m_2 + m_3 + m_4 = 1$. Then $m_5 + m_6 + m_7 \leq 3$ and $m_6 + m_7 + m_9 \leq 5$,

$$W_1(X) \leq 1 + m_5/4 + 0.22m_6 + 0.2m_7 + 0.0829m_9$$
$$+ 1.235(1 - \sum_{i=2}^{7} t_{i+1}^2 m_i - m_9/16)$$
$$< 2.1439.$$

The last inequality follows from $m_4 = 1$, $m_6 = 3$ and $m_9 = 2$.

Case 2: $W_2(X) \le 2.134$.
If $m_2 + m_3 + m_4 = 0$, i.e., no type $2, 3, 4$ items in X, then

$$W_2(X) = \min\{W_{2,1}(X), W_{2,2}(X)\} \le W_{2,2}(X) \le 2.123.$$

Else $m_2 = 1$. Then no type $1, 3, 4, 5, 6$ items in X.

$$W_2(X) = W_{2,2}(X) \le 1 + 1.8(1 - 0.65^2) = 2.0395.$$

Else $m_3 = 1$. Then no type $1, 2, 4, 5$ items in X and $m_6 + m_7 \le 3$ and $m_6 + m_7 + m_9 \le 5$,

$$W_2(X) = W_{2,2}(X) \le 1 + 0.26m_6 + 0.2m_7 + 0.0829m_9$$
$$+ 1.235(1 - 0.6^2 - 0.35^2 m_6 - m_7/9 - m_9/16)$$
$$< 2.134.$$

The last inequality follows from $m_6 = 3$ and $m_9 = 2$.
Else $m_4 = 1$. Then no type $1, 2, 3$ items in X.

$$W_2(X) \le W_{2,1}(X) \le 0 + 2.4(1 - 0.5^2) = 1.8.$$

i	$(t_{i+1}, t_i]$	$W_{3,1}(p)$	$E_{3,1}(p)$	$W_{3,2}(p)$	$E_{3,2}(p)$	$W_{4,1}(p)$	$E_{4,1}(p)$
1	$(0.7, 1]$	1	2.05	1	2.05	1	2.05
2	$(0.65, 0.7]$	1	2.37	1	2.37	0	0
3	$(0.6, 0.65]$	0	0	1	2.7778	0	0
4	$(0.5, 0.6]$	0	0	1	4	0	0
5	$(0.4, 0.5]$	1/4	1.5625	1/4	1.5625	1/4	1.5625
6	$(0.35, 0.4]$	0.26	2.123	0.26	2.123	0.26	2.123
7	$(1/3, 0.35]$	0.8/3	2.4	0.8/3	2.4	0.2/3	0.6
8	$(0.3, 1/3]$	1/9	1.235	1/9	1.235	1/9	1.235
9	$(1/4, 0.3]$	0.1338	2.141	0.0829	1.327	0.1338	2.141
10..17	$(1/5, 1/4]$	$1.99x^2$	1.99	$1.235x^2$	1.235	$1.99x^2$	1.99

Case 3: $W_3(X) \le 2.12$.
If $m_1 + m_2 + m_3 + m_4 = 0$, i.e., no type $1, 2, 3, 4$ items in X, then $m_6 + m_7 \le 4$,

$$W_3(X) = W_{3,2}(X) \le 0.26m_6 + \frac{0.8m_7}{3} + 1.5625(1 - 0.35^2 m_6 - \frac{m_7}{9}) < 2.$$

Else $m_1 = 1$ then $m_i = 0$, where $2 \le i \le 8$,

$$W_3(X) = W_{3,2}(X) \le 2.05.$$

Else $m_2 = 1$. Then no type $1, 3, 4, 5, 6$ items in X, $m_7 + m_9 \le 5$ and $m_7 \le 3$. So,
$W_3(X) = W_{3,2}(X) \le 1 + \frac{0.8m_7}{3} + 0.0829m_9 + 1.235(1 - 0.65^2 - \frac{m_7}{9} - \frac{m_9}{16}) < 2.12$.
Else $m_3 + m_4 = 1$. Then no type $1, 2, 3$ items in X.

$$W_3(X) \le W_{3,1}(X) \le 0 + 2.4(1 - 0.5^2) = 1.8.$$

Case 4: $W_4(X) = \sum_{p \in X} E_{4,1}(p)s(p)^2 \le 2.141 \sum_{p \in X} s(p)^2 \le 2.141.$
So, $\mathcal{P}(W) \le 2.1439.$

Theorem 2. *The asymptotic performance ratio of $\mathcal{A}$ for cube packing is at most 2.6852.*

Proof. For cube packing, we set parameters α_i and θ_i in the following table.

i	$1-4$	5	6	7	8	9	10	11	$12-16$
α_i	0		0	0.12	0.2	0	0.325	0.2096	0.15	0
θ_i	0		0	7	7	0	19	37	61	0
β_i^3	1		8	8	8	27	27	64	125	$(i-6)^3$

Here we set $\alpha_i = 0$ for $12 \le i \le 16$. So, their weights are defined as $1/\beta_i^3$.
We first give two tables and then use them to calculate $\mathcal{P}(W)$.

i	$(t_{i+1}, t_i]$	$W_{1,1}(p)$	$E_{1,1}(p)$	$W_{2,1}(p)$	$E_{2,1}(p)$	$W_{2,2}(p)$	$E_{2,2}(p)$
1	$(0.7, 1]$	1	2.9155	1	2.9155	1	2.9155
2	$(0.65, 0.7]$	1	3.65	1	3.65	1	3.65
3	$(0.6, 0.65]$	1	4.63	1	4.63	1	4.63
4	$(0.5, 0.6]$	1	8	0	0	1	8
5	$(0.4, 0.5]$	1/8	1.9532	1/8	1.9532	1/8	1.9532
6	$(0.35, 0.4]$	0.11	2.5656	0.1272	2.966	0.1272	2.966
7	$(1/3, 0.35]$	0.1	2.7	0.1286	3.472	0.1	2.7
8	$(0.3, 1/3]$	1/27	1.372	1/27	1.372	1/27	1.372
9	$(1/4, 0.3]$	0.025	1.6	0.04211	2.6948	0.025	1.6
10	$(1/5, 1/4]$	0.0124	1.55	0.01802	2.252	0.0124	1.55
11	$(1/6, 1/5]$	0.0068	1.4688	0.0093	2	0.0068	1.4688
12..17	$(0, 1/6]$	$1.59x^3$	1.59	$1.59x^3$	1.59	$1.59x^3$	1.59

Case 1: $W_1(X) \le 2.6852.$
If $m_1 + m_2 + m_3 + m_4 = 0$, i.e., no type $1, 2, 3, 4$ items in X, then $m_6 + m_7 \le 8$,

$$W_1(X) \le 0.11m_6 + 0.1m_7 + 1.96(1 - 0.35^3 m_6 - m_7/27) \le 2.3.$$

Else $m_1 = 1$. Then $m_i = 0$, where $2 \le i \le 8$,

$$W_1(X) \le 1 + 1.6(1 - 0.7^3) = 2.0512.$$

Else $m_2 = 1$. Then no type $1, 3, 4, 5, 6$ items in X and $m_7 \le 7$,

$$W_1(X) \le 1 + 0.1 \times 7 + 1.6(1 - 0.65^3 - 7/27) \le 2.546.$$

Else $m_3 = 1$. Then no type $1, 2, 4, 5$ items in X and $m_6 + m_7 \le 7$,

$$W_1(X) \le 1 + 0.11m_6 + 0.1m_7 + 1.6(1 - 0.6^3 - 0.35^3 m_6 - m_7/27) \le 2.5646.$$

Else $m_4 = 1$. Then $m_1 + m_2 + m_3 = 0$ and $m_5 + m_6 + m_7 \le 7$,

$$\begin{aligned} W_1(X) \le{}& 1 + m_5/8 + 0.11m_6 + 0.1m_7 \\ &+ 1.6(1 - 0.5^3 - 0.4^3 m_5 - 0.35^3 m_6 - m_7/27) \\ <{}& 2.6852. \end{aligned}$$

The last inequality follows from $m_7 = 7$ and $m_5 = m_6 = 0$.

Case 2: $W_2(X) \leq 2.6646$.

If $m_1 + m_2 + m_3 + m_4 = 0$, i.e., no type $1, 2, 3, 4$ items in X, then $m_6 + m_7 \leq 8$,

$$W_2(X) = W_{2,2} \leq 0.1272 m_6 + 0.1 m_7 + 1.96(1 - 0.35^3 m_6 - m_7/27) \leq 2.4.$$

Else $m_1 + m_2 = 1$. Then no type $1, 4, 5, 6$ items in X,

$$W_2(X) = W_{2,2}(X) = W_1(X) \leq 2.546.$$

Else $m_3 = 1$. Then no type $1, 2, 4, 5$ items in X and $m_6 + m_7 \leq 7$,

$$W_2(X) \leq 1 + 0.1272 m_6 + 0.1 m_7 + 1.6(1 - 0.6^3 - 0.35^3 m_6 - m_7/27) \leq 2.6646.$$

Else $m_4 = 1$. Then no type $1, 2, 3$ items in X and $m_6 + m_7 \leq 7$,

$$W_2(X) \leq W_{2,1}(X) \leq 0 + 0.1272 m_6 + 0.1286 m_7 +$$
$$2.6948(1 - 0.35^3 m_6 - m_7/27 - 1/8) < 2.5595.$$

The last inequality holds for $m_6 = 0$ and $m_7 = 7$.

i	$(t_{i+1}, t_i]$	$W_{3,1}(p)$	$E_{3,1}(p)$	$W_{3,2}(p)$	$E_{3,2}(p)$	$W_{4,1}(p)$	$E_{4,1}(p)$
1	$(0.7, 1]$	1	2.9155	1	2.9155	1	2.9155
2	$(0.65, 0.7]$	1	3.65	1	3.65	0	0
3	$(0.6, 0.65]$	0	0	1	4.63	0	0
4	$(0.5, 0.6]$	0	0	1	8	0	0
5	$(0.4, 0.5]$	1/8	1.9532	1/8	1.9532	1/8	1.9532
6	$(0.35, 0.4]$	0.1272	2.966	0.1272	2.966	0.1272	2.966
7	$(1/3, 0.35]$	0.1286	3.472	0.1286	3.472	0.03	0.81
8	$(0.3, 1/3]$	1/27	1.372	1/27	1.372	1/27	1.372
9	$(1/4, 0.3]$	0.04211	2.6948	0.025	1.6	0.04211	2.6948
10	$(1/5, 1/4]$	0.01802	2.252	0.0124	1.55	0.01802	2.252
11	$(1/6, 1/5]$	0.0093	2	0.0068	1.4688	0.0093	2
12..17	$(0, 1/6]$	$1.59x^3$	1.59	$1.59x^3$	1.59	$1.59x^3$	1.59

By the similar calculation with Case 1 and Case 2, we have

$$W_3(X) \leq 2.646 \text{ and } W_4(X) \leq 2.63.$$

So, $\mathcal{P}(W) < 2.6852$. (Due to page limination, we skip the details.)

5   Concluding Remarks

In this page, we reduced the gaps between the upper and lower bounds of online square packing and cube packing. But the gaps are still large. It seems possible to use computer proof as the one in [14] to get a more precise upper bound. But, the analysis becomes more complicated and more difficult than the one in [14], since we are faced to solve a two dimensional knapsack problem, rather than one dimensional knapsack problem [14]. So, how to reduce the gaps is a challenging open problem.

References

1. D.J. Brown, A lower bound for on-line one-dimensional bin packing algorithms, *Techincal report R864*, Coordinated Sci. Lab., Urbana, Illinois, 1979.
2. E.G. Coffman, M.R. Garey and D.S. Johnson, Approximation algorithms for bin packing: a survey. In Approximation Algorithms for NP-hard Problems, D. Hochbaum, Ed. PWS, Boston, MA, 1997, chapter 2.
3. D. Coppersmith, P. Paghavan, Multidimensional on-line bin packing: Algorithms and worst case analysis, *Oper. Res. Lett.* 8:17-20, 1989.
4. J. Csirik, A. van Vliet, An on-line algorithm for multidimensional bin packing, *Oper. Res. Lett.* 13: 149-158, 1993.
5. J. Csirik and G.J. Woeginger, Shelf algorithm for on-line strip packing, *Information Processing Letters* 63, 171-175, 1997.
6. L. Epstein, R. van Stee, Optimal online bounded space multidimensional packing, *SODA 2004*: 214-223.
7. L.Epstein, R. van Stee, Optimal Online Algorithms for Multidimensional Packing Problems. *SIAM J. Computing*, 35(2): 431-448, 2005.
8. L.Epstein, R. van Stee, Online square and cube packing, *Acta Inf.* 41(9): 595-606, 2005.
9. D.S. Johnson, A.J. Demers, J.D. Ullman, M. R. Garey, R. L. Graham, Worst-Case performance bounds for simple one-dimensional packing algorithms, *SIAM J. Comput.* 3(4): 299-325, 1974.
10. F.M. Liang, A lower bound for online bin packing, *Information processing letters* 10,76-79,1980.
11. C.C. Lee and D.T. Lee, A simple on-line bin-packing algorihtm, *J. ACM* 32, 562-572, 1985.
12. F.K. Miyazawa, Y. Wakabayashi, Cube packing, *Theor. Comput. Sci.* 1-3(297): 355-366, 2003.
13. P.V. Ramanan, D.J. Brown, C.C. Lee, and D. T. Lee, On-line bin packing in linear Time, *J. Algorithms* 10, 305-326, 1989.
14. S.S. Seiden, On the online bin packing problem, *J. ACM* 49, 640-671, 2002.
15. S.S. Seiden, R. van Stee, New bounds for multidimensional packing, *Algorithmica* 36(3): 261-293, 2003.
16. A. van Vliet, An improved lower bound for on-line bin packing algorithms, *Inform. Process. Lett.* 43, 277-284,1992.
17. A.C.-C. Yao, New Algorithms for Bin Packing, *J. ACM* 27, 207-227, 1980.

Competitive Online Multicommodity Routing

Tobias Harks*, Stefan Heinz, and Marc E. Pfetsch**

Konrad-Zuse-Zentrum für Informationstechnik Berlin,
Takustr. 7, 14195 Berlin, Germany
{harks,heinz,pfetsch}@zib.de

Abstract. In this paper we study online multicommodity minimum cost routing problems in networks, where commodities have to be routed sequentially. The flow of each commodity can be split on several paths. Arcs are equipped with load dependent price functions defining routing costs. We discuss a greedy online algorithm that routes each commodity by minimizing a convex cost function that only depends on the demands previously routed. We present a competitive analysis of this algorithm showing that for affine linear price functions this algorithm is $\frac{4K}{2+K}$-competitive, where K is the number of commodities. For the parallel arc case, this algorithm is optimal. Without restrictions on the price functions and network, no algorithm is competitive. Finally, we investigate a variant in which the demands have to be routed unsplittably.

1 Introduction

In this work we study the fundamental problem of sequentially routing demands in a network. We consider dynamic load dependent price functions on links. In realistic scenarios the online aspect arises due to the fact that by the time of routing a given demand, future demands are not known. We briefly outline two examples.

Open Shortest Path First (OSPF) is the most commonly used intra-domain Internet routing protocol today, see Moy [1]. Here, traffic is routed along shortest paths from source to destination with respect to weights on the links that are under the control of network operators. A default weight setting strategy is to make the weight inversely proportional to the physical link capacity as suggested by Cisco [2]. If the routing weights are interpreted as prices for reserving capacity on the corresponding link, the OSPF protocol routes demands along the cheapest path.

Minimum cost routing also arises in an inter-domain Quality of Service (QoS) market, where multiple service providers offer network resources (capacity) to enable Internet traffic with specific QoS constraints, see for example Yahaya, Suda and Harks [3,4]. In such a market, each service provider advertises prices

---

* Supported by the German research funding agency 'DFG' under the graduate program 'Graduiertenkolleg 621 (MAGSI/Berlin)'.

** Supported by the DFG Research Center MATHEON *Mathematics for key technologies* in Berlin.

T. Erlebach and C. Kaklamanis (Eds.): WAOA 2006, LNCS 4368, pp. 240–252, 2006.
© Springer-Verlag Berlin Heidelberg 2006

(weights) for resources that he wants to sell. Buying providers reserve capacity along the cheapest available path to route demand (coming from own customers) from source to destination via domains of other providers.

In this paper we investigate the *Online Multicommodity Routing Problem* (ONLINEMCRP). Here, commodities in a network have to be routed sequentially in an online fashion. The flow for each commodity can be split on several paths. The cost for each arc is defined by load dependent price functions. As far as we know, this approach has not been investigated before.

Related Work. Multicommodity routing problems have been studied in the context of traffic engineering, see Fortz and Thorup [5,6]. There, the goal is to route given demands subject to capacity constraints in order to minimize a convex load dependent penalty function. In this setting, a central planer has full knowledge of all demands, which is not the case in our approach.

Another related line of research is the investigation of efficient routing in decentralized noncooperative systems. This has been extensively studied using game theoretic concepts, cf. Roughgarden and Tardos [7], Correa, Schulz, and Stier Moses [8], and references therein. In this line of research, the efficiency of Nash equilibria are studied. Hence, rerouting of demands is allowed in this context. In our model, once a routing decision has been made this routing remains unchanged.

In the online network routing field, mainly call admission control problems have been considered. An overview article about these problems is given by Leonardi in [9]. Perhaps closest to our work is the paper by Awerbuch, Azar, and Plotkin [10], where online routing algorithms are presented to maximize throughput under the assumption that routings are irrevocable. However, Awerbuch et al. restrict the analysis to single path routing and present competitive bounds that depend on the number of nodes in the network.

Contributions. We show that no online algorithm for the ONLINEMCRP is competitive if the price functions and network are not restricted. However, for affine linear price functions we investigate a greedy online algorithm, called SEQ, and show that this algorithm is $\frac{4K}{2+K}$-competitive, where K is the number of commodities. Furthermore, we prove in this case a lower bound of $\frac{4}{3}$ on the competitive ratio for any deterministic online algorithm. If the network only consists of parallel arcs, SEQ is optimal. We also study a variant in which the demands have to be routed unsplittably. There, we prove that the offline problem is NP-hard, show that in general no competitive deterministic online algorithm exists, and present a lower bound of 2 on the competitive ratio for any deterministic online algorithm if the price functions are linear.

2   Problem Description

An instance of the *Online Multicommodity Routing Problem* (ONLINEMCRP) consists of a directed network $D = (V, A)$ and nondecreasing continuous price functions $p_a : \mathbb{R}_+ \rightarrow \mathbb{R}_+$ for each link $a \in A$. These functions define the price of reserving capacity on a link depending on the current load, see below. Furthermore, a sequence $\sigma = 1, \ldots, K$ of commodities must be routed one after the other. We assume that $K \geq 2$ and denote the set of commodities by $[K] := \{1, \ldots, K\}$. The routing decision for commodity k is *online*, that is, it only depends on the routings of commodities $1, \ldots, k-1$. Once a commodity has been routed it remains unchanged. Each commodity $k \in [K]$ has a demand $d_k > 0$ that is to be routed from its source $s_k \in V$ to its destination $t_k \in V$ after it arises.

A routing assignment, or *flow*, for commodity $k \in [K]$ is a nonnegative vector $\boldsymbol{f}^k \in \mathbb{R}_+^A$. This flow is *feasible* if for all $v \in V$

$$\sum_{a \in \delta^+(v)} f_a^k - \sum_{a \in \delta^-(v)} f_a^k = \gamma(v), \tag{1}$$

where $\delta^+(v)$ and $\delta^-(v)$ are the arcs leaving and entering v, respectively; furthermore, $\gamma(v) = d_k$ if $v = s_k$, $\gamma(v) = -d_k$ if $v = t_k$, and $\gamma(v) = 0$ otherwise. Note that splitting of demands is allowed.

Alternatively, one can consider a *path flow* for a commodity $k \in [K]$. Let $\mathcal{P}_k$ be the set of all paths from s_k to t_k in D. A path flow is a nonnegative vector $(f_P^k)_{P \in \mathcal{P}_k}$. The corresponding flow on link $a \in A$ for commodity $k \in [K]$ is then

$$f_a^k := \sum_{P \ni a} f_P^k.$$

We define $\mathcal{F}_k$ with $k \in [K]$ to be the set of vectors $(\boldsymbol{f}^1, \ldots, \boldsymbol{f}^k)$ such that $\boldsymbol{f}^i$ is a feasible flow for commodity i for $i = 1, \ldots, k$. If $(\boldsymbol{f}^1, \ldots, \boldsymbol{f}^k) \in \mathcal{F}_k$, we say that it is *feasible* for commodities $1, \ldots, k$. The entire flow for a sequence of commodities is denoted by $\boldsymbol{f} = (\boldsymbol{f}^1, \ldots, \boldsymbol{f}^K)$. Furthermore, the cost of a flow on link $a \in A$ of commodity k is defined by

$$C_a^k(f_a^1, \ldots, f_a^k) = \int_0^{f_a^k} p_a\left(\sum_{i=1}^{k-1} f_a^i + z\right) dz. \tag{2}$$

This expression can be obtained as the cost of a shortest path routing, where the demand is split into infinitesimal pieces that are routed consecutively. Hence, the integral represents the fact that an infinitesimal amount of flow increases the price for each consecutive piece. Note that $C_a^k(\cdot)$ is a convex function.

The cost for $\boldsymbol{f}^k$ is

$$C^k(\boldsymbol{f}^k) := \sum_{a \in A} C_a^k(f_a^1, \ldots, f_a^k), \tag{3}$$

and the total cost is defined by

$$C(\boldsymbol{f}) = \sum_{k=1}^K C^k(\boldsymbol{f}^k).$$

In this paper we study the greedy online algorithm SEQ that sequentially routes the requested demands with minimum cost. To this end, it solves for every $k \in [K]$ the following convex program

$$\min \quad C^k(\boldsymbol{f}^k)$$

$$\text{s.t.} \quad \sum_{a \in \delta^+(v)} f_a^k - \sum_{a \in \delta^-(v)} f_a^k = \gamma(v) \qquad \forall v \in V \qquad (4)$$

$$f_a^k \geq 0 \qquad \forall a \in A,$$

where the vectors $\boldsymbol{f}^1, \ldots, \boldsymbol{f}^{k-1}$ are fixed by solving the first $k-1$ problems. This problem can be efficiently solved within arbitrary precision in polynomial time (see Grötschel, Lovász, and Schrijver [11]). Note that SEQ always produces a feasible flow.

Using the relation

$$\frac{\partial C^k}{\partial f_a^k}(\boldsymbol{f}^k) = p_a\Big(\sum_{i=1}^k f_a^i\Big),$$

we state in the following lemma necessary and sufficient optimality conditions of the above K problems.

Lemma 1. *A feasible flow $\boldsymbol{f} = (\boldsymbol{f}^1, \ldots, \boldsymbol{f}^K)$ for the sequence $\sigma = 1, \ldots, K$ is generated by SEQ if and only if for all $k \in [K]$ the following two equivalent conditions are satisfied:*

$$i) \quad \sum_{a \in A} p_a\Big(\sum_{i=1}^k f_a^i\Big)(f_a^k - x_a^k) \leq 0 \qquad \begin{array}{l} \textit{for all feasible flows } x^k \\ \textit{for commodity } k \end{array} \qquad (5)$$

$$ii) \quad \sum_{a \in P} p_a\Big(\sum_{i=1}^k f_a^i\Big) \leq \sum_{a \in Q} p_a\Big(\sum_{i=1}^k f_a^i\Big) \qquad \begin{array}{l} \textit{for all } P, Q \in \mathcal{P}_k, \\ P \textit{ flow carrying w.r.t. } \boldsymbol{f}^k. \end{array} \qquad (6)$$

The proof is based on the first order optimality conditions and the convexity of $C^k(\cdot)$, see Dafermos and Sparrow [12].

For the sequence $\sigma = 1, \ldots, K$, an *optimal offline flow* is given by a solution $\boldsymbol{f}^\star$ of the following convex optimization problem:

$$\min \quad C(\boldsymbol{f})$$

$$\text{s.t.} \quad \sum_{a \in \delta^+(v)} f_a^k - \sum_{a \in \delta^-(v)} f_a^k = \gamma(v) \qquad \forall v \in V, \ k \in K \qquad (7)$$

$$f_a^k \geq 0 \qquad \forall a \in A, \ k \in K,$$

where $\gamma(v)$ is defined as in (1). We denote by $\text{OPT}(\sigma)$ the optimal value $C(\boldsymbol{f}^\star)$ of the above convex problem.

Using the relation

$$\frac{\partial C}{\partial f_a^k}(\boldsymbol{f}) = p_a\Big(\sum_{i=1}^K f_a^i\Big),$$

the necessary and sufficient optimality conditions of the above problem are given in the following lemma.

Lemma 2. *A feasible flow* $f = (f^1, \ldots, f^K)$ *for the sequence* $\sigma = 1, \ldots, K$ *is offline optimal if and only if for all* $k \in [K]$ *the following two equivalent conditions are satisfied:*

$$i) \quad \sum_{a \in A} p_a \Big(\sum_{i=1}^{K} f_a^i \Big) (f_a^k - x_a^k) \leq 0 \qquad \begin{array}{l} \text{for all feasible flows } x^k \\ \text{for commodity } k \end{array} \qquad (8)$$

$$ii) \quad \sum_{a \in P} p_a \Big(\sum_{i=1}^{K} f_a^i \Big) \leq \sum_{a \in Q} p_a \Big(\sum_{i=1}^{K} f_a^i \Big) \qquad \begin{array}{l} \text{for all } P, Q \in \mathcal{P}_k, \\ P \text{ flow carrying w.r.t. } f^k. \end{array} \qquad (9)$$

Note that the only difference to the optimality conditions in Lemma 1 is the summation in the price function up to commodity K instead of k. This reflects the offline aspect since all demands are known. For the proof see again Dafermos and Sparrow [12].

For a given sequence of commodities $\sigma = 1, \ldots, K$ and a solution f produced by an online algorithm ALG for σ we denote by $\text{ALG}(\sigma) = C(f)$ its cost. The online algorithm ALG is called *c-competitive* if the cost of ALG is never larger than c times the cost of an optimal offline solution. The *competitive ratio* of ALG is the infimum over all $c \geq 1$ such that ALG is c-competitive, see Borodin and El-Yaniv [13].

Remark 1. If the price functions $p_a(z)$ are constant for every arc $a \in A$, the algorithm SEQ is optimal for the offline problem. This holds because in this case the routing problems are independent from each other. In fact, each routing decision is just a shortest path problem with respect to the constant costs. Furthermore, the offline problem is a min-cost flow problem without capacity constraints. Hence, both problems can be solved more efficiently than in the general case.

Clearly, also in the case $K = 1$, the competitive ratio of SEQ is 1.

3   Competitive Analysis of SEQ

First, we show that there exists, in general, no competitive deterministic online algorithm.

Proposition 1. *If neither the network nor the price functions are restricted, there exists no competitive deterministic online algorithm for the* ONLINEMCRP.

Proof. Consider the network depicted in Figure 1. For all arcs a in the network, the price function is $p_a(z) = m \cdot z^{m-1}$ with $m > 2$. Let ALG be an arbitrary deterministic online algorithm. The first commodity has demand $d_1 = 1$ and has to be routed from node $s_1 = 1$ to node $t_1 = 4$. There are two possible paths for this commodity: path $P_1 = (1, 2, 4)$ and path $P_2 = (1, 3, 4)$. Because

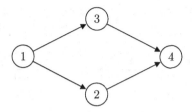

Fig. 1. Graph construction for the proofs of Proposition 1, Theorem 6, and Theorem 7

of symmetry, we can assume that ALG sends a flow of $\frac{1}{2} \le \alpha \le 1$ over path P_1 and $(1 - \alpha)$ along path P_2. Now commodity 2 arises with demand $d_2 = 1$, source $s_2 = 1$, and target $t_2 = 2$. For this demand there exists only the single path $P_3 = (1, 2)$. For this sequence σ we have the total cost

$$\text{ALG}(\sigma) = 2 \cdot \alpha^m + 2 \cdot (1 - \alpha)^m + \int_0^1 m(\alpha + z)^{m-1} \, dz$$
$$= 2 \cdot \alpha^m + 2 \cdot (1 - \alpha)^m + (\alpha + 1)^m - \alpha^m.$$

Routing the first commodity completely over path P_2 and the second over path P_3 leads to the total cost $2 \cdot 1^m + 1^m = 3 \ge \text{OPT}(\sigma)$. Letting m tend to infinity shows that in this case ALG is not competitive. □

Proposition 1 shows that to obtain competitive results, the network or the price functions have to be restricted.

3.1   Affine Linear Price Functions

Now we will show that if the price functions are affine, SEQ is $\frac{4K}{2+K}$-competitive. For affine price functions $p_a(z) = q_a \cdot z + r_a$ with $q_a \ge 0$, $r_a \ge 0$ $(a \in A)$, we have for a flow $(\boldsymbol{f}^1, \ldots, \boldsymbol{f}^k) \in \mathcal{F}_k$

$$C_a^k(\boldsymbol{f}^1, \ldots, \boldsymbol{f}^k) = q_a \Big(\sum_{i=1}^{k-1} f_a^i + \tfrac{1}{2} f_a^k \Big) f_a^k + r_a f_a^k.$$

It follows from the optimality conditions (5) that if $(\boldsymbol{f}^1, \ldots, \boldsymbol{f}^k)$ is generated by SEQ, we have

$$\sum_a \Big(q_a \sum_{i=1}^k f_a^i + r_a \Big) (f_a^k - x_a^k) \le 0, \tag{10}$$

for all feasible flows $\boldsymbol{x}^k$ for commodity k.

Theorem 1. *If the price functions of the* ONLINEMCRP *are affine, the online algorithm* SEQ *is* $\frac{4K}{2+K}$*-competitive.*

Proof. Let $\boldsymbol{f}$ be the flow generated by SEQ for the sequence σ and let $\boldsymbol{x}$ be any other feasible flow for σ. We start with the following inequality:

$$0 \le \Big(\tfrac{1}{2}\sum_{k=1}^{K} f_a^k - \sum_{k=1}^{K} x_a^k\Big)^2 = \tfrac{1}{4}\sum_{k=1}^{K}\sum_{i=1}^{K} f_a^i f_a^k - \sum_{k=1}^{K}\sum_{i=1}^{K} f_a^i x_a^k + \sum_{k=1}^{K}\sum_{i=1}^{K} x_a^i x_a^k.$$

Using the relation

$$\sum_{k=1}^{K}\sum_{i=1}^{K} f_a^i f_a^k = 2 \sum_{k=1}^{K}\Big(\sum_{i=1}^{k-1} f_a^i + \tfrac{1}{2}f_a^k\Big) f_a^k, \tag{11}$$

for the first and last sum we obtain:

$$0 \le \tfrac{1}{2}\sum_{k=1}^{K}\Big(\sum_{i=1}^{k-1} f_a^i + \tfrac{1}{2}f_a^k\Big) f_a^k - \sum_{k=1}^{K}\sum_{i=1}^{K} f_a^i x_a^k + 2 \sum_{k=1}^{K}\Big(\sum_{i=1}^{k-1} x_a^i + \tfrac{1}{2}x_a^k\Big) x_a^k.$$

Multiplying with q_a and adding over all arcs yields:

$$0 \le \sum_{a \in A} q_a \Big(\tfrac{1}{2}\sum_{k=1}^{K}\Big(\sum_{i=1}^{k-1} f_a^i + \tfrac{1}{2}f_a^k\Big) f_a^k - \sum_{k=1}^{K}\sum_{i=1}^{K} f_a^i x_a^k + 2 \sum_{k=1}^{K}\Big(\sum_{i=1}^{k-1} x_a^i + \tfrac{1}{2}x_a^k\Big) x_a^k\Big).$$

Now we add the inequality

$$0 \le \sum_{a \in A}\sum_{k=1}^{K}\Big(\tfrac{1}{2}r_a f_a^k - r_a x_a^k + 2r_a x_a^k\Big) - \tfrac{1}{K}\sum_{a \in A}\sum_{k=1}^{K} r_a f_a^k,$$

which holds because $K \ge 2$. This leads to:

$$0 \le \tfrac{1}{2}C(\boldsymbol{f}) - \sum_{a \in A}\sum_{k=1}^{K}\Big(q_a \sum_{i=1}^{K} f_a^i + r_a\Big) x_a^k + 2\,C(\boldsymbol{x}) - \tfrac{1}{K}\sum_{a \in A}\sum_{k=1}^{K} r_a f_a^k.$$

Dropping part of the second term and applying (10) yields:

$$0 \le \tfrac{1}{2}C(\boldsymbol{f}) - \sum_{a \in A}\sum_{k=1}^{K}\Big(q_a \sum_{i=1}^{k} f_a^i + r_a\Big) f_a^k + 2\,C(\boldsymbol{x}) - \tfrac{1}{K}\sum_{a \in A}\sum_{k=1}^{K} r_a f_a^k$$

$$= -\tfrac{1}{2}C(\boldsymbol{f}) + 2\,C(\boldsymbol{x}) - \tfrac{1}{2}\sum_{a \in A} q_a \sum_{k=1}^{K} f_a^k f_a^k - \tfrac{1}{K}\sum_{a \in A}\sum_{k=1}^{K} r_a f_a^k.$$

Hence,

$$C(\boldsymbol{f}) \le 4\,C(\boldsymbol{x}) - \sum_{a \in A} q_a \sum_{k=1}^{K} f_a^k f_a^k - \tfrac{2}{K}\sum_{a \in A}\sum_{k=1}^{K} r_a f_a^k$$

$$\le 4\,C(\boldsymbol{x}) - \tfrac{1}{K}\sum_{a \in A} q_a \Big(\sum_{k=1}^{K} f_a^k\Big)^2 - \tfrac{2}{K}\sum_{a \in A}\sum_{k=1}^{K} r_a f_a^k$$

$$= 4\,C(\boldsymbol{x}) - \tfrac{2}{K}\sum_{a \in A} q_a \sum_{k=1}^{K}\Big(\sum_{i=1}^{k-1} f_a^k + \tfrac{1}{2}f_a^k\Big) f_a^k - \tfrac{2}{K}\sum_{a \in A}\sum_{k=1}^{K} r_a f_a^k,$$

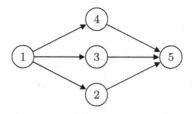

Fig. 2. Graph construction for the proof of Theorem 2

where the second inequality follows from the inequality of Cauchy-Schwarz and the last equation follows by (11). Therefore, we get $C(\boldsymbol{f}) \leq 4\,C(\boldsymbol{x}) - \frac{2}{K}C(\boldsymbol{f})$, from which the claim follows. $\qquad\square$

We do not know whether this result is tight. The best known lower bound is the following.

Theorem 2. *In case of linear cost functions no deterministic online algorithm for the* ONLINEMCRP *is c-competitive for any* $c < \frac{4}{3}$.

Proof. Consider the network displayed in Figure 2. Each arc a leaving node 1 has the same price function $p_a(z) = 4\,z$. All the other arcs (leading to node 5) have price function $p_a(z) = 0$. Let ALG be an arbitrary deterministic online algorithm. The first commodity has demand 1, which has to be routed from $s_1 = 1$ to $t_1 = 5$.

Assume the algorithm behaves like SEQ. This means that the demand gets evenly divided into three pieces: one third is routed over path $P_1 = (1, 2, 5)$, another over path $P_2 = (1, 3, 5)$, and the final third over path $P_3 = (1, 4, 5)$. We then reveal commodity 2 with demand 1 between nodes 1 and 2. For this commodity there only exists a single path $P_4 = (1, 2)$. Therefore, the cost of ALG for this sequence σ is:

$$\text{ALG}(\sigma) = \text{SEQ}(\sigma) = 3 \cdot 4 \cdot \left(\tfrac{1}{2} \cdot \tfrac{1}{3}\right) \cdot \tfrac{1}{3} + 4 \cdot \left(\tfrac{1}{3} + \tfrac{1}{2} \cdot 1\right) \cdot 1 = 4.$$

An optimal offline solution is to route half of commodity 1 over path P_2, the other half over path P_3, and commodity 2 along P_4. Therefore,

$$\text{OPT}(\sigma) = 2 \cdot 4 \cdot \left(\tfrac{1}{2} \cdot \tfrac{1}{2}\right) \cdot \tfrac{1}{2} + 4 \cdot \left(\tfrac{1}{2} \cdot 1\right) \cdot 1 = 3.$$

This leads to

$$\frac{\text{ALG}(\sigma)}{\text{OPT}(\sigma)} = \frac{4}{3}.$$

If ALG does not behave like SEQ for the first commodity, we can assume by symmetry that ALG routes $\alpha > \frac{1}{3}$ over path P_1. Hence, a demand of $1 - \alpha$ has to be routed over path P_2 and P_3. The optimal way to do this is to route $(1 - \alpha)/2$

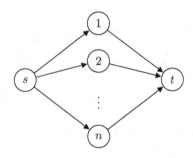

Fig. 3. Graph construction for the proof of Theorem 3

over each path. We then present commodity 2 as above. The cost of ALG for this sequence σ is

$$\mathrm{ALG}(\sigma) \geq 4 \cdot \left(\tfrac{1}{2} \cdot \alpha\right) \cdot \alpha + 2 \cdot 4 \cdot \left(\tfrac{1}{2} \cdot \tfrac{(1-\alpha)}{2}\right) \cdot \tfrac{(1-\alpha)}{2} + 4 \cdot \left(\alpha + \tfrac{1}{2} \cdot 1\right) \cdot 1 > 4.$$

since $\alpha > \tfrac{1}{3}$. Since $\mathrm{OPT}(\sigma) = 3$, we have

$$\frac{\mathrm{ALG}(\sigma)}{\mathrm{OPT}(\sigma)} > \frac{4}{3}.$$

Therefore, ALG cannot have a competitive ratio less than $\tfrac{4}{3}$.   □

As we show next, possible "good" algorithms for the ONLINEMCRP have to split the demands.

Theorem 3. *A deterministic online algorithm for the* ONLINEMCRP *that routes all demands unsplittably is not competitive, even for linear cost functions.*

Proof. Consider the network shown in Figure 3. This network contains $n + 2$ nodes and n paths from node s to node t. The price functions are $p_a(z) = 2z$ for all arcs a. Let ALG be an arbitrary deterministic online algorithm which does not split demands. We consider a single commodity with demand 1 between nodes s and t. Since ALG does not split, the cost of its routing is independent from the chosen path:

$$\mathrm{ALG}(\sigma) = 2 \cdot \left(\tfrac{1}{2} \cdot 1\right) \cdot 1 + 2 \cdot \left(\tfrac{1}{2} \cdot 1\right) \cdot 1 = 2.$$

An optimal solution splits the demand into n evenly divided pieces and sends each piece over a different path. This leads to an optimal cost of

$$\mathrm{OPT}(\sigma) = n\left(2 \cdot \left(\tfrac{1}{2} \cdot \tfrac{1}{n}\right) \cdot \tfrac{1}{n} + 2 \cdot \left(\tfrac{1}{2} \cdot \tfrac{1}{n}\right) \cdot \tfrac{1}{n}\right) = n \cdot 2 \cdot \left(\tfrac{1}{n}\right)^2 = \tfrac{2}{n}.$$

Therefore, the competitive ratio of ALG is not smaller than n. Since this holds for all $n \in \mathbb{N}$, ALG is not competitive.   □

In Section 4, we further investigate the problem variant, where splitting demand is not allowed.

3.2   Parallel Arc Case

We now consider the parallel arc case, that is, D consists of two nodes s and t and arcs from s to t only. We allow for arbitrary nondecreasing continuous price functions. Recall from Lemma 1 that f is generated by SEQ for the sequence $\sigma = 1, \ldots, K$ if and only if for all $a \in A$, $k \in [K]$ with $f_a^k > 0$:

$$p_a\left(\sum_{i=1}^{k} f_a^i\right) \leq p_{\hat{a}}\left(\sum_{i=1}^{k} f_{\hat{a}}^i\right) \qquad \text{for all } \hat{a} \in A. \tag{12}$$

By Lemma 2, a flow x solves the offline problem (7) if and only if we have for all $a \in A$ with $\sum_{k=1}^{K} x_a^k > 0$:

$$p_a\left(\sum_{k=1}^{K} x_a^k\right) \leq p_{\hat{a}}\left(\sum_{k=1}^{K} x_{\hat{a}}^k\right) \qquad \text{for all } \hat{a} \in A. \tag{13}$$

Lemma 3. *Given the sequence* $\sigma = 1, \ldots, K$, *let* $f = (f^1, \ldots, f^K)$ *be the flow generated by* SEQ *for this sequence. Define* $A_k^+ := \{a \in A : f_a^k > 0\}$ *for* $k \in [K]$. *Then,*

$$p_a\left(\sum_{i=1}^{k+1} f_a^i\right) \leq p_{\hat{a}}\left(\sum_{i=1}^{k+1} f_{\hat{a}}^i\right), \qquad \forall a \in A_k^+, \ \hat{a} \in A, \ k = 1, \ldots, K-1.$$

Proof. Let $a \in A_k^+$. First assume that $a \in A_{k+1}^+$. Then by the optimality conditions (12) for $(f^1, \ldots, f^{k+1})$ the claim follows.

Now assume $a \notin A_{k+1}^+$. Then we have for all $\hat{a} \in A$:

$$p_a\left(\sum_{i=1}^{k+1} f_a^i\right) = p_a\left(\sum_{i=1}^{k} f_a^i\right) \leq p_{\hat{a}}\left(\sum_{i=1}^{k} f_{\hat{a}}^i\right) \leq p_{\hat{a}}\left(\sum_{k=1}^{k+1} f_{\hat{a}}^i\right),$$

where the first inequality follows from the optimality condition for the flow $(f^1, \ldots, f^{k+1})$ and the second follows from the assumption that the price functions are nondecreasing. □

Theorem 4. *Consider the sequence* $\sigma = 1, \ldots, K$ *in the parallel arcs case. Let* f *be the flow generated by* SEQ *for* σ. *Then,* $C(f) \leq C(x)$ *for any feasible flow* x *for* σ, *this means,* f *is also an offline optimum.*

Proof. For the last commodity K, we have the following optimality condition:

$$p_a\left(\sum_{i=1}^{K} f_a^i\right) \leq p_{\hat{a}}\left(\sum_{k=1}^{K} f_{\hat{a}}^i\right), \ \forall a \in A_K^+, \ \hat{a} \in A. \tag{14}$$

Using Lemma 3 for $k = K - 1$ we obtain:

$$p_a\left(\sum_{i=1}^{K} f_a^i\right) \leq p_{\hat{a}}\left(\sum_{k=1}^{K} f_{\hat{a}}^i\right), \ \forall a \in A_{K-1}^+, \ \hat{a} \in A.$$

Applying Lemma 3 iteratively $K - 1$ times together with (14) yields the optimality conditions (13) for the offline optimum. □

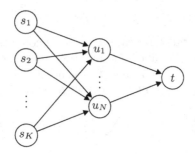

Fig. 4. Construction for the proof of Theorem 5

4   Unsplittable Routings

In this section we study the variant of the ONLINEMCRP in which demands are not allowed to be split, i.e., unsplittable routings. Such a restriction often occurs in practice, for instance in single path routing problems in telecommunication networks. It is possible to formulate a mixed integer convex program for this setting. In contrast to the splittable case, however, the offline problem is NP-hard in this case.

Theorem 5. *The offline problem for the unsplittable variant of the* ONLINE-MCRP *is NP-hard, even when the price functions are linear.*

Proof. Consider an instance of the *minimum sum of squares problem*, which is NP-complete in the strong sense (see Garey and Johnson [14]). Here, one is given nonnegative integers $d_1, \ldots, d_K$ and positive integers $N \leq K$ and J. The question is whether there exists a partition of $[K]$ into N sets $A_1, \ldots, A_N$ such that

$$\sum_{i=1}^{N} \left(\sum_{k \in A_i} d_k \right)^2 \leq J?$$

For the reduction to the offline problem, we construct a directed graph D with nodes $\{s_1, \ldots, s_K, u_1, \ldots, u_N, t\}$ and the following arcs: For each $k \in [K]$ and $i \in [N]$ we have an arc (s_k, u_i) with price function 0. For each $i \in [N]$ we add an arc $a = (u_i, t)$ with price function $p_a(z) = 2z$; see Figure 4. Furthermore, for $k \in [K]$ there are demands d_k between s_k and t.

We now claim that there exists an unsplittable solution to the offline problem of value at most J if and only if the answer to the minimum sum of squares problem is positive. To see this, first assume that $A_1, \ldots, A_N$ is the wanted partition. Then if $k \in A_i$, we route commodity k along u_i to t. Using (11), we obtain the following costs:

$$2 \sum_{i=1}^{N} \sum_{k \in A_i} \left(\sum_{\substack{j \in A_i \\ j < k}} d_j + \tfrac{1}{2} d_k \right) d_k = \sum_{i=1}^{N} \sum_{k \in A_i} \sum_{j \in A_i} d_k \, d_j = \sum_{i=1}^{N} \left(\sum_{k \in A_i} d_k \right)^2.$$

This proves the forward direction of the claim. Conversely, assume that there exists an unsplittable flow of value J. For $i = 1, \ldots, N$, let A_i be the set of indices k whose corresponding demands are routed over the arc (u_i, t). Again the cost is given as above, which shows that there exits a solution to the minimum sum of squares problem. $\square$

When the price functions are constant, both the unsplittable variants of (4) and (7) are min-cost flow problems and hence, polynomial time solvable.

Theorem 6. *In general there exists no competitive deterministic online algorithm for the unsplittable variant of the* ONLINEMCRP.

Proof. Given the network shown in Figure 1, where each arc a has a price function $p_a(z) = m \cdot z^{m-1}$ with $m > 2$. Let ALG be an arbitrary deterministic online algorithm for the considered problem. First, we reveal a commodity with demand $d_1 = 1$, source $s_1 = 1$, and target $t_1 = 4$. Without loss of generality, we can assume that ALG uses path $P_1 = (1, 2, 4)$ to route this demand. Commodity 2 is released with demand $d_2 = 1$, source $s_2 = 1$, and target $t_2 = 2$. For this commodity there only exists a single path $P_2 = (1, 2)$. Hence, for this sequence σ, ALG yields the cost

$$\text{ALG}(\sigma) = 2 \cdot 1^m + \int_0^1 m(1 + z)^{m-1}\, dz = 2 + (1 + 1)^m - 1^m = 1 + 2^m.$$

The optimal cost is $\text{OPT}(\sigma) = 3$ which is achieved by routing commodity 1 over path $P_3 = (1, 3, 4)$ and commodity 2 along path P_2. Therefore, for m going to infinity it follows that ALG is not competitive. $\square$

Theorem 7. *If we consider only linear price functions, no deterministic online algorithm has a competitive ratio less than 2 for the unsplittable variant of the* ONLINEMCRP.

Proof. Consider the network shown in Figure 1, where each link a is equipped with the same price function $p_a(z) = 2\,z$. Let ALG be an arbitrary deterministic online algorithm. We first reveal commodity 1 with demand $d_1 = 1$, source $s_1 = 1$, and target $t_1 = 4$. This request can be routed over path $P_1 = (1, 2, 4)$ or over path $P_2 = (1, 3, 4)$. W.l.o.g. assume that ALG chooses path P_1. Now we release two more commodities from node 1 to 2 and from node 2 to 4, respectively. Both have a demand of 1. Since for each of the last two commodities there exists only a single path, the assignment by ALG for this sequence σ leads to a cost of

$$\text{ALG}(\sigma) = 2 \cdot 2 \cdot (\tfrac{1}{2} \cdot 1) \cdot 1 + 2 \cdot (1 + \tfrac{1}{2} \cdot 1) \cdot 1 + 2 \cdot (1 + \tfrac{1}{2} \cdot 1) \cdot 1 = 8.$$

An optimal routing is achieved by routing commodity 1 along path P_2, and commodity 2 and 3 over their single paths. Since the optimal cost for σ is $\text{OPT}(\sigma) = 4$, the competitive ratio of ALG is at least 2. $\square$

5   Final Comments and Future Research

In practice, routings have to consider capacities, which we ignored in our approach. In this case, however, one can easily construct examples in which any online algorithm does not even produce a feasible solution. Further requirements in practice include path length restrictions and survivability issues.

In the future, we plan to investigate the competitiveness of SEQ for nonlinear prices functions. It is also an open issue whether the competitiveness bound in Theorem 1 is tight and whether the optimality results in Theorem 4 can be extended.

References

1. Moy, J.: OSPF: Anatomy of an Internet Routing Protocol. Addison-Wesley (1999)
2. Cisco: OSPF   Design   Guide.   Documentation   available   at http://www.cisco.com/en/US/tech/tk365 (2006)
3. Yahaya, A., Suda, T.: iREX: Inter-domain QoS Automation using Economics. In: Proceedings of IEEE CCNC. (2006)
4. Yahaya, A., Harks, T., Suda, T.: iREX: Efficient Inter-domain QoS Automation using Economics. In: Proceedings of IEEE Globecom. (2006)
5. Fortz, B., Thorup, M.: Optimizing OSPF/IS-IS weights in a changing world. IEEE JSAC **20** (2002) 756–767
6. Fortz, B., Thorup, M.: Increasing internet capacity using local search. Computational Optimization and Applications **29** (2004) 13–48
7. Roughgarden, T., Tardos, E.: How bad is selfish routing? Journal of the ACM **49** (2002) 236–259
8. Correa, J.R., Schulz, A.S., Stier Moses, N.E.: Selfish routing in capacitated networks. Math. Oper. Res. **29** (2004) 961–976
9. Fiat, A., Woeginger, G.J., eds.: Online Algorithms: The State of the Art. Volume 1442 of Lecture Notes in Computer Science. Springer (1998)
10. Awerbuch, B., Azar, Y., Plotkin, S.: Throughput-competitive on-line routing. In: 34th Annual Symposium on Foundations of Computer Science (FOCS) 1993, Palo Alto, IEEE (1993) 32–40
11. Grötschel, M., Lovász, L., Schrijver, A.: Geometric Algorithms and Combinatorial Optimization. 2nd edn. Volume 2 of Algorithms and Combinatorics. Springer-Verlag, Heidelberg (1993)
12. Dafermos, S., Sparrow, F.: The traffic assignment problem for a general network. J. Res. Natl. Bur. Stand., Sect. B **73** (1969) 91–118
13. Borodin, A., El-Yaniv, R.: Online Computation and Competitive Analysis. Cambridge University Press (1998)
14. Garey, M.R., Johnson, D.S.: Computers and Intractability. A Guide to the Theory of NP-Completeness. W. H. Freeman and Company, New York (1979)

The k-Allocation Problem and Its Variants

Dorit S. Hochbaum[1,*] and Asaf Levin[2]

[1] Department of Industrial Engineering and Operations Research and Walter A. Haas School of Business, University of California, Berkeley
`hochbaum@ieor.berkeley.edu`
[2] Department of Statistics, The Hebrew University Jerusalem, Israel
`levinas@mscc.huji.ac.il`

Abstract. In the process of reviewing and ranking projects by a group of reviewers, each reviewer is assumed to review a partial list of projects, up to k projects. Each individual reviewer then ranks and compares all pairs of k projects. The k-allocation problem is to determine the allocation of up to k projects to each reviewer within the expertise set of the reviewer so that the resulting union of reviewed projects has certain desirable properties. One property of the k-allocation is to have all pairs of projects compared by at least one reviewer. This we call the k-complete problem.

In cases when the property of k-complete cannot be achieved, one might settle for other properties. One such basic requirement is that each pair of projects is comparable via a *ranking path* which is a sequence of pairwise rankings of projects implying a comparison of all pairs on the path. A k-allocation with a ranking path between each pair is the *connectivity-k-aloc*. Since the robustness of relative comparisons deteriorates with the length of the ranking path, another property is that between each pair of projects there will be at least one ranking path that has at most two hops or q hops for fixed values of q. Another property that increases robustness of the ranking is to find a k-allocation so there are at least p disjoint ranking paths between each pair.

We model all these problems as graph problems and show that the CONNECTIVITY-k-ALOC problem is polynomially solvable using matroid intersection, the k-complete problem is NP-hard unless $k = 2$, and all other considered variants of the k-allocation properties problem are NP-complete for all values of $k \geq 2$. We provide approximation algorithms for an optimization problem related to the k-complete problem.

Keywords: Approximation algorithms, allocation problem, maximum coverage problem.

1 Introduction

In a typical group decision making scenario, a group of individuals rank a set of projects. In the group decision making setup discussed here the total number of

---

* Research supported in part by NSF awards No. DMI-0085690 and DMI-0084857.

T. Erlebach and C. Kaklamanis (Eds.): WAOA 2006, LNCS 4368, pp. 253–264, 2006.

projects n is too large to be assigned to each reviewer. Moreover, each reviewer has his/her own expertise set and the allocated projects must lie within the expertise set. Therefore, each reviewer reviews a partial list of projects that includes up to k projects in order to balance the work load. The allocation of up to k projects per reviewer within their expertise set is said to be a k-allocation.

One major challenge is to come up with an aggregate ranking that reflects the opinions of all reviewers and is fair and representative. There is a large body of literature that address this challenge reviewed in [19] (see also Kemeny and Snell [23], Brans and Vincke [4], Bartholdi, Tovey and Trick [3], Keener [22], Fuller and Carlsson [14], and Fernandez and Olemdo [12]). The aggregate ranking is affected by the assignment of projects to individual reviewers. For example, an assignment whereby one sub-group of reviewers evaluates one subset of projects, and the remaining reviewers evaluate the remaining projects, renders an overall ranking impossible as the two subsets of projects are not comparable. In spite of its importance, the allocation of projects and its impact on the quality of the resulting aggregate ranking is an aspect of group decision often overlooked. Cook et al. [8] are the only researchers that have explicitly addressed this issue.

Our purpose here is to address the allocation of projects to reviewers. In a typical scenario the reviewing work has to be allocated to different reviewers in such a way that the subset of projects allocated to each reviewer is contained in their expertise set and the number of projects allocated to each reviewer does not exceed k. This is for instance the case in NSF review panels. We study in the full version an extension of this model where the number of projects assigned reviewer j depends on the reviewer and is equal k_j.

There are a number of properties that might be associated with the allocation process. At the most basic level it is required that each project is reviewed by at least one reviewer. A much stronger requirement is that each pair of projects is evaluated and compared by at least one reviewer. Other types of properties are elaborated on below.

A convenient formalism for the problems and the associated models is as graph representation. The input to the problem is an undirected complete graph $G = (V, E)$ defined on the set of projects V and all possible pairs (edges) E, an integer number k and a collection of node subsets representing the expertise set of each of L reviewers, $S_1, \ldots, S_L \subseteq V$. In order to maintain a reasonable and balanced workload, each reviewer is assigned at most k projects out of the set of possible projects, S_j. So a feasible allocation consists, for each $j \in \{1, 2, \ldots, L\}$, of a subset $V_j \subseteq S_j$ such that $|V_j| \leq \min\{|S_j|, k\}$. Each reviewer is able to make a direct comparison between each pair of projects he/she reviews. The set of pairs compared by each reviewer forms a clique (a complete subgraph) of size $|V_j|$, C_{V_j} (i.e., C_{V_j} is the edge set of a clique over the node set V_j). For a given feasible allocation, the set of covered projects is $\cup_{j=1}^{L} V_j$, and the set of compared project pairs is $\cup_{j=1}^{L} C_{V_j}$. The properties of the graph of the edges covered by this union of cliques are closely related to the quality of the ranking decision that can be achieved.

Let the *review graph* of covered projects and compared project pairs be $G^R = (V^R, E^R)$ where $V^R = \cup_{j=1}^L V_j$ and $E^R = \cup_{j=1}^L C_{V_j}$. The graph G^R is a multi-graph – that is, there could be multiple edges between some pairs of nodes (because more than one reviewer reviews this pair). A pair of projects $i, j \in V^R$ is said to be *directly compared* if edge $[i, j] \in E^R$. For a directly compared pair there is input from at least one reviewer on the extent of preference of one project to the other. This relative rank comparison is typically expressed in an additive form or in a multiplicative form. A detailed discussion on intensity of preferences and the additive versus the multiplicative forms of preferences is provided in [19]. We will use throughout the additive form in which p_{ij} expresses by how much the rank of i exceeds the rank of j . So $p_{ji} = -p_{ij}$ and the magnitude of p_{ij} is the *intensity* of the preference of i to j. Each (undirected) edge $[i, j]$ in the graph G^R is formed of a pair of (directed) arcs (i, j), (j, i) with the associated values p_{ij} and p_{ji}.

Although not all pairs may be directly compared by a given allocation, we can deduce a relative ranking of two projects i and j if there exists a sequence of directly compared pairs: $[i, i_1], [i_1, i_2], \ldots, [i_{p-1}, j] \in E^R$. Such sequence corresponds to a path in the graph G^R and the *implied ranking* of this path is $p_{ij} = \sum_{q=0}^{p-1} p_{i_q, i_{q+1}}$ where $i = i_0$, $j = i_p$. We call this path a *ranking path* of *length* p. A direct comparison is then a ranking path of length 1. Since the process of evaluating projects and comparing them is not accurate, an implied ranking by a long path may be impacted by cumulative errors in the comparisons. This effect may be mitigated by multiple ranking paths between given pairs or by having the ranking paths of bounded length. The presence of multiple ranking paths between all pairs correspond to the increased edge, or node, *connectivity* of the review graph.

As an illustration of these concepts consider the graph G^R in Figure 1. In this graph we take $k = 2$, and therefore each reviewer reviews only a pair of projects. The endpoints of each edge form the allocation to one reviewer. The intensities of the preferences are given as p_{ij} for $i < j$. In this graph project 1 and 2 are reviewed by two different reviewers. There are four implied ranking paths between projects 1 and 5 of intensities 2, 4, .5 and -2. Among those the value .5 is the intensity of a direct comparison.

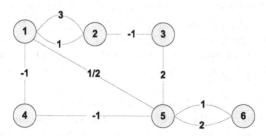

Fig. 1. An illustration review graph G^R. The numbers along the edges are the intensity p_{ij} for $i < j$.

Preliminaries and notations. For a graph H we denote by n_H and m_H the number of nodes and edges, respectively, in H. For an integer ℓ, an *ℓ-subset* is a subset of ℓ nodes.

The standard definitions of edge-connectivity and node-connectivity are as follows: A connected graph $H = (U, F)$ is *p-edge connected* if the removal of up to $p - 1$ edges from F results in a connected graph. A connected graph $H = (U, F)$ is *p-node connected* if the removal of up to $p - 1$ nodes from U results in a connected graph. We also define a new connectivity measure of the review graph that we call *reviewer-connectivity* defined as follows: A review graph G^R is a *p-reviewer connected* if the removal of all the edges corresponding to at most $p - 1$ reviewers from G^R results in a connected graph. We say that a pair of projects has p reviewer disjoint ranking paths between them, if the removal of at most $p - 1$ reviewers from the review graph keep the two projects in the same connected component of the resulting graph.

A polynomial time algorithm A for a minimization problem (maximization problem) is a *ρ-approximation algorithm* if it always returns a feasible solution whose objective value is at most (at least) ρ times the optimum.

A list of properties. We now list a set of properties for a desirable allocation of projects.

1. A basic requirement is the ability to compare each pair of projects once the review process is done. To that end we require that every pair of projects is comparable via a ranking path. In terms of the review graph G^R this goal is to find an allocation so that G^R is connected. The CONNECTIVITY-k-ALOC problem is to find a feasible allocation so that G^R is connected.

2. If an appropriate allocation exists, then it is desirable that all pairs of projects should be directly comparable. Cook et al. [8] recently studied the problem of maximizing the number of directly comparable projects pairs. The problem is therefore to determine the subsets V_j so that the union of the edges in the complete graphs (or cliques) induced on V_j, C_{V_j}, $|\bigcup_{j=1}^{L} C_{V_j}|$, is maximum. The goal in the MAX k-COMPLETE COVERAGE problem is to select sets $V_j \subseteq S_j$ of size at most k, that maximize the number of edges, $|\bigcup_{j=1}^{L} C_{V_j} \cap E|$. We call this problem the MAX k-COMPLETE COVERAGE problem. The k-COMPLETE problem is the decision problem of deciding whether a given complete graph G has an optimal solution of the MAX k-COMPLETE COVERAGE problem that equals the number of edges in G, $\binom{n_G}{2}$. So the MAX k-COMPLETE COVERAGE problem is a more general problem than the k-COMPLETE problem, in that it is an optimization problem, rather than a decision problem and in that the problem is defined on an undirected graph $G = (V, E)$ which is not necessarily complete.

3. p-EDGE CONNECTIVITY: In order to increase the reliability of the implied rankings it is desirable that there will be more than a single ranking path between each pair. To that end we require that there are at least p edge disjoint ranking paths in G^R between each pair of projects. I.e., the removal of at most $p - 1$ pairwise comparisons by a given reviewer results a connected review graph. In

the example in Figure 1 there are four ranking paths between 1 and 5 with only three of them edge-disjoint. In this graph nodes 3 and 4 have only two edge disjoint paths between them. So this allocation is a solution to the $p = 2$ edge disjoint requirement. The second review graph shown in Figure 2 is 3-edge connected. The associated problem is to maximize the number of pairs of projects that have at least p edge disjoint ranking paths between them. This associated problem is called the (k, p)-EDGE CONNECTIVITY ALLOCATION problem denoted as $(k, p) - ECon$.

4. p-REVIEWER CONNECTIVITY: Similarly to the p-edge connectivity problem, in order to increase the reliability of the implied rankings, we ask that that the review graph will be p reviewer connected. I.e., the removal of at most $p - 1$ reviewers results in a connected review graph. The motivation for studying this problem is that the implied rankings of some pairs depend on very few reviewers and this situation can skew the results. The associated problem is to maximize the number of pairs of projects that have at least p reviewer disjoint ranking paths between them. This associated problem is called the (k, p)-REVIEWER CON- NECTIVITY ALLOCATION problem denoted as $(k, p) - RCon$. We note that when $k = 2$ the $(2, p)$-ECon and the $(2, p)$-RCon problems are equivalent problems. However, for larger values of k, the notion of reviewer connectivity is different from that of edge connectivity.

The 2-REVIEWER CONNECTIVITY AUGMENTATION problem is the associated optimization problem defined as follows. The input is a feasible solution to the CONNECTIVITY-k-ALOC problem, and integer numbers q and $k \geq 2$. The goal is to augment the solution to the CONNECTIVITY-k-ALOC problem using at most q additional reviewers, where each of these has an expertise set equals to the set of all projects and we can assign at most k projects for each additional reviewer. The goal is to maximize the number of pairs of projects such that the resulting review graph (constructed by adding the q additional reviewers to the solution of the CONNECTIVITY-k-ALOC problem) has two reviewer-disjoint paths between them. The motivation for studying this problem is the fact that in some cases there are few reviewers that can review the whole set of projects (these reviewers might be the panel members) though their expertise level in each subject is smaller than the one of the regular reviewers. Therefore, we would like to have a feasible solution to the CONNECTIVITY-k-ALOC problem using a high level expert in each of the projects, and to use the panel members in a method to increase the robustness of the resulting ranking, by providing a 2-reviewer disjoint paths between some pairs of projects.

5. p-NODE CONNECTIVITY: Similarly to the p-edge connectivity problem, in order to increase the reliability of the implied rankings, we ask that there are at least p node disjoint ranking paths in G^R between each pair of projects. I.e., the removal of at most $p - 1$ projects results in a connected review graph. The motivation for this is that low node connectivity indicates that the implied rankings of some pairs depend on very few projects and can skew the results. In the example in Figure 1 the review graph is only 1 node connected, as there are no two node disjoint paths between node 1 and node 6. Therefore, the implied

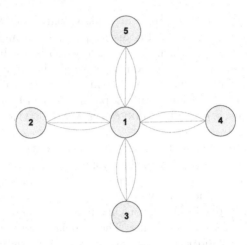

Fig. 2. A review graph G^R with star topology

ranking of nodes 1 and 6 depends only on the relative strength of project 5. When project 5 is particularly strong, then the implied ranking of projects 1 and 6 may not be meaningful as the extent of the differentiation between them is dominated by the strength of project 5. This example can be magnified in a star review graph such as the one shown in Figure 2. The review graph in this example is a 1-node connected. The associated problem is to maximize the number of pairs of projects that have at least p node disjoint ranking paths between them. This associated problem is called the (k, p)-NODE CONNECTIVITY ALLOCATION problem denoted as $(k, p) - NCon$.

6. q-HOP: As the length of the ranking path increases, the robustness and reliability of the implied ranking decreases. It is therefore desirable to limit the length of the ranking paths. For each pair of projects we require the existence of at least one ranking path that has length of at most q hops. That means that the ranking path has at most q edges. When all projects are directly comparable the allocation provides a 1-hop review graph. The graph in Figure 1 is a 3-hop review graph and the graph in Figure 2 is a 2-hop review graph. The associated problem called (k, q)-HOP problem is to maximize the number of pairs of projects that have at least one q-hop ranking path between them.

We note that the CONNECTIVITY-k-ALOC problem can be solved by solving any of the following problems: The $(k, 1)$-ECon, the $(k, 1)$-RCon, the $(k, 1)$-NCon, and the $(k, |V| - 1)$-hop. This is so because in all these optimization problems the goal function value of the optimal solution equals $\binom{|V|}{2}$ if and only if the CONNECTIVITY-k-ALOC instance is feasible.

The graph model. The problem of allocating evaluation tasks to reviewers can be cast as the H-graph k-clique cover problem defined as follows. Given a set V, L sets $S_1, \ldots, S_L \subseteq V$, and an integer k, find subsets $V_1, \ldots, V_L$, with

$V_i \subseteq S_i$, and $|V_i| \leq k$, so that the review multi-graph $G^R = (\cup_{i=1}^L V_i, \cup_{i=1}^L C_{V_i})$ has property H. We are interested in the following properties H. For example, G^R is a complete graph as in the k-COMPLETE problem.

Related Research. The Perron-Frobenius Theorem states the algebraic conditions that guarantee a positive unique solution eigenvector $\mathbf{r}$ to the system $\mathbf{Ar} = \lambda \mathbf{r}$. This subject is related to aggregate ranking when the matrix $\mathbf{A}$ represents the pairwise comparisons between all pairs of objects. Further, each column of the matrix can be viewed as the ranking provided by one reviewer. The eigenvector $\mathbf{r}$, if exists, is a principal eigenvector – corresponding to the largest eigenvalue. This theorem and the generated principal eigenvector $\mathbf{r}$ have been used for decades to generate an aggregate ranking from a matrix of rankings that can be viewed as provided by different reviewers. The condition to the existence of such eigenvector is that the matrix $\mathbf{A}$ is *irreducible*. (For a statement of the theorem see e.g. [22].) The irreducibility of the matrix is equivalent to the connectivity of G^R. In spite of the importance of the irreducibility property, condition for the allocation of evaluation tasks so as to achieve a matrix with this property have not been studied.

The k-COMPLETE problem and the MAX k-COMPLETE COVERAGE problem were recently studied by Cook et al. [8] who gave integer programming formulations and a branch-and-bound (exponential time) algorithm for solving these problems as well as a heuristic algorithm.

The (k,q)-HOP problem was addressed recently by Park and Newman [24] in ranking college football teams. They include in a graph a *directed* arc from i to j if (football) team i wins against team j. They conclude an implied win if there is a directed path with q hops, where the weight of this implied win decreases exponentially with q. The algorithm they developed is based on the diminished importance of the implied paths as a function of the number of hops. This model is different from ours in that the graph is determined as an *outcome* of the evaluation (playing the games) process which determines the directions of the arcs. In our model the topology (of the undirected graph) is determined by the k-allocation and only the intensities are determined by the evaluations.

The problem of finding a minimum set of edges that must be added to a given (simple) subgraph so that the resulting (simple) graph is 2-edge connected has been studied previously. Eswaran and Tarjan [10] provided a sufficient and necessary conditions, and a linear time algorithm to construct an optimal solution is given in [21,25]. In order to extend the linear time algorithm of [21,25] for multigraphs, we shrink each pair of nodes with parallel edges between them, where we traverse the pairs of nodes in a BFS manner. So this preprocessing takes a linear time, and then we apply the algorithm of [21,25]. Therefore, if $k = 2$ and the number of edges that can be augmented to G^R is large enough so that it is possible to make the whole review graph a 2-edge connected graph (or 2-reviewer connected graph), then such a feasible solution can be found in linear time.

We are using here two auxiliary problems: the maximum coverage problem and the densest k-subgraph problem. The MAXIMUM COVERAGE PROBLEM is defined on a given a collection of elements and a collection $\mathcal{F}$ of subsets of the element set. The objective is to select up to L members of $\mathcal{F}$ that cover a maximum number of elements. The MAXIMUM COVERAGE WITH CARDINALITY CONSTRAINTS PROBLEM (MCCC) is a variant of the maximum coverage problem where $\mathcal{F}$ is partitioned into L sub-collections $\mathcal{F}_1, \mathcal{F}_2, \ldots, \mathcal{F}_L$, and the constraint restricting the choice of at most L subsets from $\mathcal{F}$ is replaced by a set of constraints enforcing for each j the choice of a single member of $\mathcal{F}_j$.

The DENSEST k-SUBGRAPH PROBLEM is defined on an undirected graph $G = (V, E)$ and an integer number k. The goal is to find a subset $V' \subseteq V$ of at most k nodes so as to maximize the number of edges in the induced subgraph of G over V'. This problem is known to be NP-hard and the current best known approximation algorithm for this problem has an approximation ratio of $O(n^{-1/3+\delta})$ for a positive fixed number δ [11]. Moreover, there is an $O(\frac{k}{n})$-approximation algorithm for this problem (see for example [2]).

Paper overview. In Section 2 we show that CONNECTIVITY-k-ALOC problem is polynomially solvable via intersection of two matroids. In Section 3 we study two related decision and optimization problems: the k-COMPLETE problem and the MAX k-COMPLETE COVERAGE problem. We first determine the complexity of the MAX k-COMPLETE COVERAGE problem as a function of k: we show that for $k = 2$ the MAX 2-COMPLETE COVERAGE is polynomially solvable, whereas for any fixed value of k greater than 2 the k-COMPLETE is NP-complete, and hence the MAX k-COMPLETE COVERAGE problem is NP-hard. Next, we consider several approximation algorithms for the MAX k-COMPLETE COVERAGE problem: A trivial $\frac{1}{k-1}$-approximation for even values of k and $\frac{1}{k}$-approximation for odd values of k. The trivial algorithm applies to all values of k (even if k is not fixed); we then show that we can apply the greedy algorithm which was devised by Chekuri and Kumar [7]. This greedy algorithm is a $\frac{1}{2} \cdot \rho$-approximation algorithm, where ρ is the approximation ratio of an approximation algorithm for the densest k-subgraph problem (i.e., $\rho = O(n^{-1/3+\delta})$ for a fixed value of $\delta > 0$, or $\rho = O(\frac{k}{n})$). Next, it is shown that for a fixed value of k the MAX k-COMPLETE COVERAGE problem can be reduced to MCCC. Therefore, for fixed value of k, we can apply a recent algorithm by Ageev and Sviridenko [1] with an approximation ratio of $1 - \left(1 - \frac{1}{r}\right)^r$ where r can be as large as $O(n^{k-2})$ (so this approximation ratio is about $1 - \frac{1}{e} = 0.63212$). This algorithm of [1] has high complexity as it requires to solve a linear program with number of variables as large as the number of all possible k-subsets. The study of the MAX k-COMPLETE COVERAGE problem is concluded with a discussion in the full version of several extensions and modifications. In Section 4 we study the (k, p)-ECon, the (k, p)-RCon and the (k, p)-NCon problems. We prove that the $(k, 2)$-RCon is NP-complete for all fixed values of $k \geq 2$. We show that the (k, p)-NCon is NP-complete for all values of p and k such that $p \geq 2$ and $k \geq 2$, and we prove that the (k, p)-ECon problem is NP-complete is $k \geq 2$ and $2k - 2 \geq p \geq k$. In Section 5 we show that the 2-REVIEWER CONNECTIVITY AUGMENTATION problem is polynomially solvable using a dynamic programming algorithm. In Section 6, we prove that the (k, q)-HOP problem is NP-complete for all fixed values of $q \geq 2$ and $k \geq 2$. Omitted proofs will appear in the full version.

2   The CONNECTIVITY-k-ALOC Problem

Here we state that the CONNECTIVITY-k-ALOC problem is polynomially solvable as an instance of an intersection of two matroids. The proof is provided in the Appendix.

Theorem 1. *The* CONNECTIVITY-k-ALOC *problem is solvable in polynomial time* $O((|P| + |R|)^3)$. *If the number of reviewers is fixed then the problem is solvable in* $O((|P| + |R|) \log(|P| + |R|))$ *time.*

3   The k-COMPLETE and the MAX k-COMPLETE COVERAGE Problems

3.1   Complexity Classification

The following theorem proves that the MAX 2-COMPLETE COVERAGE is polynomially solvable via a matching procedure, whereas the k-COMPLETE for each fixed value of k ($k \geq 3$) is NP-complete. The proof of the theorem is given in the appendix.

Theorem 2. *The* MAX 2-COMPLETE COVERAGE *problem is solvable in time* $O(\min\{m_G L^{1.5}, L^{2.5}/\log(m_G + L)\})$. *The* k-COMPLETE *is NP-hard for all fixed values of k such that $k \geq 3$.*

3.2   Approximation Algorithms

Since the MAX k-COMPLETE COVERAGE problem is NP-hard, we turn our attention to approximation algorithms.

The trivial algorithm. The so-called trivial algorithm is a generalization of the matching procedure used to solve the MAX 2-COMPLETE COVERAGE problem. Given an instance of MAX k-COMPLETE COVERAGE we construct a bipartite graph $B = (A_1; A_2, E_B)$. For each subset S_j there is a corresponding node in A_1 denoted by v_{S_j}, and for each edge e of G there is a corresponding node in A_2 denoted by u_e. There is an edge $(v_{S_j}, u_e) \in E_B$ if both endpoints of e belong to S_j.

A b-matching in the bipartite graph B is a set of edges M that has up to b_i edges adjacent to node i. For each node i in A_1, $b_i = \lfloor \frac{k}{2} \rfloor$, and for each node j in A_2, $b_j = 1$. A maximum b-matching has a maximum number of edges among all b-matchings. From the maximum b-matching, M, we generate a feasible solution to the MAX k-COMPLETE COVERAGE by setting, for each S_j, the subset V_j is the one consisting of the endpoints of its matched edges $\{e_q \in A_2 | (s_j, e_q) \in M\}$ in the b-matching.

Theorem 3. *The trivial approximation algorithm is a $\left(\frac{1}{k-1}\right)$-approxi mation algorithm for even values of k and a $\left(\frac{1}{k}\right)$-approximation algorithm for odd values of k. The complexity of the algorithm is $O(nmL\log n)$.*

Applying the greedy algorithm when k is not fixed. We denote by ρ the approximation factor of an approximation algorithm for the densest k-subgraph problem (so $\rho = O(n^{-1/3+\delta}) < 1$, or $\rho = O(\frac{k}{n})$). When k is a fixed constant then $\rho = 1$ (by testing all possible k-subsets of nodes).

The greedy algorithm iteratively picks subsets that cover, each in turn, the maximum number of uncovered elements. In each step the subset picked must be a subset of S_j such that the algorithm did not select earlier another subset of S_j. Chekuri and Kumar [7] proved that this algorithm is a $\frac{1}{2}$-approximation algorithm. If in each step of the greedy algorithm, instead of picking the subset that covers the maximum number of uncovered elements (among the subsets of S_j such that the algorithm did not select earlier another subset of S_j), the algorithm picks a subset that covers at least β times the maximum number of uncovered elements (note that $\beta \leq 1$), then Chekuri and Kumar showed that the resulting approximation ratio is $\frac{1}{2}\beta$.

Theorem 4. *For a ρ-approximation algorithm for the densest k-subgraph problem, then there is a $\frac{1}{2}\rho$-approximation algorithm for the MAX k-COMPLETE COVERAGE problem. Hence there is an $O(\min\{n^{-1/3+\delta}, \frac{k}{n}\}) < 1$ approximation algorithm for the MAX k-COMPLETE COVERAGE problem. Moreover, if there is a ρ'-approximation algorithm for the MAX k-COMPLETE COVERAGE problem, then there is also a ρ'-approximation algorithm for the densest k-subgraph problem.*

Transforming the MAX k-COMPLETE COVERAGE problem into MCCC when k is fixed. For each S_j we write down the list of all subsets of S_j that have exactly k elements. Denote by $\mathcal{F}_j$ the resulting family of k-subsets of S_j, and if $|S_j| < k$ we let $\mathcal{F}_j = \{S_j\}$. Denote $\mathcal{F} = \bigcup_j \mathcal{F}_j$. The MAX k-COMPLETE COVERAGE problem is to choose one set from each $\mathcal{F}_j$ such that the number of covered edges with endpoints in a common set is maximized. The resulting problem is an instance of the MAXIMUM COVERAGE WITH CARDINALITY CONSTRAINTS PROBLEM. The size of this instance is polynomial if we assume that k is fixed. With this reduction and the approximation algorithm in [1] we have,

Theorem 5. *When k is a fixed constant, there is a $(1 - \frac{1}{e})$-approximation algorithm for the MAX k-COMPLETE COVERAGE problem.*

4   The (k,p)-RCon, the (k,p)-NCon and the (k,p)-ECon Problems

Theorem 6. *The $(k,2)$-RCon problem is NP-complete for all fixed values of k such that $k \geq 2$. The (k,p)-NCon problem is NP-complete for all fixed values of p and k such that $p \geq 2$ and $k \geq 2$.*

It remains to consider the complexity status of the (k, p)-ECon problem. We first note that if each reviewer has an expertise set that contains at least k projects and $p \le k - 1$, then the review graph is p-edge connected if and only if it is 1-edge connected (i.e., if it is a feasible solution to the CONNECTIVITY-k-ALOC problem). We next consider the case where $p \ge k$, and we prove the following.

Proposition 1. *The (k, p)-ECon problem is NP-complete for all values of $k \ge 2$ such that $2k - 2 \ge p \ge k$.*

5 The 2-REVIEWER CONNECTIVITY AUGMENTATION Problem

In this section we show how to solve in polynomial time the 2-REVIEWER CONNECTIVITY AUGMENTATION problem. We define a *block* to be a non-trivial (i.e., with at least two nodes) maximal node set such that between each pair of nodes in this set there are at least 2-reviewer disjoint paths.

Lemma 1. *Given an optimal solution such that there is a pair of additional reviewers r, r' where r review projects p_1 and p_2, and r' reviews p_3 and p_4 (among perhaps other projects), then p_1, p_2, p_3, and p_4 belong to a common block.*

Our algorithm **Augment_DP** for solving the 2-REVIEWER CONNECTIVITY AUGMENTATION problem is based on the above lemma and uses dynamic programming.

Theorem 7. *Algorithm* **Augment_DP** *solves the* 2-REVIEWER CONNECTIVITY AUGMENTATION *problem in $O(n^5 k^2 q^2)$ time.*

6 The (k, q)-HOP Problem

Theorem 8. *The (k, q)-HOP problem is NP-complete for all fixed values of q and k such that $q \ge 2$ and $k \ge 2$.*

References

1. A. A. Ageev and M. I. Sviridenko, "Pipage rounding: a new method of constructing algorithms with proven performance guarantee," *Journal of Combinatorial Optimization*, **8**, 307–328, 2004.
2. Y. Asahiro, K. Iwama, H. Tamaki and T. Tokuyama, "Greedily finding a dense subgraph," *J. Algorithms*, **34**, 203–221, 2000.
3. J. J. Bartholdi, C. A. Tovey and M. A. Trick, "The computational difficulty of manipulating an election," *Social Choice and Welfare*, **6**, 227–241, 1989.
4. J. P. Brans and Ph. Vincke, "A preference ranking organization method," *Management Science*, **31**, 647–656, 1985.
5. C. Brezovec, G. Cornuejols and F. Glover, "Two algorithms for weighted matroid intersection," *Mathematical Programming*, **36**, 39-53, 1986.
6. C. Brezovec, G. Cornuejols and F. Glover, "A matroid algorithm and its application to the efficient solution of two optimization problems on graphs," *Mathematical Programming*, **42**, 471-487, 1988.

7. C. Chekuri and A. Kumar, "Maximum coverage problem with group budget constraints and applications," in *Proceedings of APPROX 2004*, 72–83, 2004.

8. W.D. Cook, B. Golany, M. Kress, M. Penn and T. Raviv. Optimal allocation of proposals to reviewers to facilitate effective ranking. *Management Science*, **51**, 655–661, 2005.

9. D. Dor and M. Tarsi, "Graph decomposition is NP-complete: a complete proof of Holyer's conjecture," *SIAM Journal on Computing*, **26**, 1166–1187, 1997.

10. K. P. Eswaran and R. E. Tarjan, "Augmentation problems," *SIAM Journal on Computing*, **5**, 653-665, 1976.

11. U. Feige, G. Kortsarz and D. Peleg, "The Dense k-Subgraph Problem," *Algorithmica*, **29**, 410-421, 2001.

12. E. Fernandez and R. Olemdo, "An agent model based on ideas of concordance and discordance for group ranking problems," *Decision Support Systems*, **39**, 429–443, 2005.

13. G. N. Frederickson and M. A. Srinivas, "Algorithms and data structures for an expanded family of matroid intersection problems," *SIAM Journal on Computing*, **18**, 112-138, 1989.

14. R. Fuller and Ch. Carlsson, "Fuzzy multiple criteria decision making: recent developments," *Fuzzy sets and systems*, **78**, 139–153, 1996.

15. H. N. Gabow and R. E. Tarjan, "Almost optimum speed-ups of algorithms for bipartite matching and related problems," in *Proceedings of STOC 1988*, 514–527, 1988.

16. H. N. Gabow and R. E. Tarjan, "Faster scaling algorithms for network problems," *SIAM Journal of Computing*, **18**, 1013–1036, 1989.

17. M. R. Garey and D. S. Johnson, *Computers and Intractability*, W.H. Freeman and Co., New York, 1979.

18. D. S. Hochbaum and B. Chandran, "Further below the flow decomposition barrier of maximum flow for bipartite matching and maximum closure." Manuscript, UC Berkeley April 2004.

19. D. S. Hochbaum and A. Levin, "Methodologies for the group rankings decision." Manuscript, UC Berkeley April 2005, to appear *Management Science*.

20. I. Holyer, "The NP-completeness of some edge-partition problems," *SIAM Journal on Computing*, **10**, 713–717, 1981.

21. T. Hsu and V. Ramachandran, "On finding smallest augmentation to biconnect a graph," *SIAM Journal on Computing*, **22**, 889-912, 1993.

22. J. P. Keener, "The Perron-Frobenius theorem and the rating of football teams," *SIAM review*, **35**, 80–93, 1993.

23. J. G. Kemeny and J. L. Snell, "Preference ranking: An axiomatic approach," In *Mathematical models in the social sciences*, Boston, Ginn, 9–23, 1962.

24. J. Park, and M.E.J. Newman, "A network-based ranking system for US college football," *Journal of Statistical Mechanics: Theory and Experiment*, (Oct. 31, 2005). Abstract available at http://www.iop.org/EJ/abstract/1742-5468/2005/10/P10014.

25. A. Rosenthal and A. Goldner, "Smallest augmentations to biconnect a graph," *SIAM Journal on Computing*, **6**, 55-66, 1977.

26. A. Schrijver, "Combinatorial optimization polyhedra and efficiency", Springer-Verlag, Berlin, 2003.

An Experimental Study of the Misdirection Algorithm for Combinatorial Auctions

Jörg Knoche* and Piotr Krysta**

Dept. of Computer Science, University of Dortmund

Abstract. Single-minded combinatorial auctions (CA) are auctions in which a seller wants to sell diverse kinds of goods and each of the potential buyers, also called bidders, places a bid on a combination, i.e., a subset of the goods. There is a severe computational limitation in CA, as the problem of computing the optimal allocation is NP-hard and even hard to approximate. There is thus interest in polynomial time approximation algorithms for this problem. Recently, many such approximation algorithms were designed, among them greedy and local search based algorithms. One of these is a so-called misdirection algorithm combining both approaches and using a non-standard, misdirected, local search approach with neighborhood of size 2. This algorithm has the best known provable approximation ratio for the problem in terms of the sizes of bids. Its analysis, however, is quite complicated. We study this algorithm and its variants on typical instances designed for CAs. On question is if larger neighborhood helps – the question that seems quite difficult to address theoretically at the moment, taking into account already complex analysis for size 2 neighborhood. We also study experimentally other aspects of the misdirection algorithm, and finally present a comparison to other approximation algorithms.

1 Introduction

There has been an increasing interest in the recent years in so-called combinatorial auctions (CA). These are auctions where a seller wants to sell diverse kinds of goods and the potential buyers, called bidders, place bids on the combinations, i.e., subsets of goods. Such auctions were suggested for auctioning, e.g., spectrum licenses, landing slots or computational resources, see [23] for a survey.

When the auction concerns many related kinds of goods, combinatorial auctions are particularly well suited as they allow buyers to express their valuations on combinations of goods, which should lead to more economically efficient allocations. One of the main obstacles of dealing with combinatorial auctions is the

---

* Department of Computer Science, Dortmund University, Baroper Str. 301, 44221 Dortmund, Germany. E-mail: `joerg.knoche@cs.uni-dortmund.de`. The author is partially supported by DFG grant Kr 2332/1-2 within Emmy Noether program.
** Department of Computer Science, Dortmund University, Baroper Str. 301, 44221 Dortmund, Germany. E-mail: `piotr.krysta@cs.uni-dortmund.de`. The author is supported by DFG grant Kr 2332/1-2 within Emmy Noether program.

T. Erlebach and C. Kaklamanis (Eds.): WAOA 2006, LNCS 4368, pp. 265–278, 2006.

computational hardness of the problem of determining the optimal allocation for a given collection of buyers. This problem, called also winner determination, is known to be NP-hard and even hard to approximate [15].

For simplicity, we assume that each bidder desires only a single subset of goods and places a positive valuation only on this particular set of goods (any superset obviously values the same for this bidder). Such kind of restricted bidders, called single-minded, has been introduced by Lehmann et al. in their seminal paper [15], and has since been intensively investigated, e.g., [2,17,7,14].

In this paper we are interested in the winner determination problem in single-minded CAs, well known in discrete optimization as the set packing problem. Given a family of subsets of a given universe, each subset with a prescribed value, this problem asks for a maximum value set packing, i.e., a pairwise-disjoint subfamily of the subsets. Here, each single subset models a single bidder desiring that subset and its value is the bidder's valuation. The disjointness constraints corresponds to the fact that the seller cannot sell the same good to two bidders.

Since the set packing problem is NP-hard and even hard to approximate, polynomial time approximation algorithms are of interest. In fact, many approximation algorithms have been designed for this problem, see, e.g., [11] for a survey. We will be interested in approximating this problem in terms of d, which is the size of the largest set in the input family. Having the value of d small in terms of combinatorial auctions (CA) means that the bidder's preference sets are of size at most d, which clearly is the natural assumption for the bidders.

It is known that even unweighted, i.e., when values of the sets are unit, set packing problem with sets of size at most d is NP-hard to approximate to within $O(d/\log d)$ [12]. We cannot, thus, expect a much better than merely d approximation factor for this problem in polynomial time. It is quite easy to obtain a precisely d-approximation algorithm via the greedy method. In fact, a d-approximation with a running time $O(|\mathcal{S}|^2)$ was given by Hochbaum [13], and a $|\mathcal{S}|^{O(1/\epsilon)}$-time $(d-1+\epsilon)$-approximation for any fixed $\epsilon > 0$, by Bafna et al. [4] and Arkin & Hassin [3], where $\mathcal{S}$ is the given family of sets.

It turns out that even reducing the constant in front of d is a challenging problem. Chandra and Halldórsson [8] succeed to give a $\frac{2}{3}(d+1)$-approximation for this problem, at the expense of running time of $\Omega(|\mathcal{S}|^d)$, which is not polynomial if d is not a fixed constant. Berman [5] has improved this ratio to $\frac{1}{2}(d+1)$ but the running time is also $\Omega(|\mathcal{S}|^d)$. The first known better than d approximation in polynomial time is due to Berman and Krysta [6], who gave a factor $\frac{2}{3}d$ approximation for this problem in time roughly $O(|\mathcal{S}|^2)$. This last result is also the best known to date approximation ratio for the considered problem.

Berman and Krysta [6] consider in fact a slightly more general problem, that is, a maximum weighted independent set problem in a $(d+1)$-claw-free graph. Their algorithm is based on the local search method with a local misdirected [1]

---

[1] Misdirection is meant to mean here that while performing one step of the local search, we locally optimize a different (misdirected) objective than the original one.

objective and neighborhood of size 2 [2]. The theoretical analysis of this algorithm in [6] is quite technical and complex. Thus, for instance, a natural question of extending this analysis to neighborhood of size 3 seems quite challenging. This is the place where experimental analysis may help. This, and other questions and properties of the misdirection algorithm that seem difficult, or even impossible to address theoretically, are the subject of this experimental paper. We also study other greedy algorithms and an LP-based randomized rounding algorithm.

Outline. The rest of this paper is organized as follows. Section 2 has formal definitions, description of the algorithms and instances. Section 3 includes the experimental analysis of the misdirection algorithm. Section 4 compares all algorithms by running time and approximation factor.

2   Problem Definition, Algorithms and Instances

We formally define the set packing problem. We are given a finite set of goods U with $|U| = m$, and a family of subsets of U denoted by $\mathcal{S} \subseteq 2^U$. A given set $S \in \mathcal{S}$ models a bidder, and thus we identify set S with the set of bidders, and $|\mathcal{S}| = n$. Let also each $S \in \mathcal{S}$ have an associated weight $w(S) \in \mathbb{R}_+$, modeling the valuation of bidder S. The set packing problem (winner determination in CAs) asks for finding a packing, that is, a subfamily $\mathcal{S}' \subseteq \mathcal{S}$ such that any two distinct sets $S, T \in \mathcal{S}'$ are disjoint, $S \cap T = \emptyset$, and the total weight of $\mathcal{S}'$, $w(\mathcal{S}') = \sum_{S \in \mathcal{S}'} w(S)$, is maximized. We will assume in this paper that $d \in \mathbb{N}_+$ is the maximum size of any bid, i.e., $\max\{|S| : S \in \mathcal{S}\} = d$. Set packing problem is known to be NP-hard to approximate to within a factor of $O(d/\log d)$ [12].

2.1   Description of the Algorithms

We describe below the approximation algorithms for set packing we will use.

Greedy-1: This is algorithm Greedy-1 from paper [14]. It first sorts the sets in $\mathcal{S}$ by non-increasing values of $w(S)/\sqrt{|S|}$, and then goes through all sets in this order and puts them into the solution maintaining the feasibility of the packing. Since our input data is represented as a 0/1 matrix A such that $A_{e,S} = 1$ iff $e \in S$, implementing Greedy-1, the time to compute $|S|$ for each $S \in \mathcal{S}$ is taken into account. Greedy-1 is know to be $\sqrt{m}$-approximate for set packing [15].

Greedy-2: The same as Greedy-1 above, but the sorting is with respect to non-increasing values of $w(S)$. Note, that this algorithm is a bit faster than Greedy-1, since we do not need to compute $|S|$ for $S \in \mathcal{S}$. Hochbaum [13] shows that it is d-approximate for set packing.

Greedy-3: The same as Greedy-1 above, but the sorting is with respect to non-increasing values of $w(S)/|S|$. We did not put this algorithm into our

---

[2] We use the term neighborhood to denote a parameter ℓ in the misdirection algorithm, but in fact the "real" neighborhood size is roughly $O(|\mathcal{S}|^{\ell})$.

diagrams – see further explanations below. In theory Greedy-3 has an approximation ratio of d (in fact [14] shows ratio $d + 1$ for Greedy-3 on a more general problem, but the analysis of [14] slightly modified shows ratio d for set packing).

Misdirect-noGreed: This is the basic misdirection algorithm of Berman and Krysta [6]. It starts with an empty solution and performs all possible local exchanges. While this algorithm is defined in [6] for a slightly more general than set packing problem we redefine it here for the latter problem. Before we describe it we need some notation. Let $\mathcal{P}, \mathcal{R} \subseteq 2^U$ be two given set families. We define $N(\mathcal{P}, \mathcal{R}) = \{R \in \mathcal{R} : \exists P \in \mathcal{P} \text{ such that } P \cap R \neq \emptyset\}$. We also define $\mathcal{P} \triangleleft \mathcal{R} = (\mathcal{P} \setminus N(\mathcal{R}, \mathcal{P})) \cup \mathcal{R}$. Observe that if family $\mathcal{P}$ and family $\mathcal{R}$ is a packing then so is family $\mathcal{P} \triangleleft \mathcal{R}$. Let also $w^\alpha(\mathcal{P}) = \sum_{P \in \mathcal{P}}(w(P))^\alpha$, for a given $\alpha > 1$. Algorithm Misdirect-noGreed, also called $\ell - Imp^\alpha$, with neighborhood size $\ell = 2$ and value of $\alpha = 1.71$ (from [6]), is as follows.

> Algorithm $2 - Imp^\alpha$
> $\mathcal{P} \leftarrow \emptyset$
>   **while** there exists pair of sets $\{S, T\} \subseteq \mathcal{S}$ that improves $w^\alpha(\mathcal{P})$ **do**
> $\mathcal{P} \leftarrow \mathcal{P} \triangleleft \{S, T\}$

This algorithm can be described in words simply as follows: start from an empty packing, and as long as there is a pair of sets that on adding to the current packing (and removing all conflicting sets) improves the misdirected objective $w^\alpha(\cdot)$, perform such a local exchange. This algorithm may not have polynomial running time, and it is shown in [6] how to make it polynomial – see algorithm Misdirect below. Berman and Krysta prove an upper bound of (roughly) $\frac{2}{3}d$ on the approximation ratio of algorithm $2 - Imp^{1.71}$. More precisely, they prove that $\alpha = 1.71$ is the value of α giving the best approximation ratio of $\frac{2}{3}d$.

Misdirect: This is the original misdirection algorithm described in paper [6]. First it runs algorithm Greedy-2. Let $\mathcal{P}$ be the output greedy packing. Then, rescale the weights so that $w(\mathcal{P}) = k \cdot |\mathcal{S}| = k \cdot n$, for some fixed $k \in \mathbb{N}_+$, and run algorithm Misdirect-noGreed after replacing the function w with $\overline{w}(S) = \lfloor w(S)^\alpha \rfloor^{1/\alpha}$, starting with initial solution $\mathcal{P}$. Berman and Krysta show that the running time of this modified algorithm is bounded by $O(k^\alpha (dn)^{2+\alpha})$, and its approximation ratio is at most $\frac{k}{k-1} \cdot \frac{2}{3} \cdot d$, for any choice of $k \in \mathbb{N}_+$.

RandRound: This is the most typical approximation algorithm, see [18,19,22], for packing problems like set packing. It first solves the linear programming relaxation of the set packing problem and then performs the standard randomized rounding. An iteration may not produce a feasible packing and that is why we repeat the randomized rounding step 750 times and take the best output solution. Please note that we do not optimize the number of iterations of RandRound and take it into account just to compare with the above (combinatorial) algorithms, and our comparison is fair – see further sections. Srinivasan [21] proved that RandRound has an $O(d)$-approximation for set packing, where, in particular the constant in the ratio $O(d)$ is larger than 1.

2.2   Description of the Instances

Vohra / de Vries: These instances are described by Zurel and Nisan [24], and by de Vries and Vohra [23]. The description below follows [24]. These instances are called in our experiments *prob.i.m.n.d*, where $i = 1, 2, 3, 4$ according to the definitions below, and numbers m, n and d are as defined previously. (Some of the instances may not have the last part d in *prob.i.m.n.d* defined.)

1. Random: For each bid, pick the number of goods randomly from $\{1, 2, \ldots, m\}$. Randomly choose that many goods without replacement. Pick the bid weight (valuation) randomly from $[0, 1]$. Then, $m \in \{100, \ldots, 400\}$, $n \in \{500, \ldots, 1000\}$.
2. Weighted Random: The same as for Random, but the bid weight is picked from $[0, \text{number of goods in bid}]$. Then, $m \in \{100, \ldots, 400\}$, and $n \in \{500, \ldots, 2000\}$.
3. Uniform: For each bid, pick a constant number of goods randomly from $\{1, 2, \ldots, m\}$. Randomly choose that many goods without replacement. Pick the bid weights randomly from $[0, 1]$. Then, $m \in \{25, \ldots, 100\}$, $n \in \{50, \ldots, 1100\}$, and the bid size $d \in \{3, 8, 11\}$.
4. Decay: For each bid, give it one random good. Then repeatedly add a new random good with probability γ until that good was not added or the bid contains all m goods. Pick the bid valuations randomly from $[0, \text{number of goods in bid}]$. Also, $m \in \{50, \ldots, 200\}$, $n \in \{50, \ldots, 200\}$, and the probability $\gamma \in [5\%, 95\%]$.

CATS: These are instances generated by the CATS program described in paper [16] by Leyton-Brown, Pearson and Shoham. The used distributions are arbitrary, paths, regions and scheduling. We have used the standard parameters to generate these instances, and only the number of bids and goods was varying. For more precise description of this instance generator see [16] and the web page http://cats.stanford.edu/. These instances are referred in our paper to as *name.m.n*, where *name* $\in \{arb, paths, reg, sched\}$ and the names $\{arb, paths, reg, sched\}$ correspond to the ones above.

Fujishima / Sandholm: These are the instances described in [1], which can be found at web page http://user.it.uu.se/~tein/cmb/index.html. They contain instances generated according to random, uniform, decay, binomial and exponential distribution. The distributions used in our experiments are uniform, binomial with 1500 bids and exponential. From each distribution the first 5 instances have been taken. We refer to those instances as *name.nr*, where *name* $\in \{exp, uni, bin\}$ and *nr* is the number of the instance. Following the cited paper, we keep m, n fixed to some specific values.

Uniform ([20]): Draw the same number of randomly chosen items for each bid. Pick an integer valuation from $[500, 1500]$ and multiply by the number of commodities. The number of goods and bids are fixed to $m = 100$ and $n = 500$.

Binomial ([9]): The probability distribution for a bid requesting j goods out of m goods in the market is $f(j) = p^j \cdot (1 - p)^{m-j} \cdot \binom{m}{j}$ with $p = 0.2$. An integer valuation is drawn from 500 to 1500 and multiplied by j. The number of goods and bids here are $m = 150$ and $n = 1500$.

Exponential ([9]): The probability distribution is defined as $f_e(j) = c \cdot e^{-j/5}$ (c is implicitly defined by $\sum_{j=1}^{m} f_e(j) = 1$, where m as before is the number of goods). The valuation is an integer, rectangularly drawn from $[500, 1500]$ and multiplied by the number of requested goods j. Again, we fix $m = 30$ and $n = 3000$.

Random: These instances were randomly generated by us with a fixed number of goods m, bids n, and goods per bid (the d value). We choose randomly n subsets (bids) of size d out of m goods (possibly with repetitions). Then for each generated bid its weight is randomly chosen from $[0, 1]$ and multiplied by d. We call these instances $Randomx$, where x is the serial number of the instance.

Test-Setup: All algorithms are implemented in Java 1.4.2 and run on AthlonXP 1900 MHz machine with 768 MB RAM under WindowsXP with Service Pack 1. In all tested instances the optimal solution was found by using CPLEX 6.5.2.

3   Analyzing the Misdirection Algorithm

3.1   Proved Versus Achieved Approximation

A natural question after one succeeds to prove a bound on the approximation ratio is how rough this bound is as compared to one obtained on typical instances. Indeed, also in the case of the misdirection algorithm $2 - Imp^{1.71}$ with the starting greedy solution this bound turns out to be rough. This, of course, is not surprising, but just confirms the known phenomenon that most likely there are only few worst-case, untypical instances. For some data, see Figure 1.

instance	prob.1.100.1050.3	prob.3.75.50.8	prob.3.100.50.3	Random1	bin150	bin1506	uni31
d	3	8	3	5	47	48	3
proved-apx.	2,000	5,166	2,000	3,229	30,351	30,996	2,000
achieved-apx.	1,086	1,352	1,037	1,000	1,007	1,0	1,091

instance	prob.3.100.100.3	prob.3.100.1300.3	prob.3.75.100.13	exp50	exp51	exp52	exp53
d	3	3	13	4	4	4	4
proved-apx.	2,000	2,000	8,395	2,606	2,606	2,606	2,606
achieved-apx.	1,000	1,096	1,214	1,055	1,038	1,026	1,056

Fig. 1. The proved approximation ratio is just $\frac{2}{3}d$, and the achieved ratio is calculated by comparing to the optimal solution.

3.2   Larger Neighborhood

We investigate here one of our main questions concerning algorithm Misdirect, namely if it is worth investing time to find theoretical analysis showing better factors when $\ell \geq 3$. Let us first consider a tight example in [6] for the ratio of $\frac{2}{3}d$

of $2 - Imp^\alpha$ when $d = 3$. This is given by two examples in Lemma 2.1 and 2.2 in [6]. These two lemmas together imply Lemma 2.3 in [6] stating that for $d = 3$ the ratio is at least $(\sqrt{5d^2 - 8d + 4} + 2 - d)/2 = \frac{2}{3}d = \beta \cdot d = 2$, where $\beta^\alpha = \frac{1}{2}$, which implies $\beta = \frac{2}{3}$ and $\alpha \approx 1.71$, which is the best theoretical value of α for $d = 3$ found in [6]. (Note, taking $\alpha \approx 1.71$ gives approximation factor $\frac{2}{3}d$ for *all* values of d. For other values of d slightly better values of α are found [6].)

It can, however, easily be checked that the locally optimal solutions w.r.t. $\ell = 2$ in these two examples from [6] are not locally optimal anymore when $\ell = 3$ and parameter β from [6] fulfills $\frac{1}{3} < \beta^\alpha < 1$. We, therefore, see that the size 3 neighborhood indeed helps, but certainly this is not enough evidence. What about typical instances ?

We judge here the improvement in the weight of the solution when we change the neighborhood size ℓ in algorithm Misdirect from 2 to 4 on CA instances. We have selected the instances in Figure 2 (a) only from type *prob.i.m.n.d* (additional number in the brackets is just the serial number). The reason being that these are quite small instances and the running time increases rapidly when ℓ is raised. Obviously, we did not run the tests for larger ℓ if we reached optimum, i.e., 100% earlier. For each value of $\ell \in \{3, 4\}$ we used one value of α for all tested instances, namely the best found.

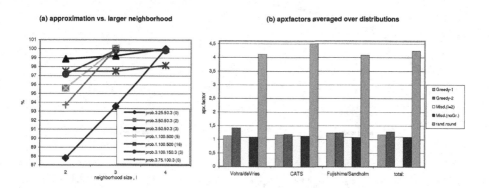

Fig. 2. (a) 100% on the vertical axis corresponds to the weight of an optimal solution. (b) shows approximation factors of all algorithms averaged over all instances.

Conclusions: It seems it is worth to try to analyze $\ell - Imp^\alpha$ theoretically for larger values of ℓ. Even in the typical instances we selected we observe an improvement in the approximation ratio ranging from 1% to about 10%. This seems not much, but note, that our instances are average "typical" ones. On the other hand it is plausible that there are better improvements possible on other larger instances as well, but so far our limiting factor for such tests was rapidly increasing running time of our implementations.

Testing which α is best for $\ell \geq 3$ we observed that smaller values give better results for larger neighborhoods. Our diagram in Fig. 2 (a) shows the best solution found over all α tested (between 1 and 2 with step size 0.01), but we found one value of $\alpha = 1.05$ which gives these best results for $\ell = 3$. Same value of α gives best results for most of the instances when $\ell = 4$, but we could not test many of them because of high run time. For a majority of instances an α smaller than 1.1 seemed to be best. But α-values up to 1.2 may also give better results depending on the instance. For $\ell = 2$ we found that on roughly 30% of the instances smaller values of α in $(1.0, 1.2]$ were better than 1.71.

3.3   Misdirected Versus Standard Local Search

In this section we compare the standard local search, that is with $\alpha = 1$, with the misdirected local search, that is with $\alpha > 1$, both with neighborhood of size 2. More precisely, we compare $2 - Imp^1$ with $2 - Imp^{1.71}$. Both used algorithms start with an empty initial solution (Misdirect-noGreed).

The diagrams in Fig. 3 show the increase of the solution's original weight, $w(S)$, at every local exchange for the respective algorithms. The weight of the final solution output by $2 - Imp^{1.71}$ algorithm is 100% on the vertical axis.

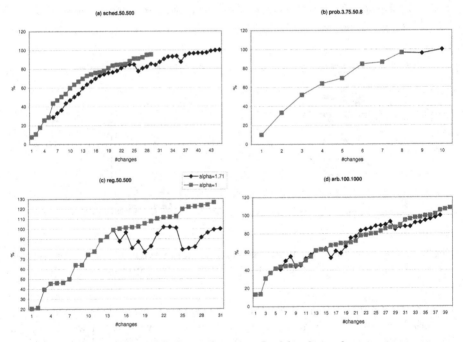

Fig. 3. Misdirected vs. standard local search

Conclusions: One can observe that of course in case of $2 - Imp^{1.71}$ there are many jumps down, which in many cases lead to better locally optimal solutions in the future exchanges.

Berman and Krysta [6] show an example on which the misdirected algorithm avoids some bad local optima that would lead to an approximation factor of d, which is the case for the standard local search, instead of factor of $\frac{2}{3}d$ for the misdirected algorithm. In fact, this example is quite specific one. A question is if such a behavior also occurs for typical instances. Indeed, we found some instances where this is the case, see Figure 3 (a) and (b). There are also instances in which the misdirected solution was worse – see Figure 3 (c) and (d).

The major kind of behavior that we observed is, however, the fact that in about 80% of all the instances we tested the two curves for $2 - Imp^{1.71}$ and for $2 - Imp^1$ split at some exchange, earlier or later, and at the very point of the split, the weight of the solution of $2 - Imp^{1.71}$ jumps down. As examples, see Figure 3. Also in these majority of instances the $2 - Imp^{1.71}$-curve is below the $2 - Imp^1$-curve. We also observe that the remaining 20% of the instances where this behavior does not occur, do not come from one specific type of instances. We have not observed a clear correlation between the jump down and the fact that the final Misdirect solution was better than that of standard local search.

3.4   How Fast Is the Local Optimum Reached?

We have taken into account here how many local exchanges are needed to reach the final locally optimal solution for the Misdirect-noGreed algorithm.

In Figure 4 we draw the quality of the solution (w.r.t. the original weight $w(\cdot)$) as a function on the number of local exchanges. We see that within one class of instances there are easier instances, where there are small number of high jumps, and there are harder instances, where we have many small jumps. One striking observation is that for Fujishima/Sandholm instances, see Fig. 4 (a) the instances have been clearly divided into easy, medium, and hard. In particular easy are instances generated w.r.t. binomial distribution, medium – ones generated with exponential distribution, and hard – generated with uniform distribution. Observe, also, that in the case of "medium", exponential instances there are many jumps down, which may suggest that in those cases there are many bad local optima and they are avoided by Misdirect-noGreed.

A similar picture can be obtained for Misdirect-noGreed with neighborhood of size 3 in Figure 4 (d), where the easiest instances are the ones of Vohra/de Vries – see also Fig. 4 (c) for those instances and neighborhood of size 2.

Considering Misdirect-noGreed with $\ell = 2$ on Vohra/de Vries instances we found that there are also all levels of difficulty, see Fig. 4 (c).

Finally, we observed that in the case of Misdirect, increasing k which is used in scaling the weights does not increase its running time on tested instances (though a bound of $O(k^\alpha (dn)^{2+\alpha})$ on the running time in [6] suggests the opposite). To explain this we found instances where the value of $\overline{w}(S)^\alpha$ increases by more than 1 in the exchanges, and not just by 1 as assumed in the $O(k^\alpha (dn)^{2+\alpha})$ bound.

4   Comparing All the Algorithms

This section is devoted to the comparison of all the algorithms that we tested, that is the two greedy algorithms and two misdirection algorithms, and the randomized rounding algorithm.

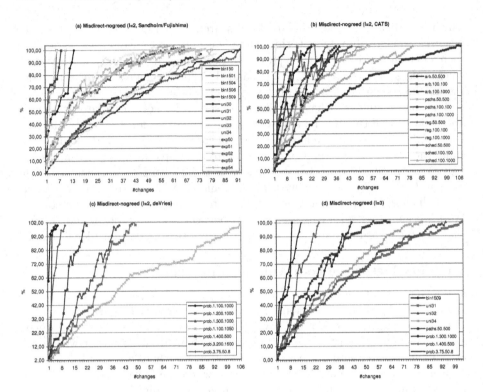

Fig. 4. How fast is the local optimum reached by Misdirect-noGreed for all distributions?

4.1 Running Times

In terms of running time it is no surprise that the greedy algorithms are much faster than the misdirection ones. For a comparison see Fig. 5, where we averaged over 13 instances for the Vohra/de Vries distributions, 12 instances of the CATS distributions and 15 instances of the Fujishima/Sandholm distributions.

Conclusions: We see that typically, the greedy algorithms are faster than the misdirection ones by at least a factor of 100. We also see that the Fujishima/Sandholm instances are most time demanding for all algorithms, and the CATS instances are somehow least time demanding for all algorithms.

Note that when calculating the running time of RandRound in Fig. 5 we are fair and only take into account the randomized rounding iterations and disregard the time for solving the LP relaxation. The reason for this is that we solve the LPs exactly by CPLEX, but for such packing LPs there are faster (approximate) LP solvers, e.g., [10].

For RandRound we observe that our 750 iterations of randomized rounding phase lead to running time higher than that of greedy algorithms (Fig. 5), but the approximation factors achieved are much worse (Fig. 6, Fig. 2 (b)).

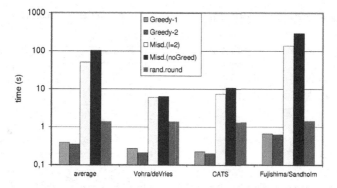

Fig. 5. Running times of all algorithms averaged over the instances

We have also found two instances from the Vohra/de Vries distribution where all the algorithms have a larger running time than on the other instances. Both have a small d value of 3. Thus, possibly, the running time increases, when the number of bids is the same but the d value is smaller. Finally we observed that obtaining optimal solution with CPLEX was 10 to 100 times slower than running a greedy algorithm. Except one type of the Fujishima/Sandholm instances, namely exponential, where CPLEX was about 4 times faster than greedy.

4.2 Approximation Factors

We first describe the diagrams for comparing the approximation factors. Fig. 6 (a), (b) shows the approximation factors for every algorithm depending on the number of goods. Each data point is an average over 3-10 instances. Fig. 6 (c), (d) shows the approximation factors for every algorithm depending on the number of bids, where each data point is an average over 3-5 instances. Finally, Fig. 2 (b) shows the approximation factors averaged over all instances of a distribution. The Vohra/de Vries value is averaged over 13 instances. The CATS value is averaged over 12 (3 of each kind of distribution), and the Fujishima/Sandholm value–over 15 instances (5 of each distribution type).

Conclusions: We see that Misdirect has the best approximation factors which as we know is also the case in the theoretically proven results. This, however, is with the expense of much higher running times – see the previous subsection. Also, we did not put here Greedy-3 into the diagrams, because we observed that Greedy-1 always had better approximation ratio than Greedy-3 (there are only very few instances of the Fujishima-Sandholm type where the ratio of Greedy-3 is better only by 0.02%). This is somehow interesting since in theory Greedy-1 has ratio roughly $\sqrt{m}$ (which is $\sqrt{30} \approx 5.47$ for those instances), but Greedy-3 has ratio d (which is 4 for those instances).

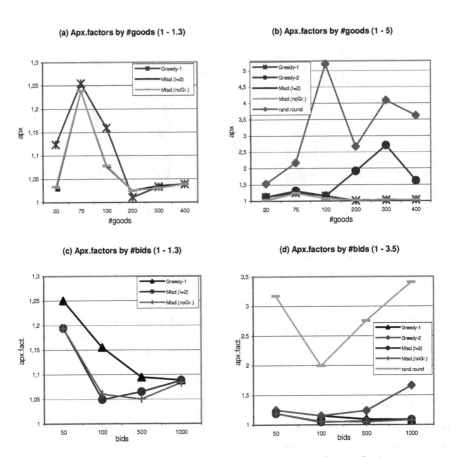

Fig. 6. Approximation factors of all algorithms averaged over the instances

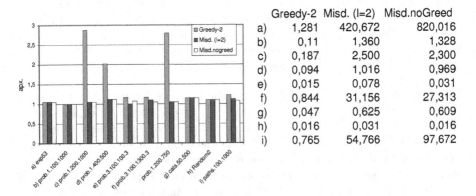

	Greedy-2	Misd. (l=2)	Misd.noGreed
a)	1,281	420,672	820,016
b)	0,11	1,360	1,328
c)	0,187	2,500	2,300
d)	0,094	1,016	0,969
e)	0,015	0,078	0,031
f)	0,844	31,156	27,313
g)	0,047	0,625	0,609
h)	0,016	0,031	0,016
i)	0,765	54,766	97,672

Fig. 7. Approximation factors of Misdirect with/without Greedy and of Greedy itself. The table on the right is the running time (s) for those instances.

4.3 How Profitable Is Using Greedy Inside Misdirect?

Fig. 7 shows that using greedy solution as the starting point for the misdirection algorithm does not lead to much better approximation factors. Also, the running times are not much different between Misdirect with and without Greedy.

Final conclusion: The best algorithm suggested by our experiments is Greedy-1 if we want a good trade-off for both running time and approximation quality. It also wins when the simplicity and easy implementation are the concerns. Greedy-1 is fastest, and gives in most instances the best ratio and in all instances ratios worse than slowest Misdirect, by at most 8%.

5   Conclusion and Future Work

Our experiments suggest that it should be interesting to try to theoretically prove better ratios for Misdirect with larger neighborhoods. The theoretical bounds on the running time of Misdirect appeared to be quite rough on typical instances. We also observed some interesting behavior on how Misdirect avoids some local optima. Among the tested algorithms Greedy-1 turns out to be best if we want a fast algorithm, good approximation ratios and simple implementation. For the future, we plan to also test other algorithms for multi-packing problems, and conducting experiments on larger instances.

Acknowledgment. We would like to thank Piotr Berman for some useful discussions on experimental analysis.

References

1. A. Andersson, M. Tenhunen, F. Ygge. Integer Programming for Combinatorial Auction Winner Determination. *Proc. 4th Int. Conf. on Multiagent Systems (IC-MAS)*, 2000.
2. A. Archer, C.H. Papadimitriou, K. Talwar, and É. Tardos. An approximate truthful mechanism for combinatorial auctions with single parameter agents. In *Proc. of 14th SODA*, 2003.
3. E. M. Arkin, and R. Hassin. On local search for weighted k-set packing. In the *Proc. ESA '97*, LNCS **1284**, Springer, 1997.
4. V. Bafna, B. Narayana, R. Ravi. Nonoverlapping local alignments (Weighted independent sets of axis-parallel rectangles). *Discr. Applied Math.*, **71**, 41–53, 1996.
5. P. Berman. A $d/2$ Approximation for Maximum Weight Independent Set in d-Claw Free Graphs. *Nordic J. Computing*, **7**(3), pp. 178–184, 2000.
6. P. Berman and P. Krysta. Optimizing misdirection. In *Proc. 14th SODA*, 2003.
7. P. Briest, P. Krysta, and B. Vöcking. Approximation Techniques for Utilitarian Mechanism Design. In *Proc. 37th ACM STOC*, 2005.
8. B. Chandra and M. M. Halldórsson. Greedy local improvement and weighted packing approximation. In *Proc. SODA*, 1999.
9. Y. Fujishima, K. Leyton-Brown, and Y. Shoham. Taming the computational complexity of combinatorial auctions: Optimal and approximate approaches. In *Proc. 16th Int. Joint Conference on Artificial Intelligence (IJCAI)*, pp. 548–553, 1999.

10. N. Garg and J. Könemann. Faster and Simpler Algorithms for Multicommodity Flow and Other Fractional Packing Problems. In *Proc. 39th IEEE FOCS*, 1998.

11. M.M. Halldórsson. A survey on independent set approximations. In *Proc. APPROX '98*, Springer LNCS 1444, pp. 1–14, 1998.

12. E. Hazan, S. Safra and O. Schwartz. On the Hardness of Approximating k-Dimensional Matching. In *APPROX*, 2003.

13. D. S. Hochbaum. Efficient bounds for the stable set, vertex cover, and set packing problems. *Discr. Applied Math.*, **6**, pp. 243–254, 1983.

14. P. Krysta. Greedy Approximation via Duality for Packing, Combinatorial Auctions and Routing. In *Proc. of 30th MFCS*, 2005.

15. D. Lehmann, L. O'Callaghan, and Y. Shoham. Truth revelation in approximately efficient combinatorial auctions. In *Proc. 1st ACM Conference on Electronic Commerce (EC)*, 1999.

16. K. Leyton-Brown, M. Pearson, and Y. Shoham. Towards a Universal Test Suite for Combinatorial Auction Algorithms. In *Proc. 2nd ACM Conference on Electronic Commerce (EC)*, 2000.

17. A. Mu'alem and N. Nisan. Truthful Approximation Mechanisms for Restricted Combinatorial Auctions. In *Proc. 18th AAAI Conf. on Artificial Intelligence*, 2002.

18. P. Raghavan. Probabilistic Construction of Deterministic Algorithms: Approximating Packing Integer Programs. *J. Comput. Syst. Sci.*, **37(2)**, 130-143, 1988.

19. P. Raghavan and C.D. Thompson. Randomized rounding: a technique for provably good algorithms and algorithmic proofs. *Combinatorica*, **7**: 365–374, 1987.

20. T. W. Sandholm. An algorithm for optimal winner determination in combinatorial auctions. *Proc. 16th Int. Joint Conf. on Artificial Intelligence (IJCAI)*, 542–547, 1999.

21. A. Srinivasan. A extension of the Lovász Local Lemma and its applications to integer programming. *In Proc. 7th ACM-SIAM SODA*, 1996.

22. A. Srinivasan. Improved Approximation Guarantees for Packing and Covering Integer Programs, *SIAM J. Computing*, Vol. 29, 648–670, 1999.

23. S. de Vries and R. Vohra. Combinatorial Auctions: A Survey. *INFORMS J. Computing*, **15(3)**, pp. 284–309, 2003.

24. E. Zurel and N. Nisan. An Efficient Approximate Allocation Algorithm for Combinatorial Auctions. In the *Proc. EC*, 2001.

Reversal Distance for Strings with Duplicates: Linear Time Approximation Using Hitting Set

Petr Kolman[1,*] and Tomasz Waleń[2,**]

[1] Charles University in Prague
Faculty of Mathematics and Physics
Department of Applied Mathematics
kolman@kam.mff.cuni.cz
[2] Warsaw University
Faculty of Mathematics, Informatics and Mechanics
walen@mimuw.edu.pl

Abstract. In the last decade there has been an ongoing interest in string comparison problems; to a large extend the interest was stimulated by genome rearrangement problems in computational biology but related problems appear in many other areas of computer science. Particular attention has been given to the problem of *sorting by reversals* (SBR): given two strings, A and B, find the minimum number of reversals that transform the string A into the string B (a *reversal* $\rho(i,j)$, $i < j$, transforms a string $A = a_1 \ldots a_n$ into a string $A' = a_1 \ldots a_{i-1} a_j a_{j-1} \ldots a_i a_{j+1} \ldots a_n$).

Primarily the problem has been studied for strings in which every symbol appears exactly once (that is, for permutations) and only recently attention has been given to the general case where duplicates of the symbols are allowed. In this paper we consider the problem k-SBR, a version of SBR in which each symbol is allowed to appear up to k times in each string, for some $k \geq 1$. The main result of the paper is a $\Theta(k)$-approximation algorithm for k-SBR running in time $O(n)$; compared to the previously known algorithm for k-SBR, this is an improvement by a factor of $\Theta(k)$ in the approximation ratio, and by a factor of $\Theta(k)$ in the running time. Crucial ingredients of our algorithm are the suffix tree data structure and a linear time algorithm for a special case of a disjoint set union problem.

Keywords: Approximation algorithms, String comparison, Sorting by reversals, Minimum common string partition, Suffix trees.

1 Introduction

In the last decade there has been an ongoing interest in string comparison problems. To a large extent the interest was stimulated by genome rearrangement problems in computational biology but related problems appear in many

---

* Supported by project 1M0021620808 (ITI) of Ministry of Education of the Czech Republic.
** Supported by the Polish Scientific Research Committee (KBN) under grant GR-1946.

T. Erlebach and C. Kaklamanis (Eds.): WAOA 2006, LNCS 4368, pp. 279–289, 2006.

other areas of computer science, in data compression or text processing to name a few. One of the important problems is to measure the similarity of two strings. Particular attention has been given to the problem of *sorting by reversals* (SBR): given two strings, A and B, find the *reversal distance* of A and B, which is the minimum number of reversals that transform the string A into the string B. A *reversal* $\rho(i,j)$, $1 \leq i < j \leq n$, is an operation that transforms a string $A = a_1 \ldots a_n$, into a string $A' = a_1 \ldots a_{i-1} a_j a_{j-1} \ldots a_i a_{j+1} \ldots a_n$ (that is, the reversal $\rho(i,j)$ reverses the order of symbols in the substring $a_i \ldots a_j$ of A). In the case of signed strings, each symbol is given a sign $+$ or $-$, and the reversal operation also flips the sign of each symbol in the reversed substring.

Primarily the problem has been studied for strings in which every symbol appears exactly once (that is, for permutations); even in this setting the problem is NP-hard for unsigned permutations [2] and, surprisingly, the problem is in P for signed permutations [10]. Only recently attention has been given also to the general case where duplicates of the symbols are allowed. We denote by k-SBR the version of SBR in which each symbol is allowed to appear up to k times in each string, for some $k \geq 1$. Christie and Irving [4] prove that unsigned SBR is NP-hard for binary strings and Chen et al. [3] show that 2-SBR is NP-hard. The best approximation ratio for the general signed SBR is $O(\log n \log^* n)$ (following from the work of Cormode and Muthukrishnan [6]); there are $O(1)$-approximation algorithms for signed 2-SBR and 3-SBR [3,5,9]. Kolman [11] describes a greedy-like $O(k^2)$-approximation algorithm for k-SBR running in $O(kn)$ time. Most of the above mentioned algorithms exploit the close relationship between the minimum common string partition problem (see below for definition) and the problem of sorting by reversals: they find an approximation for the static problem MCSP and turn it into a solution for SBR; this is also the approach that we take in this paper. For an overview of other related results and for more details about the relation between MCSP and SBR, we refer to the paper [11].

The main results of this paper are $\Theta(k)$-approximation algorithms for k-MCSP and k-SBR running in time $O(n)$; compared to the previously known algorithms for k-MCSP and k-SBR, this is an improvement by a factor of $\Theta(k)$ in the approximation ratio, and by a factor of $\Theta(k)$ in the running time.

On a high level, the algorithm works as follows: given the strings A and B, the algorithm turns them into an instance of the minimum hitting set problem and, exploiting special properties of the instance, it computes an approximation of the minimum hitting set which is in turn transformed into an approximate solution for k-MCSP; a solution for k-SBR is obtained from a solution of the relevant k-MCSP problem by the standard technique mentioned above. Crucial ingredients of the algorithm are a linear time procedure for construction of a suffix tree [7] and a linear time algorithm for a special case of a disjoint set union problem [8].

1.1   Notation

We stick to the notation used in the previous paper on k-SBR [11]. For a (signed or unsigned) string $P = a_1 \ldots a_n$, we denote by $-P$ the result of reversal $\rho(1,n)$

of P (e.g., for $P = +a + b - d$, we have $-P = +d - b - a$; for $P = abd$, we have $-P = dba$). We say that two (signed or unsigned) strings $A = a_1a_2 \ldots a_n$ and $B = b_1b_2 \ldots b_n$ are *identical*, $A = B$, if $a_i = b_i$ for each $i \in 1, \ldots, n$ (in the case of signed strings, $a_i = b_i$ involves also the equality of the signs), and they are *congruent*, $A \cong B$, if $A = B$ or $A = -B$ (note that for the sake of notational simplicity we overload the sign $\cong$ so that it has a slightly different meaning for signed and unsigned strings).

Throughout the paper we assume that the symbols are represented by integers from the set $\Sigma = \{1, 2, \ldots, n\}$. We also assume that each symbol appears the same number of times in A and B (for the signed version, we count together the occurrences of a symbol with positive and negative signs). Clearly, this is a necessary and sufficient condition for A and B to have a finite reversal distance. We call such strings *related*.

The length of a string A is denoted by $|A|$. A *duo* is a string of length two. A *partition* of a string A is a sequence $\mathcal{P} = (P_1, P_2, \ldots, P_m)$ of strings whose concatenation is equal to A, that is, $P_1P_2 \ldots P_m = A$. The strings P_i are called the *blocks* of $\mathcal{P}$ and their number is the *size* of the partition. Given a partition $\mathcal{P} = (P_1, P_2, \ldots, P_m)$, if $l = \sum_{j=1}^{i} |P_j|$ for some $i \in \{1, 2, \ldots, m-1\}$, we say that the pair $l, l+1$ is a *break* of the partition $\mathcal{P}$ and a_la_{l+1} is a *broken duo* of the partition $\mathcal{P}$.

For two strings A and B, we say that S is a *common substring with respect to the relation* $=$ if S is a substring of A and a substring of B; we say that S is a *common substring with respect to the relation* $\cong$, if S is a substring of A and there exists a substring R of B such that $S \cong R$, or S is a substring of B and there exists a substring R of A such that $S \cong R$. When not necessary, we will often avoid specifying the relation and will talk only about a common substring.

To *cut* a duo a_ia_{i+1} of a block $P = a_j \ldots a_k$ of a partition of A, for some $j \leq i < k$, means to replace the block P in the partition by two blocks $P_1 = a_j \ldots a_i$ and $P_2 = a_{i+1} \ldots a_k$. For a string $C = c_1, \ldots, c_n$, we denote by $duos(C)$ the set of duos of the string C, that is, $duos(C) = \{c_ic_{i+1} \mid 1 \leq i \leq n-1\}$.

SBR is closely related to the minimum common string partition problem. Given a partition $\mathcal{P} = (P_1, \ldots, P_m)$ of a string A and a partition $\mathcal{Q} = (Q_1, \ldots, Q_m)$ of a string B, we say that the pair $\pi = (\mathcal{P}, \mathcal{Q})$ is a *common partition* of A and B with respect to the relation $\mathsf{Rel} \in \{=, \cong\}$, if there exists a permutation σ on $1, \ldots, m$ such that for each $i \in 1, \ldots, m$, $(P_i, Q_{\sigma(i)}) \in \mathsf{Rel}$. The *minimum common string partition problem* (MCSP) is to find a common partition of A, B with the minimum size, denoted by $\mathsf{MCSP}(A, B)$. The restricted version of MCSP, where each letter occurs at most k times in each input string, is denoted by k-MCSP. Similarly as for SBR, there is a signed and an unsigned variant of the problem. In *unsigned* MCSP, the input consists of two unsigned strings, and the relation $=$ is used; in *signed* MCSP, the input consists of two signed strings and the relation $\cong$ is used. For unsigned strings, we define yet another variant of the problem, *reversed* MCSP (RMCSP), in which the (unsigned) strings are compared by the relation $\cong$. Chen et al. [3] observed that for any two related signed strings A and B, the sizes of the optimal solutions of MCSP and SBR differ only by a constant multiplicative factor.

An analogous observation applies for related unsigned strings and the problems reversed MCSP and SBR; we refer to the paper [11] for further details.

The rest of the paper is organized as follows. Section 2 is devoted to a simple algorithm for k–MCSP that is based on the Hitting Set problem. In Section 3 we describe how to modify the algorithm to get an $O(k)$ approximation for k-MCSP. In Section 4 we deal with the running time of the algorithm and we show how to implement the algorithm in linear time, using the suffix tree data structure. Finally, Section 5 describes how to modify the algorithm so that it works also for the signed and reversed variants of MCSP and thus, for signed and unsigned SBR.

2   Common Partition Via Hitting Set

In *Minimum Hitting Set Problem*, we are given a set U and a collection $\mathcal{S}$ of subsets of U, that is, $\mathcal{S} = \{S_1, \ldots, S_k\}$ such that $S_i \subseteq U$ for $i = 1, \ldots, k$. The task is to find a *minimum hitting set* for $\mathcal{S}$ which is a smallest set $H \subseteq U$ such that $H \cap S_i \neq \emptyset$ for each $i \in 1, \ldots, k$. Minimum Hitting Set problem is equivalent to Minimum Set Cover [1].

We are going to use an algorithm for Minimum Hitting Set Problem as a procedure for MCSP. The idea behind the algorithm is simple. Given the strings A and B and a string X such that the number of occurrences of X in A is larger (or smaller, resp.) than the number of occurrences of X in B, we know that even in the minimum common partition of A and B at least one duo in (an occurrence of) X in A (or in B, resp.) must be broken. The algorithm aims at "hitting" (that is, cutting) all substrings of A and B that have a different number of occurrences. This motivates the following definition.

For two strings A and X, let $\#\mathsf{substr}(A, X)$ be the number of all occurrences of the substring X in the string A. For a partition $\mathcal{P} = (P_1, P_2, \ldots, P_m)$ and a string X, we denote by $\#\mathsf{blocks}(\mathcal{P}, X)$ the number of blocks $P_i = X$ in $\mathcal{P}$.

Algorithm HS

input: strings A, B
construct an instance $(U, \mathcal{S})$ of the Hitting Set problem:
   $U \leftarrow duos(A) \cup duos(B)$
   $T \leftarrow \{X \in \Sigma^* \mid \#\mathsf{substr}(A, X) \neq \#\mathsf{substr}(B, X)\}$
   $\mathcal{S} \leftarrow \{duos(X) \mid X \in T\}$
solve (approximately) the Minimum Hitting Set problem:
   $\Phi \leftarrow$ a hitting set for $(U, \mathcal{S})$
transform the hitting set into a common partition:
   $\mathcal{A}, \mathcal{B} \leftarrow$ for each duo $xy \in \Phi$, cut all occurrences of xy in the strings A, B
output: $(\mathcal{A}, \mathcal{B})$

Lemma 1. *The partition* $(\mathcal{A}, \mathcal{B})$ *computed by the algorithm* HS *is a common partition of the strings A and B.*

Proof. The proof is by contradiction. Suppose that there exists a block $X \in \mathcal{A}$ such that $\#\mathsf{blocks}(\mathcal{A}, X) \neq \#\mathsf{blocks}(\mathcal{B}, X)$; if there are several such blocks, take as X the longest one. Since the block X is not cut by any duo from Φ we have $duos(X) \cap \Phi = \emptyset$, and since Φ is a correct answer for the Hitting Set problem, it holds that $duos(X) \notin S$. We conclude that $\#\mathsf{substr}(A, X) = \#\mathsf{substr}(B, X)$. We aim to get a contradiction by inferring an equality for $\#\mathsf{blocks}(\mathcal{A}, X)$ and $\#\mathsf{blocks}(\mathcal{B}, X)$.

Exploiting the fact that X is not cut by any duo from Φ, it is possible to calculate the numbers $\#\mathsf{blocks}(\mathcal{A}, X)$ and $\#\mathsf{blocks}(\mathcal{B}, X)$ by the following formula (by $X \sqsubseteq Y$ we denote that X is a substring of Y and by $X \sqsubset Y$ that X is a proper substring of Y):

$$\#\mathsf{blocks}(\mathcal{A}, X) = \#\mathsf{substr}(A, X) - \sum_{Y \sqsubseteq A, X \sqsubset Y} \#\mathsf{substr}(Y, X) \cdot \#\mathsf{blocks}(\mathcal{A}, Y)$$

$$\#\mathsf{blocks}(\mathcal{B}, X) = \#\mathsf{substr}(B, X) - \sum_{Y \sqsubseteq B, X \sqsubset Y} \#\mathsf{substr}(Y, X) \cdot \#\mathsf{blocks}(\mathcal{B}, Y)$$

By our choice, X is the longest block with $\#\mathsf{blocks}(\mathcal{A}, X) \neq \#\mathsf{blocks}(\mathcal{B}, X)$ (informally, a "wrong" block); therefore for all strings Y satisfying $X \sqsubset Y$ we have $\#\mathsf{blocks}(\mathcal{A}, Y) = \#\mathsf{blocks}(\mathcal{B}, Y)$. We conclude that $\#\mathsf{blocks}(\mathcal{A}, X) = \#\mathsf{blocks}(\mathcal{B}, X)$, which is a contradiction.

Lemma 2. *The algorithm HS finds a $2k$-approximation of the minimum common partition (if an exact procedure for a minimum hiting set is available).*

Proof. Consider any common partition $\mathcal{A}', \mathcal{B}'$ of A and B. Then, every duo in a minimum hitting set for the instance (U, S) must appear as a broken duo in $\mathcal{A}'$ or $\mathcal{B}'$. That is, (half of) the size of the minimum hitting set is a lower bound on the size of the minimum common partition. Observing that the algorithm cuts at most k duos for each duo in the set Φ, the claim follows.

Observe that by replacing the optimal procedure for Minimum Hitting Set by an α-approximation procedure, the algorithm HS finds a $2k\alpha$-approximation of the minimum common partition.

Unfortunately, Minimum Hitting Set problem is hard to approximate; to achieve a good approximation ratio, we need to investigate special properties of the instance (U, S). This is the subject of the next section.

3 $O(k)$-Approximation Ratio for MCSP

Let $(\mathcal{A}_o, \mathcal{B}_o)$ denote a minimum common partition of strings A and B (if there are several minimum common partitions, we choose any of them); we say that the breaks in $\mathcal{A}_o$ and $\mathcal{B}_o$ are the *optimal breaks*. There are $2|\mathcal{A}_o| - 2$ optimal breaks. We say that a substring $X = a_i \dots a_j$ (resp., $X = b_i \dots b_j$) *goes over* an optimal break if there exists an optimal break $l, l+1$ in $\mathcal{A}_o$ (resp., in $\mathcal{B}_o$) such that $i \leq l < j$.

Recall the definition of the set $T = \{X \in \Sigma^* \mid \#\mathsf{substr}(A, X) \neq \#\mathsf{substr}(B, X)\}$; informally, T is the set of all wrong substrings. Note that in the instance of the Hitting Set problem, most of the substrings in T are redunadant. To be more specific, if $X, Y \in T$ and X is a proper substring of Y, then we can remove Y from the set T and a hitting set for $\{duos(X) \mid X \in T \setminus \{Y\}\}$ will still be a hitting set for $\mathcal{S}$. Using this observation it is possible to substantially reduce the size of the set $\mathcal{S}$. In particular, the relation $\sqsubseteq$ induces a partial order on the set T; let $T_{\min} \subseteq T$ be the set of all minimal elements of T, with respect to the relation $\sqsubseteq$. Then $T_{\min}$ satisfies the desired property

(P) if $X, Y \in T$, and X is a proper substring of Y, then $Y \notin T_{\min}$,

and, at the same time, a hitting set for the set $\mathcal{S}' = \{duos(X) \mid X \in T_{\min}\}$ is a hitting set for $\mathcal{S}$.

Lemma 3. *If $X \in T_{\min}$ then there exists an occurrence of X in A or in B that goes over an optimal break.*

Proof. Consider a string $X \in T_{\min}$ and suppose that no occurrence of X in A and B goes over an optimal break. Then every occurrence of X in A or B is a substring of some block in the minimum common partition $(\mathcal{A}_o, \mathcal{B}_o)$. Since $\mathcal{A}_o$ and $\mathcal{B}_o$ consists of the same multiset of blocks and no occurrence of X goes over an optimal break, we have $\#\mathsf{substr}(A, X) = \#\mathsf{substr}(B, X)$. This implies $X \notin T$, which is a contradiction.

Using the lemma, we assign to each string in $T_{\min}$ an optimal break. In particular, for $X \in T_{\min}$, let $f(X)$ denote the optimal break that an occurrence of X in A or in B goes over; if there is more than one such optimal break, we choose an arbitrary one.

Example: For $A = abaab$, $B = ababa$, the minimum common partition is $(aba, ab), (ab, aba)$, $ba \in T_{min}$ and $f(ba) =$ the break $2, 3$ in the partition of B.

Lemma 4. *If $X, Y \in T_{\min}$, $X = x_1, \ldots, x_l$ and $f(X) = f(Y)$, then $duos(Y) \cap \{x_1 x_2, \ x_{l-1} x_l\} \neq \emptyset$.*

Proof. Since X and Y go over the same optimal break, their overlap has size at least two. Moreover, since X is not a proper substring of Y and vice versa (by property (P) and the assumptions of the lemma), the claim follows (cf. Figure 1).

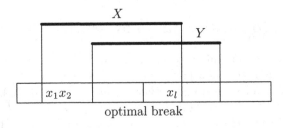

optimal break

Fig. 1. Illustration of Lemma 4

The consequence of Lemma 4 is the following. Let $\mathcal{A}$ be a partition of A and $\mathcal{B}$ be a partition of B and let $X = x_1 \ldots x_l$ be a common substring of $\mathcal{A}$ and $\mathcal{B}$ such that $X \in T_{\min}$. Then, by cutting all occurrences of $x_1 x_2$ and $x_{l-1} x_l$ in $\mathcal{A}$ and $\mathcal{B}$ we "hit" (that is, we cut) also (a duo in) each string from $T_{\min}$ that goes over the optimal break $f(X)$. Thus, if we choose for each optimal cut one string from T_{min} that goes over it (if there is any such string for the cut; if there is no such string, we ignore this cut) and put together the first and the last duos of each such string, then we get a hitting set for $T_{\min}$ of size at most twice the size of the minimum hitting set. Of course, we do not know the optimal breaks so we have to construct the hitting set in a different way. The following algorithm does it:

Algorithm FAST HS

input: strings A, B
    compute a set T' such that $T_{\min} \subseteq T'$ and T' is of size $O(n)$
    $\Phi \leftarrow \emptyset$
    $\mathcal{A} \leftarrow (A), \mathcal{B} \leftarrow (B)$
    for each $X \in T'$ in order of increasing length **do**
        if $duos(X) \cap \Phi = \emptyset$ **then**
            add the first and last duo of X to Φ
            cut all occurrences of the first and last duo of X in the partitions $\mathcal{A}, \mathcal{B}$
output: $(\mathcal{A}, \mathcal{B})$

Lemma 5. *If a string X passes the test $duos(X) \cap \Phi = \emptyset$ in the above algorithm, then $X \in T_{\min}$.*

Proof. Suppose, for a contradiction, that X passed the test yet $X \notin T_{min}$. Let Φ' denote the set Φ just before processing the string X. The assumption $X \notin T_{min}$ implies that there exists a string $X' \in T_{\min}$ such that X' is a proper substring of X. Since $|X'| < |X|$, the string X' has been processed before the string X and therefore $duos(X') \cap \Phi' \neq \emptyset$. Moreover, since $duos(X') \subseteq duos(X)$, it holds that $duos(X) \cap \Phi' \neq \emptyset$, and therefore X cannot pass the test, which is a contradiction.

Theorem 1. *The algorithm* FAST HS *computes a 4k-approximation of the minimum common partition of A and B.*

Proof. If X_1, X_2 are two different strings for which the set Φ was increased then, by Lemma 4, $f(X_1) \neq f(X_2)$. Thus, the set Φ was increased at most $|\mathcal{A}_o| + |\mathcal{B}_o| - 2$ times and therefore the final set Φ contains at most $2 \cdot (|\mathcal{A}_o| + |\mathcal{B}_o| - 2)$ duos.

Since we are dealing with an instance of k-MCSP, each duo from the set Φ introduces at most k cuts. It follows that

$$|\mathcal{A}| \leq k \cdot 2 \cdot (|\mathcal{A}_o| + |\mathcal{B}_o| - 2) + 1 \leq 4k \cdot |\mathcal{A}_o| \ .$$

Remark: The approximation ratio applies even if we measure the size of a common partition not by the number of blocks but by the number of breaks.

Lower bound. Let $A = ba\{ab\}^{k-1}$ and $B = \{ab\}^k$. Then the set Φ consists of two duos $\{aa, ab\}$ and the partition computed by the algorithm FAST HS has size $k+1$ while the minimum common partition has size 3. Thus, the apporximation ratio of the algorithm FAST HS is $\Omega(k)$.

4   Linear Running Time

We are going to describe how to implement the algorithm in linear time. The linear implementation heavily uses the *suffix tree* data structure and the fact that a suffix tree of a string of length m can be constructed in time $O(m)$ for constant size alphabets [12] and even for integer alphabets [7].

We start with the construction of the set T'. Let $ and # be two characters that do not appear in A. We compute the suffix tree τ of the string $C = A\$B\#$. Recall that each leaf of the tree τ corresponds to a suffix of C. We mark by A each leaf of τ that corresponds to a suffix starting in the substring A of C, and we mark by B each leaf of τ that corresponds to a suffix starting in the substring B of C. For each node v of τ we compute the number $numA(v)$ of leaves in the subtree of v marked by A and the number $numB(v)$ of leaves in the subtree of v marked by B; this requires time $O(n)$, for strings A, B of length n. For a node v of τ, let $s(v)$ denote the concatenation of the labels of the edges between the root and the node v and, for $v \neq root$, let $s'(v)$ denote the concatenation of $s(parent(v))$ with the first character of the label of the edge $(parent(v), v)$. If $s'(v)$ does not contain the characters $ and # we say that v is a *proper* node. Observe that for each proper node v, $numA(v) = \#\mathsf{substr}(A, s'(v))$ and $numB(v) = \#\mathsf{substr}(B, s'(v))$. Thus, if $numA(v) \neq numB(v)$ we know that $s'(v) \in T$. Once we have the suffix tree τ and the values $numA(v)$ and $numB(v)$ for all vertices, we easily compute a set

$$T' = \{s'(v) \mid v \text{ is a proper node and } numA(v) \neq numB(v)\}$$

by traversing the tree τ in, say, breadth first search order. The set T' can be computed in $O(n)$ running time. It is also easy to observe that, the size of T' is bounded by $O(n)$ (since the suffix tree consist of $O(n)$ nodes). We also note that for each string $X \in T_{\min}$ there is a proper node v such that $s'(v) = X$ and $numA(v) \neq numB(v)$ which guarantees that $T_{\min} \subseteq T'$.

To give an example, consider strings $A = abaab$ and $B = ababa$. The suffix tree of the string $C = A\$B\#$ is given in Figure 2 and the relevant sets are as follows:

$$T' = \{aa, aba, abaa, abab, ba, baa, bab\}$$
$$T_{\min} = \{aa, ba\}$$
$$\Phi = \{aa, ba\}$$
$$\mathcal{A} = (ab, a, ab)$$
$$\mathcal{B} = (ab, ab, a)$$

To finish the description of the fast implementation of the algorithm, it remains to describe how to maintain the set Φ, how to test the condition

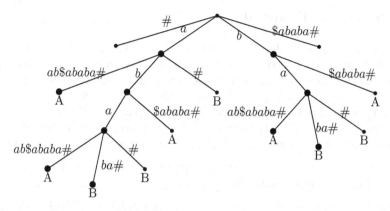

Fig. 2. Suffix tree τ of the string $C = abaab\$ababa\#$. The larger dots denote the proper nodes.

$duos(X) \cap \Phi \neq \emptyset$ and how to realize the cuts. We employ a data structure for the *set–splitting problem* [8]. In this problem, we are given a set consisting of the integers $\{1, \ldots, m\}$ and the task is to perform an intermixed sequence of the following two operations:

- $split(i)$ – splits the set containing i into two sets, one with all integers smaller than i and the other with all integers greater than or equal to i,
- $find(i)$ – returns the smallest integer in the set containing i.

Gabow and Tarjan [8] describe a data structure that requires $O(1)$ amortized time for each operation. In our setting, we maintain for each partition $\mathcal{A}$ and $\mathcal{B}$ a separate data structure that stores information about cuts in that partition. Initially, each structure consists of only one set, the set $\{1, \ldots, n\}$. Each time when we add a duo cd to Φ we perform the cuts of the partitions $\mathcal{A}$ and $\mathcal{B}$ as follows:

for each occurrence of the duo cd in $\mathcal{A}$ **do**
$\quad A.split(j+1)$, where j is the position of the current occurrence cd in A (i.e.,
$\quad a_j a_{j+1} = cd$)
for each occurrence of the duo cd in $\mathcal{B}$ **do**
$\quad B.split(j+1)$, where j is the position of the current occurrence cd in B (i.e.,
$\quad b_j b_{j+1} = cd$)

Since every duo appearing in A and B is processed at most once by the algorithm the total number of *split* operations is at most $O(n)$.

For an occurrence $a_i \ldots a_j = X$ (resp., $b_i, \ldots, b_j = X$) of the substring $X \in T'$, it holds that $duos(X) \cap \Phi = \emptyset$ if and only if $A.find(i) = A.find(j)$ (resp., $B.find(i) = B.find(j)$). This provides a way for testing the condition $duos(X) \cap \Phi \neq \emptyset$ in constant time.

Theorem 2. *The above implementation of the algorithm* FAST HS *runs in linear time.*

5   Sorting by Reversals

One can easily modify the algorithms HS and FAST HS to work also for the relation $\cong$, for both signed and unsigned strings. We redefine #substr(A, S) so that it counts occurrences of both S and $-S$ in A; the definitions of the sets T, T' and T_{min} remain unchanged. The new definition of #substr requires a small change in the computation of the set T': we compute a suffix tree of the string $C = A\#B\$(-A)\#(-B)\$$ (the brackets are only used to denote the scope of the reversal operation). We also need a slight change in Lemma 3 and Lemma 4:

Lemma 3a. If $X \in T_{min}$ then there exists an occurrence of X or $-X$ in A or in B that goes over an optimal break.

Lemma 4a. If $X, Y \in T_{min}$, $X = x_1, \ldots, x_l$ and $f(X) = f(Y)$, then $duos(Y) \cap \{x_1 x_2, \ x_{l-1} x_l, \ -(x_1 x_2), \ -(x_{l-1} x_l)\} \neq \emptyset$.

Finally, whenever the original algorithm cuts duos xy, the modified algorithm also cuts duos $-(xy)$. This increases the approximation ratio by a factor of two.

Theorem 3. *The algorithm* FAST HS *computes in linear time* $\Theta(k)$-*approximation for signed, unsigned and reversed k-MCSP and for signed and unsigned k-SBR.*

6   Conclusion

We presented $\Theta(k)$-approximation algorithms for signed and unsigned k-MCSP and k-SBR, running in time $O(n)$. A challenging open question is whether it is possible to get a nontrivial approximation ratio independent of the parameter k (or at least less dependent, say an approximation ratio $O(\log k)$).

References

1. G. Ausiello, A. D'Atri, and M. Protasi. Structure preserving reductions among convex optimization problems. *Journal of Computer and System Sciences*, 21(1):136–153, 1980.
2. A. Caprara. Sorting permutations by reversals and Eulerian cycle decompositions. *SIAM Journal on Discrete Mathematics*, 12(1):91–110, 1999.
3. X. Chen, J. Zheng, Z. Fu, P. Nan, Y. Zhong, S. Lonardi, and T. Jiang. Assignment of orthologous genes via genome rearrangement. *IEEE/ACM Transactions on Computational Biology and Bioinformatics*, 2(4):302–315, 2005.
4. D. A. Christie and R. W. Irving. Sorting strings by reversals and by transpositions. *SIAM Journal on Discrete Mathematics*, 14(2):193–206, 2001.
5. M. Chrobak, P. Kolman, and J. Sgall. The greedy algorithm for the minimum common string partition problem. *ACM Transactions on Algorithms*, 1(2):350–366, 2005.
6. G. Cormode and S. Muthukrishnan. The string edit distance matching problem with moves. In *Proceedings of the 13th Annual ACM-SIAM Symposium On Discrete Mathematics (SODA)*, pages 667–676, 2002.

7. M. Farach. Optimal suffix tree construction with large alphabets. In *Proceedings of the 38th Annual Symposium on Foundations of Computer Science (FOCS)*, pages 137–143, 1997.
8. H. N. Gabow and R. E. Tarjan. A linear-lime algorithm for a special case of disjoint set union. *Journal of Computer and System Sciences*, 30(2):209–221, 1985.
9. A. Goldstein, P. Kolman, and J. Zheng. Minimum Common String Partition Problem: Hardness and Approximations. *The Electronic Journal of Combinatorics*, 12(1), 2005.
10. S. Hannenhalli and P. A. Pevzner. Transforming cabbage into turnip: polynomial algorithm for sorting signed permutations by reversals. *Journal of the ACM*, 46(1):1–27, 1999.
11. P. Kolman. Approximating reversal distance for strings with bounded number of duplicates. In *Proceedings of the 30th International Symposium on Mathematical Foundations of Computer Science (MFCS)*, volume 3618 of *Lecture Notes in Computer Science*, pages 580–590, 2005.
12. P. Weiner. Linear pattern matching algorithms. In *14th IEEE Symposium on switching and automata theory*, pages 1–11, 1973.

Approximating the Unweighted k-Set Cover Problem: Greedy Meets Local Search

Asaf Levin

Department of Statistics, The Hebrew University, Jerusalem, Israel
levinas@mscc.huji.ac.il

Abstract. In the unweighted set-cover problem we are given a set of elements $E = \{e_1, e_2, \ldots, e_n\}$ and a collection $\mathcal{F}$ of subsets of E. The problem is to compute a sub-collection $SOL \subseteq \mathcal{F}$ such that $\bigcup_{S_j \in SOL} S_j = E$ and its size $|SOL|$ is minimized. When $|S| \leq k$ for all $S \in \mathcal{F}$ we obtain the unweighted k-set cover problem. It is well known that the greedy algorithm is an H_k-approximation algorithm for the unweighted k-set cover, where $H_k = \sum_{i=1}^{k} \frac{1}{i}$ is the k-th harmonic number, and that this bound on the approximation ratio of the greedy algorithm, is tight for all constant values of k. Since the set cover problem is a fundamental problem, there is an ongoing research effort to improve this approximation ratio using modifications of the greedy algorithm. The previous best improvement of the greedy algorithm is an $\left(H_k - \frac{1}{2}\right)$-approximation algorithm. In this paper we present a new $\left(H_k - \frac{196}{390}\right)$-approximation algorithm for $k \geq 4$ that improves the previous best approximation ratio for all values of $k \geq 4$. Our algorithm is based on combining local search during various stages of the greedy algorithm.

1 Introduction

In the WEIGHTED SET-COVER PROBLEM we are given a set of elements $E = \{e_1, e_2, \ldots, e_n\}$ and a collection $\mathcal{F}$ of subsets of E, where $\cup_{S \in \mathcal{F}} S = E$ and each $S \in \mathcal{F}$ has a positive cost c_S. The goal is to compute a sub-collection $SOL \subseteq \mathcal{F}$ such that $\bigcup_{S \in SOL} S = E$ and its cost $\sum_{S \in SOL} c_S$ is minimized. Such a sub-collection of subsets is called a *cover*. When we consider instances of the WEIGHTED SET-COVER such that each S_j has at most k elements ($|S| \leq k$ for all $S \in \mathcal{F}$), we obtain the WEIGHTED k-SET COVER PROBLEM. The UNWEIGHTED SET COVER PROBLEM and the UNWEIGHTED k-SET COVER PROBLEM are the special cases of the WEIGHTED SET COVER and of WEIGHTED k-SET COVER, respectively, where $c_S = 1 \ \forall S \in \mathcal{F}$.

It is well known (see [2]) that a greedy algorithm is an H_k-approximation algorithm for the weighted k-set cover, where $H_k = \sum_{i=1}^{k} \frac{1}{i}$ is the k-th harmonic number, and that this bound is tight even for the unweighted k-set cover problem (see, [12,15]). For unbounded values of k, Slavík [19] showed that the approximation ratio of the greedy algorithm for the unweighted set cover problem is $\ln n - \ln \ln n + \Theta(1)$. Feige [5] proved that unless $NP \subseteq DTIME(n^{polylog\ n})$ the unweighted set cover problem cannot be approximated within a factor $(1-\epsilon) \ln n$,

T. Erlebach and C. Kaklamanis (Eds.): WAOA 2006, LNCS 4368, pp. 290–301, 2006.

for any $\epsilon > 0$. Raz and Safra [18] proved that if $P \neq NP$ then for some constant c, the unweighted set cover problem cannot be approximated within a factor $c \log n$. This result shows that the greedy algorithm is an asymptotically best possible approximation algorithm for the weighted and unweighted set cover problem (unless $NP \subseteq DTIME(n^{polylog\ n})$). The unweighted k-set cover problem is known to be NP-complete [13] and MAX SNP-hard for all $k \geq 3$ [3,14,16]. Another algorithm for the weighted set cover problem by Hochbaum [10] has an approximation ratio that depends on the maximum number of subsets that contain any given item (the local-ratio algorithm of Bar-Yehuda and Even [1] has the same performance guarantee). See Paschos [17] for a survey on these results.

In spite of the above bad news Goldschmidt, Hochbaum and Yu [7] modified the greedy algorithm for the unweighted set cover and showed that the resulting algorithm has a performance guarantee of $H_k - \frac{1}{6}$. Halldórsson [8] presented an algorithm based on local search that has an approximation ratio of $H_k - \frac{1}{3}$ for the unweighted k-set cover, and a $(1.4 + \epsilon)$-approximation algorithm for the unweighted 3-set cover. Duh and Fürer [4] further improved this result and presented an $(H_k - \frac{1}{2})$-approximation algorithm for the unweighted k-set cover. We will base our algorithm on the algorithm of Duh and Fürer [4], and therefore we will review their algorithm and results in Section 3. All of these improvements [7,8,4] are based on running the greedy algorithm until each new subset covers at most t new elements (where $t = 2$ in [7] and larger values of t in [8,4]) and then switch to another algorithm.

Regarding approximation algorithms for the weighted k-set cover problem within a factor better than H_k, a first improvement step was given by Fujito and Okumura [6] who presented $H_k - \frac{1}{12}$-approximation algorithm for the k-set cover problem where the cost of each subset is either 1 or 2. More recently, Hassin and Levin [9] provided an $\left(H_k - \frac{k-1}{8k^9}\right)$-approximation algorithm.

The MAXIMUM SET PACKING PROBLEM is the following related problem: We are given a set of elements $E = \{e_1, e_2, \ldots, e_n\}$ and a collection $\mathcal{F}$ of subsets of E, where $\cup_{S \in \mathcal{F}} S = E$, and the goal is to compute a maximum size set packing, i.e., a sub-collection $\mathcal{F}' \subseteq \mathcal{F}$ of disjoint subsets. Hurkens and Schrijver [11] proved that a local-search algorithm for the maximum set packing problem where each subset in $\mathcal{F}$ has at most k elements, is a $\left(\frac{2}{k} - \epsilon\right)$-approximation algorithm. Therefore, this local-search algorithm has a better performance guarantee than the greedy selection rule that returns any maximal sub-collection. The greedy selection rule has an approximation ratio of $\frac{1}{k}$.

We first observe that w.l.o.g., there is an optimal solution to the set cover problem such that each element is covered by exactly one subset of the optimum: Let an optimal solution to the problem consist of a collection of sets S_j^*, $j \in J^*$, with $\cup_{j \in J^*} S_j^* = E$. We now construct another optimal solution formed of element-disjoint sets S_j' where $S_j' \subseteq S_j^*$ for all $j \in J^*$. To do that, we assign each element $e \in E$ to the smallest index set S_j^*, $j \in J^*$ that contains e. We modify the instance by adding the sets S_j' for all j to the collection $\mathcal{F}$. If the algorithm or the optimal solution decides to pick such a set S_j', we interpret this as picking the set S_j^*. Henceforth, any optimal solution will be considered to have this disjointness property, so each $e \in E^*$ belongs to exactly one set S_j^*. We consider an optimal solution OPT that satisfies the disjointness property.

We define a *j-set* to be a set with j elements. We fix an optimal solution OPT, and we say that a k-set is an *optimal k-set* if it is contained in OPT.

Paper overview: In Section 2 we review the greedy algorithm for the unweighted minimum k-set cover problem, and its analysis. In Section 3 we review the semi-local optimization algorithm of [4]. In Section 4 we present our improved algorithm. We analyze its performance in Section 5, i.e., we show that our algorithm is an $\left(H_k - \frac{196}{390}\right)$-approximation algorithm for the unweighted k-set cover problem where $k \geq 4$, improving the earlier $\left(H_k - \frac{1}{2}\right)$-approximation algorithm of [4]. We conclude in Section 6 by discussing open questions.

2   The Greedy Algorithm

In this section we review the greedy algorithm for the unweighted k-set cover problem and the proof of its performance guarantee.

The greedy algorithm starts with an empty collection of subsets in the solution and no item being covered. Then, it iterates the following procedure until all items are covered:

Let w_S be the number of uncovered items in a set $S \in \mathcal{F}$, and the current *ratio of S* is $r_S = \frac{1}{w_S}$. Let S^* be a set such that r_{S^*} is minimized. The algorithm adds S^* to the collection of subsets of the solution, defines the items of S^* as covered, and assigns a *price* of r_{S^*} to all the items that are now covered but were uncovered prior to this iteration (i.e., the items that were first covered by S^*).

Johnson [12], Lovász [15] and Chvátal [2] showed that the greedy algorithm is an H_k-approximation algorithm for the unweighted k-set cover.

Chvátal's proof is the following: first note that the cost of the greedy solution equals the sum of prices assigned to the items. Second, consider a set S that belongs to an optimal solution OPT. Then, OPT pays 1 for S. When the i-th item of S is covered by the greedy algorithm, the algorithm could choose S as a feasible set with a current ratio of $\frac{1}{|S|-i+1}$. Therefore, the price assigned to the this item is at most $\frac{1}{|S|-i+1}$. It follows that the total price assigned to the items of S is at most $\sum_{i=1}^{|S|} \frac{1}{|S|-i+1} = \sum_{i'=1}^{|S|} \frac{1}{i'} \leq H_k$, and therefore, the approximation ratio of the greedy algorithm is at most H_k.

3   The Semi-local Optimization Algorithm

Duh and Fürer [4] suggested the following procedure to approximate the unweighted 3-set cover problem. In a pure local improvement step, we replace a number of sets with fewer sets to form a new cover with a reduced cost. To define a semi-local step, they observed that once the 3-sets are selected the remaining instance can be solved optimally in polynomial time. Thus a local change in the

3-sets allows any global changes in the 2-sets and 1-sets and such a change is called a *semi-local change*.

They allowed the algorithm to remove one 3-set and insert at most a pair of 3-sets if one of the following happens: either the total cost is reduced, or the total cost remains the same and the number of 1-sets in the resulting solution is reduced (thus the total cost is the primary objective whereas the number of 1-sets is a secondary objective). This results in the approximation algorithm for the unweighted 3-set cover of [4]. They showed that this is a $\frac{4}{3}$-approximation algorithm. More precisely, the following proposition was proved in [4].

Proposition 1. *Assume that an optimal solution for the unweighted 3-set cover instance has b_1, b_2, and b_3 1-sets, 2-sets and 3-sets, respectively. Then the solution that the semi-local optimization algorithm returns, costs at most $b_1 + b_2 + \frac{4}{3}b_3$. Moreover, the number of 1-sets in the solution that the algorithm returns, is at most b_1.*

In order to extend their result to larger values of k, they suggested the following algorithm:

1. **Greedy Phase.** For $j = k$ down to 6 do:
   greedily choose a maximal collection of j-sets.
2. **Restricted Phase.** For $j = 5$ down to 4 do:
   choose a maximal collection of j-sets with the restriction that the choice of these j-sets does not increase the number of 1-sets.
3. **Semi-local Optimization Phase.** Run the semi-local optimization algorithm on the remaining instance.

Note that the following question is answered within polynomial time during the Restricted phase: Does the addition of a j-set S to the current solution increase the number of 1-sets in the resulting solution (returned by the algorithm)? Duh and Fürer proved that this algorithm is an $\left(H_k - \frac{1}{2}\right)$-approximation, and they also showed that this bound is tight for the semi-local optimization algorithm.

4 The Algorithm

In this section we present our modification of the semi-local optimization algorithm where we use a local-search algorithm during the phase where each new set covers exactly four previously uncovered elements.

Algorithm A

1. **Greedy Phase.** For $j = k$ down to 6 do:
   greedily choose a maximal collection of disjoint j-sets (each covering exactly j new elements).
2. **Restricted Phase.** Choose a maximal collection of disjoint 5-sets with the restriction that the choice of these 5-sets does not increase the number of 1-sets.

3. **Local-Search Phase.** Choose a collection of disjoint 4-sets such that the choice of these 4-sets does not increase the number of 1-sets and this collection has a local maximum size. The requirement of local maximum size means that removing a 4-set from this collection does not allow us to add at least a pair of 4-sets (without increasing the number of 1-sets).

4. **Semi-Local Optimization Phase.** Run the semi-local optimization algorithm on the remaining instance.

In Phase 3 we are using local-search whose neighborhood is defined by removing one 4-set and inserting at least a pair of 4-sets as long as the number of 1-sets in the returned solution does not increase. The use of this local-search procedure is motivated by the approximation algorithm of [11] for the maximum set packing problem. This improved phase is the corner stone on which our improved approximation ratio is based.

Each iteration takes polynomial time because checking whether the number of 1-sets in the resulting solution increases, takes polynomial time. Therefore, Algorithm A is a polynomial time algorithm that returns a feasible solution. Therefore, we establish the following lemma:

Lemma 1. *For every value of k, Algorithm A returns a feasible solution in polynomial time.*

In the next section we analyze the performance guarantee of Algorithm A.

5   The Analysis of Algorithm A

In this section we analyze the performance guarantee of Algorithm A. We consider an optimal solution OPT, and bound the performance guarantee of A. Recall that we assume that OPT is a partition of the element set E. We now further characterize the structure of OPT.

Lemma 2. *W.l.o.g., each set of OPT is a k-set.*

Proof. Assume that the claim does not hold on an instance I. We create a new instance I' such that the optimal solution OPT' for I' costs k times the cost of OPT, and the solution returned by A on I' costs more than k times the solution returned by algorithm A on I, and we will conclude that if there is a bad example for the algorithm there is a bad example for the algorithm that shows the same approximation ratio such that the property of the lemma holds.

To do so we first take k disjoint copies of the instance I. Clearly, the optimal solution $O\hat{P}T$ for this new instance costs exactly k times the cost of OPT, and it is a union of k copies of OPT. Then, we add new elements to $O\hat{P}T$'s existing sets so that each set in this sub-collection is a k-set. Note that the number of the new elements is divisible by k. Last, we add new k-sets of these new elements, such that the algorithm picks these new k-sets (of the new elements) in its first steps, and then continue like it acts on I on each of the k copies of I. Therefore, OPT' costs exactly k times the cost of OPT (we can use the sets of $O\hat{P}T$ that we increased), however the cost of the solution returned by A on I' is strictly larger than k times the cost of the solution returned by A on I. □

Next, we allocate a price for each element in the following way:

- For an element that is covered by an i-set during Phases 1, 2 and 3 of Algorithm A, we allocate a price of $\frac{1}{i}$.
- We consider special 2-sets and 3-sets that are named *sibling 2-sets* defined as follows (see [4] for introduction of this term): a 2-set or a 3-set S chosen by the semi-local optimization phase such that one of the elements in S is the last uncovered element of an optimal k-set (this element is called the *primary element*) and the remaining elements of S belong to a common optimal k-set (i.e., to the same set in the optimal solution). The elements of a sibling 2-set that are not primary are called *secondary elements*. A sibling 2-set is the result of the fact that the Semi-Local Optimization phase does not create a new singleton, and therefore, if an optimal k-set has $k-1$ covered elements at the end of Phase 3 out of which at least one is covered during Phases 2 or 3, then the last element belongs to at least a 2-set (and is not a singleton). We allocate a cost of $\alpha = \frac{4}{5}$ for the primary element of a sibling 2-set, and for each of its secondary elements we allocate a cost of $1 - \alpha = \frac{1}{5}$.
- For the other elements that we cover during Phase 4, we assign at most a unit price for each selected set such that the following holds (such an allocation of prices is feasible according to Proposition 1):
  • For each three elements that belong to a common k-set of OPT, are covered during Phase 4, and do not intersect with a sibling 2-set, we assign a total price of $\frac{4}{3}$.
  • For each pair of elements that belong to a common k-set of OPT, are covered during Phase 4, and do not intersect with a sibling 2-set, we assign a total price of one unit.

By the allocation of the prices and Proposition 1, we conclude the following lemma:

Lemma 3. *The cost of the solution returned by Algorithm A is at most the total price of all the elements.*

Next, we define a *bad set*. Given an optimal k-set S, if $k \geq 5$, then S is a *bad set* if at the end of Phase 1 S has exactly five uncovered elements from which exactly one element is covered during Phase 3 and one of the following holds: Either exactly one element of S is covered during Phase 2 and none of the three remaining elements belongs to a sibling 2-set, or none of the elements of S is covered during Phase 2 and exactly one element of S belongs to a sibling 2-set. If $k = 4$, then S is a bad set if exactly one of its elements is covered during Phase 3. An optimal k-set that is not bad is a *good set*.

The outline of the proof of our improved approximation ratio is as follows: in Lemma 4 we will prove that the total price of an optimal set is better than $H_k - \frac{1}{2}$ if the optimal set is good and it equals $H_k - \frac{1}{2}$ for bad sets. Afterwards, in Lemma 5 we will show that there is a constant proportion of the optimal sets that have to be good sets. Combining the two results together we will establish our improved approximation ratio.

Lemma 4. *Assume that $k \geq 4$. The total price assigned to an optimal bad k-set is at most $\rho_b = H_k - \frac{1}{2}$. The total price assigned to an optimal good k-set is at most $\rho_g = H_k - \frac{16}{30}$.*

Proof. Let S be an optimal k-set. Denote by $price(S)$ the total price assigned to the elements of S. First assume that S is a bad set. If $k \geq 5$, then the j-th covered element from S during Phase 1 is assigned a price of at most $\frac{1}{k-j+1}$, the element that is covered during Phase 2 is assigned a price of $\frac{1}{5}$ (if it exists), the element that belongs to a sibling 2-set is assigned a price of $\frac{1}{5}$ (if it exists), the element that is covered during Phase 3 is assigned a price of $\frac{1}{4}$, and the remaining three elements are assigned a total price of at most $\frac{4}{3}$. Hence, $price(S) \leq \sum_{i=6}^{k} \frac{1}{i} + \frac{1}{5} + \frac{1}{4} + \frac{4}{3} = H_k - \frac{1}{2} = \rho_b$. If $k = 4$, then S has a single element covered during Phase 3 that pays a price of $\frac{1}{4}$ and the three remaining elements are assigned a total price of at most $\frac{4}{3}$. So again $price(S) \leq H_k - \frac{1}{2} = \rho_b$.

It remains to prove the second part of the lemma regarding the total price of a good set. So assume that S is a good set. We denote by N_g the number of elements of S that remains uncovered at the end of Phase 1. We denote by N_r (N_l) the number of elements of S that are covered during Phase 2 (Phase 3). Our proof is based on a detailed case analysis.

First assume that $k = 4$. Then, the Greedy phase and the Restricted phase do not select sets, and therefore $N_g = 4$ and $N_r = 0$.

- Assume that $N_l = 4$. Then, each element of S is covered during Phase 3 and pays a price of $\frac{1}{4}$. Therefore, $price(S) = 1 < H_4 - \frac{16}{30} = \rho_g$.
- Assume that $N_l = 3$. Then, each element of S that is covered during Phase 3 pays a price of $\frac{1}{4}$, and the remaining element pays a price of at most $\frac{4}{5}$ (this is because since no singletons are created, this remaining element either belongs to a sibling 2-set and then it pays at most $\frac{4}{5}$, or it belongs to a 3-set and in this case it pays $\frac{1}{3}$). Therefore, $price(S) \leq \frac{3}{4} + \frac{4}{5} = \frac{93}{60} = \frac{125}{60} - \frac{32}{60} = H_4 - \frac{16}{30} = \rho_g$.
- Assume that $N_l = 2$. Then, each element of S that is covered during Phase 3 pays a price of $\frac{1}{4}$. The two remaining elements pay a total price of at most 1. Thus, $price(S) \leq \frac{3}{2} < \rho_g$.
- Assume that $N_l = 1$. Then, the element of S that is covered during Phase 3, pays a price of $\frac{1}{4}$. Since S is a good set, it contains at least one element that belongs to a sibling 2-set that pays $\frac{1}{5}$ (since $N_l = 1$, it is not the primary element). The other two elements of S have total price of at most 1. Therefore, $price(S) \leq \frac{1}{4} + \frac{1}{5} + 1 = \frac{87}{60} < \frac{93}{60} = \rho_g$.
- Assume that $N_l = 0$. Since S is not added to the solution during Phase 3, it must contain an element that belongs to a sibling 2-set, and pays a price of $\frac{1}{5}$. The other three elements pay a total price of at most $\frac{4}{3}$ (this is also the case if some of them belong to sibling 2-sets). Therefore, $price(S) \leq \frac{1}{5} + \frac{4}{3} = \frac{92}{60} < \rho_g$

It remains to consider the case where $k \geq 5$. First note that by the greedy selection rule during the greedy phase, we conclude that $N_g \leq 5$. Moreover, the j-th covered element from S during the greedy phase (for $1 \leq j \leq k - 5$) is assigned a price of at most $\frac{1}{k-j+1}$. Moreover, since the Restricted phase and the

Local-Search phase do not create new singletons, we conclude that if $N_g \geq 2$, then the maximum price of an element of S is at most $\frac{4}{5}$.

- Assume that $N_g \leq 2$. Then, the $k - 4$-th, the $k - 3$-rd, and the $k - 2$-nd covered elements from S are covered during Phase 1, and therefore assigned a price of at most $\frac{1}{6}$ for each. The last two elements of S are assigned a total price of at most $\frac{4}{5} + \frac{1}{4}$ (this is the case where one of them is covered during Phase 3 and the last element is from sibling 2-set, and the other cases cause a smaller cost). Therefore, $price(S) \leq H_k - H_5 + \frac{3}{6} + \frac{4}{5} + \frac{1}{4} = H_k - H_5 + \frac{31}{20} = H_k - \frac{137}{60} + \frac{93}{60} = H_k - \frac{44}{60} < \rho_g$.
- Assume that $N_g = 3$. Then, the $k - 4$-th and the $k - 3$-rd covered elements from S are covered during Phase 1, and therefore assigned a price of at most $\frac{1}{6}$ for each.
  - If $N_r + N_l = 0$, then the last three elements of S are covered during Phase 4, and pay a total price of at most $\frac{4}{3}$. Therefore, $price(S) \leq H_k - H_5 + \frac{2}{6} + \frac{4}{3} = H_k - \frac{137}{60} + \frac{5}{3} = H_k - \frac{37}{60} < \rho_g$.
  - If $N_r + N_l = 1$, then the last two elements of S are covered during Phase 4, and pay a total price of at most 1. The $k - 2$-nd element of S is covered during either Phase 2 or Phase 3, and so it pays a price of at most $\frac{1}{4}$. Therefore, the last three elements of S pay a total price of at most $\frac{5}{4} < \frac{4}{3}$ and again $price(S) < \rho_g$.
  - If $N_r + N_l = 2$, then the last uncovered element pays at most $\frac{4}{5}$ (if it belongs to a sibling 2-set, and otherwise it pays less). The $k - 2$-nd and the $k - 1$-st covered elements from S are covered during either Phase 2 or Phase 3, and therefore each of these is assigned a price of at most $\frac{1}{4}$. Again the last three elements of S pay at most $\frac{4}{5} + \frac{2}{4} < \frac{4}{3}$, and therefore $price(S) < \rho_g$.
  - If $N_r + N_l = 3$, then each of the last three elements of S pays a price of at most $\frac{1}{4}$, and in total they pay less than $\frac{4}{3}$. Therefore, $price(S) < \rho_g$.
- Assume that $N_g = 4$. Then, the $k - 4$-th covered element from S is covered during Phase 1, and therefore pays a price of at most $\frac{1}{6}$. Among the last four elements of S there is at least one element that pays at most $\frac{1}{4}$. To see this fact note that if none of the elements of S belong to a set that is chosen during Phase 2 or Phase 3, then S has an element that belongs to a sibling 2-set (otherwise, we add S to the solution during Phase 3 contradicting the maximality of the collection that we choose during Phase 3), and in each of these cases the element pays at most $\frac{1}{4}$. The other three elements pay at most $\max\{\frac{4}{3}, \frac{4}{4} + \frac{2}{4}\} = \frac{4}{3}$. Therefore, $price(S) \leq H_k - H_5 + \frac{1}{6} + \frac{1}{4} + \frac{4}{3} = H_k - \frac{137}{60} + \frac{105}{60} = H_k - \frac{32}{60} = \rho_g$.
- Assume that $N_g = 5$.
  - Assume that $N_r = N_l = 0$. By the maximality of the sets chosen during Phase 3, we conclude that S has at least two elements that belong to sibling 2-sets, and therefore each of these pays $\frac{1}{5}$. The other three elements of S pay at most $\frac{4}{3}$. Therefore, $price(S) \leq H_k - H_5 + \frac{2}{5} + \frac{4}{3} = H_k - \frac{137}{60} + \frac{104}{60} = H_k - \frac{33}{60} < \rho_g$.
  - Assume that $N_r = 1$ and $N_l = 0$. The element of S that is covered during Phase 2, pays a price of $\frac{1}{5}$. By the maximality of the sets chosen

during Phase 3, we conclude that S has an element that belongs to a sibling 2-set and pays $\frac{1}{5}$. The remaining three elements pay at most $\frac{4}{3}$. Therefore, $price(S) \leq H_k - H_5 + \frac{1}{5} + \frac{1}{5} + \frac{4}{3} = H_k - \frac{33}{60} < \rho_g$.

- Assume that $N_r \geq 2$. The elements of S that are covered during Phase 2 pay a price of $\frac{1}{5}$ each. The last three elements pay a total price of at most $\frac{4}{3}$. Therefore, $price(S) \leq H_k - H_5 + \frac{2}{5} + \frac{4}{3} = H_k - \frac{33}{60} < \rho_g$.

- Assume that $N_r \leq 1$ and $N_l = 1$. Since S is a good set, we conclude that either $N_r = 1$ and S has an element that belongs to a sibling 2-set, or S has at least two elements that belong to sibling 2-sets. The element of S that is covered during Phase 2 (if it exists) pays a price of $\frac{1}{5}$, the element of S that is covered during Phase 3 pays a price of $\frac{1}{4}$, and each element of S that belongs to a sibling 2-set pays $\frac{1}{5}$. The two last remaining elements have a total price of at most 1. Therefore, $price(S) \leq H_k - H_5 + \frac{1}{5} + \frac{1}{4} + \frac{1}{5} + 1 = H_k - \frac{137}{60} + \frac{99}{60} = H_k - \frac{38}{60} < \rho_g$.

- Assume that $N_r \leq 1$ and $N_l = 2$. By the maximality of the sets that we choose during Phase 2, we conclude that if $N_r = 0$ then S has an element that belongs to a sibling 2-set and pays $\frac{1}{5}$. Therefore, S has an element that pays $\frac{1}{5}$ (this is the one that is covered during Phase 2, or the one that belongs to a sibling 2-set). Each of the elements of S that is covered during Phase 3, pays a price of $\frac{1}{4}$. The two remaining elements pay a total price of 1. Therefore, $price(S) \leq H_k - H_5 + \frac{1}{5} + \frac{2}{4} + 1 = H_k - \frac{137}{60} + \frac{102}{60} = H_k - \frac{35}{60} < \rho_g$.

- Assume that $N_r \leq 1$ and $N_l = 3$. By the maximality of the sets that we choose during Phase 2, we conclude that if $N_r = 0$, then S has an element that belongs to a sibling 2-set and pays $\frac{1}{5}$. Therefore, S has an element that pays $\frac{1}{5}$ (this is the one that is covered during Phase 2, or the one that belongs to a sibling 2-set). Each of the elements of S that is covered during Phase 3, pays a price of $\frac{1}{4}$. The remaining element pays at most $\frac{4}{5}$. Therefore, $price(S) \leq H_k - H_5 + \frac{1}{5} + \frac{3}{4} + \frac{4}{5} = H_k - \frac{137}{60} + \frac{105}{60} = H_k - \frac{32}{60} = \rho_g$.

- Assume that $N_r \leq 1$ and $N_l = 4$. By the maximality of the sets that we choose during Phase 2, we conclude that if $N_r = 0$, then S has an element that belongs to a sibling 2-set and pays $\frac{1}{5}$. Therefore, S has an element that pays $\frac{1}{5}$ (this is the one that is covered during Phase 2, or the one that belongs to a sibling 2-set). Each of the elements of S that is covered during Phase 3, pays a price of $\frac{1}{4}$. Therefore, $price(S) \leq H_k - H_5 + \frac{1}{5} + \frac{4}{4} = H_k - \frac{137}{60} + \frac{72}{60} = H_k - \frac{65}{60} < \rho_g$.

$\square$

Denote by n_b the number of bad sets in OPT and by n_g the number of good sets in OPT.

Lemma 5. $n_b \leq 12 n_g$.

Proof. Consider a bad set S in OPT. At the beginning of Phase 3, S has four uncovered elements such that none of these belong to a sibling 2-set. Since S is a bad set we cover exactly one of its elements during Phase 3. Consider a set S'

chosen in Phase 3. Then, there is a good set $S'' \in OPT$ such that $S'' \cap S' \neq \emptyset$. To see this note that otherwise during Phase 3 we could replace S' by the bad sets it intersects (each such set has four elements that consist of a 4-set that we could add to the solution without increasing the number of singletons). Since we did not apply this step, we conclude that at least one of its intersecting sets from OPT is a good set.

A good set $S \in OPT$ can intersect at most four sets that we choose during Phase 3. These four sets can intersects at most 12 other sets of OPT. These 12 sets might be bad sets. Therefore, the claim follows.   □

Theorem 1. *Algorithm A is a* $\left(H_k - \frac{196}{390}\right)$-*approximation algorithm for the unweighted k-set cover problem.*

Proof. By Lemma 1, the algorithm returns a feasible solution in polynomial time. It remains to establish its approximation ratio.

$$
\begin{aligned}
A &\leq n_g \cdot \rho_g + n_b \cdot \rho_b \\
&= n_g \cdot \left(H_k - \frac{16}{30}\right) + n_b \cdot \left(H_k - \frac{1}{2}\right) \\
&\leq (n_g + n_b) \cdot \left[\frac{1}{13} \cdot \left(H_k - \frac{16}{30}\right) + \frac{12}{13} \cdot \left(H_k - \frac{1}{2}\right)\right] \\
&= OPT \cdot \left[\frac{1}{13} \cdot \left(H_k - \frac{16}{30}\right) + \frac{12}{13} \cdot \left(H_k - \frac{1}{2}\right)\right] \\
&= OPT \cdot \left(H_k - \frac{196}{390}\right),
\end{aligned}
$$

where the first inequality follows by Lemma 3, the first equation follows by Lemma 4, the second inequality follows by Lemma 5, the second equation follows because the cost of OPT is exactly $n_b + n_g$, and the last equation follows by simple algebra.   □

6   Concluding Remarks

In this paper we addressed the fundamental problem of unweighted k-set cover problem, and introduced an improvement over the previously best known algorithm for all values of k such that $k \geq 4$. Although we obtain a small improvement over the algorithm of Duh and Fürer [4], we think that our analysis is not tight and the approximation ratio of our algorithm can be improved. Improving the analysis of our Algorithm A is left for future research.

In this paper we showed that incorporating a local-search procedure in various stages of the greedy algorithm instead of only where each set has at most three uncovered elements, provides a better approximation ratio. We conjecture that incorporating local-search procedures in each greedy phase decreases the approximation ratio further. Such an algorithm replaces the Greedy phase by the following phase:

Improved phase: For $j = k, k - 1, k - 2, \ldots, 6$ do: apply local-search to choose an approximated maximum size collection of j-sets (each covering exactly j new elements).

It is easily noted that using the Improved phase instead of the Greedy phase in Algorithm A does not harm the approximation ratio of the resulting algorithm. We leave the analysis of this improved algorithm for future research.

References

1. R. Bar-Yehuda and S. Even, "A linear time approximation algorithm for the weighted vertex cover problem," *Journal of Algorithms*, **2**, 198-203, 1981.
2. V. Chvátal, "A greedy heuristic for the set-covering problem," *Mathematics of Operations Research*, **4**, 233-235, 1979.
3. P. Crescenzi and V. Kann, "A compendium of NP optimization problems", http://www.nada.kth.se/theory/problemlist.html, 1995.
4. R. Duh and M. Fürer, "Approximation of k-set cover by semi local optimization," *Proc. STOC 1997*, 256-264, 1997.
5. U. Feige, "A threshold of $\ln n$ for approximating set cover", *Journal of the ACM*, **45**, 634-652, 1998.
6. T. Fujito and T. Okumura, "A modified greedy algorithm for the set cover problem with weights 1 and 2," *Proc. ISAAC 2001*, 670-681, 2001.
7. O. Goldschmidt, D. S. Hochbaum and G. Yu, "A modified greedy heuristic for the set covering problem with improved worst case bound," *Information Processing Letters*, **48**, 305-310, 1993.
8. M. M. Halldórsson, "Approximating k set cover and complementary graph coloring," *Proc. IPCO 1996*, 118-131, 1996.
9. R. Hassin and A. Levin, " A better-than-greedy approximation algorithm for the minimum set cover problem," *SIAM J. Computing*, **35**, 189-200, 2006.
10. D. S. Hochbaum, "Approximation algorithms for the weighted set covering and node cover problems," *SIAM Journal on Computing*, **11**, 555-556, 1982.
11. C. A. J. Hurkens and A. Schrijver, "On the size of systems of sets every t of which have an SDR, with an application to the worst-case ratio of heuristics for packing problems", *SIAM Journal on Discrete Mathematics*, **2**, 68-72, 1989.
12. D. S. Johnson, "Approximation algorithms for combinatorial problems," *Journal of Computer and System Sciences*, **9**, 256-278, 1974.
13. R. M. Karp, "Reducibility among combinatorial problems," Complexity of computer computations (R.E. Miller and J.W. Thatcher, eds.), Plenum Press, New-York, 1972, 85-103.
14. S. Khanna, R. Motwani, M. Sudan and U. V. Vazirani, "On syntactic versus computational views of approximability," *SIAM Journal on Computing*, **28**, 164-191, 1998.
15. L. Lovász, "On the ratio of optimal integral and fractional covers," *Discrete Mathematics*, **13**, 383-390, 1975.
16. C. H. Papadimitriou and M. Yannakakis, "Optimization, approximation and complexity classes," *Journal of Computer System Sciences*, **43**, 425-440, 1991.

17. V. T. Paschos, "A survey of approximately optimal solutions to some covering and packing problems," *ACM Computing Surveys*, **29**, 171-209, 1997.
18. R. Raz and S. Safra, "A sub-constant error-probability low-degree test, and sub-constant error-probability PCP characterization of NP", *Proc. STOC 1997*, 475-484, 1997.
19. P. Slavík, "A tight analysis of the greedy algorithm for set cover," *Journal of Algorithms*, **25**, 237-254, 1997.

Approximation Algorithms for
Multi-criteria Traveling Salesman Problems*

Bodo Manthey[1],[**] and L. Shankar Ram[2]

[1] Yale University, Department of Computer Science
manthey@cs.yale.edu
[2] ETH Zürich, Institut für Theoretische Informatik
shankar.lakshminarayanan@ag.ch

Abstract. In multi-criteria optimization, several objective functions are to be optimized. Since the different objective functions are usually in conflict with each other, one cannot consider only one particular solution as optimal. Instead, the aim is to compute so-called Pareto curves. Since Pareto curves cannot be computed efficiently in general, we have to be content with approximations to them.

We are concerned with approximating Pareto curves of multi-criteria traveling salesman problems (TSP). We provide algorithms for computing approximate Pareto curves for the symmetric TSP with triangle inequality (Δ-STSP), symmetric and asymmetric TSP with strengthened triangle inequality ($\Delta(\gamma)$-STSP and $\Delta(\gamma)$-ATSP), and symmetric and asymmetric TSP with weights one and two (STSP$(1,2)$ and ATSP$(1,2)$).

We design a deterministic polynomial-time algorithm that computes $(1 + \gamma + \varepsilon)$-approximate Pareto curves for multi-criteria $\Delta(\gamma)$-STSP for $\gamma \in [\frac{1}{2}, 1]$. We also present two randomized approximation algorithms for multi-criteria $\Delta(\gamma)$-STSP achieving approximation ratios of $\frac{2\gamma^3 + \gamma^2 + 2\gamma - 1}{2\gamma^2} + \varepsilon$ and $\frac{1+\gamma}{1+3\gamma-4\gamma^2} + \varepsilon$, respectively. Moreover, we design randomized approximation algorithms for multi-criteria $\Delta(\gamma)$-ATSP (ratio $\frac{1}{2} + \frac{\gamma^3}{1-3\gamma^2} + \varepsilon$ for $\gamma < 1/\sqrt{3}$), STSP$(1,2)$ (ratio $4/3$) and ATSP$(1,2)$ (ratio $3/2$).

The algorithms for $\Delta(\gamma)$-ATSP, STSP$(1,2)$, and ATSP$(1,2)$ as well as one algorithm for $\Delta(\gamma)$-STSP are based on cycle covers. Therefore, we design randomized approximation schemes for multi-criteria cycle cover problems by showing that multi-criteria graph factor problems admit fully polynomial-time randomized approximation schemes.

1 Introduction

In many practical optimization problems, there is not only one single objective function to measure the quality of a solution, but there are several such functions.

---

* A full version of this work is available at http://arxiv.org/abs/cs/0606040.

** Supported by the Postdoc-Program of the German Academic Exchange Service (DAAD). On leave from Saarland University. Work done in part at the University of Lübeck supported by DFG research grant RE 672/3 and at Saarland University.

T. Erlebach and C. Kaklamanis (Eds.): WAOA 2006, LNCS 4368, pp. 302–315, 2006.

Consider for instance buying a car: We (probably) want to buy a cheap car that is fast and has a good gas mileage. How do we decide which car is the best one for us? Of course, with respect to any single criterion, making the decision is easy. But with multiple criteria involved, there is no natural notion of a best choice. The aim of *multi-criteria optimization* (also called multi-objective optimization or Pareto optimization) is to cope with this problem. To transfer the concept of a best choice to multi-criteria optimization, the notion of *Pareto curves* was introduced (cf. Section 1.1 and Ehrgott [12]). A Pareto curve is a set of solutions that can be considered optimal.

However, for most optimization problems, Pareto curves cannot be computed efficiently. Thus, we have to be content with approximations to them.

The *traveling salesman problem* (TSP) is one of the best-known combinatorial optimization problems [16]. An instance of the TSP is a complete graph with edge weights, and the aim is to find a Hamiltonian cycle (also called a tour) of minimum weight. Since the TSP is NP-hard [14], we cannot hope to always find an optimal tour efficiently. For practical purposes, however, it is often sufficient to obtain a tour that is close to optimal. In such cases, we require *approximation algorithms*, i.e., polynomial-time algorithms that compute such near-optimal tours.

While the approximability of several variants of the single-criterion TSP has been studied extensively in the past decades, not much is known about the approximability of multi-criteria TSP. The classical TSP is about a traveling salesman who has to visit a certain number of cities and return back home in a shortest tour. "Real" saleswomen and salesmen do not face such a simple situation. Instead, while arranging their tours, they have to bear in mind several objectives that are to be optimized. For instance, the distance travelled and the travel time should be minimized while the journey should be as cheap as possible. This gives rise to multi-criteria TSP, for which we design approximation algorithms in this paper.

1.1 Preliminaries

Graphs and Optimization Problems. Let $G = (V, E)$ be a graph (directed or undirected) with edge weights $w : E \to \mathbb{N}$. We define the weight of a subgraph $G' = (V', E')$ of G or a subset E' of the edges of G as the sum of the weights of its edges: $w(G') = w(E') = \sum_{e \in E'} w(e)$. For $k \in \mathbb{N}$, we define $[k] = \{1, 2, \ldots, k\}$.

TSP in general is the following optimization problem: Given a graph with edge weights, find a Hamiltonian cycle, i.e., a cycle that visits every vertex of the graph exactly once, of minimum weight. In this paper, we are concerned with several variants of the TSP, which are defined below. In case of undirected graphs, we speak of the symmetric TSP (STSP), while in case of directed graphs, we refer to the problem as the asymmetric TSP (ATSP).

Definition 1 (TSP). *An instance of* **Δ-STSP** *is an undirected complete graph* $G = (V, E)$ *with edge weights* $w : E \to \mathbb{N}$ *that fulfill triangle inequality, i.e.,* $w(\{u, v\}) \leq w(\{u, x\}) + w(\{x, v\})$ *for all distinct vertices* $u, v, x \in V$.

For $\gamma \in [\frac{1}{2}, 1]$, $\boldsymbol{\Delta(\gamma)}$**-STSP** *is the restriction of* Δ-STSP *to instances that satisfy γ-strengthened triangle inequality, i. e.,* $w(\{u, v\}) \leq \gamma \cdot (w(\{u, x\}) + w(\{x, v\}))$ *for all distinct vertices* u, v, x.

STSP(1, 2) *is the special case of* Δ-STSP *where only one and two are allowed as edge weights, i. e.,* $w : E \to \{1, 2\}$.

$\boldsymbol{\Delta}$**-ATSP**, $\boldsymbol{\Delta(\gamma)}$**-ATSP**, *and* **ATSP(1, 2)** *are analogously defined except that the graphs are directed.*

In all variants, Hamiltonian cycles of minimum weight are sought.

For $\gamma = 1$, $\Delta(\gamma)$-STSP becomes Δ-STSP and $\Delta(\gamma)$-ATSP becomes Δ-ATSP. As γ gets smaller, the edge weights become more and more structured. For $\gamma = 1/2$, all edge weights are equal. The γ-strengthened triangle inequality can also be considered as a data-dependent bound [7]: Given an instance of metric TSP, we compute the minimum γ such that the instance fulfills γ-strengthened triangle inequality. If $\gamma < 1$, then we obtain a better performance guarantee for our approximate solution than with triangle inequality alone.

Throughout the paper, a matching always means a perfect matching, i. e., a set of edges such that every vertex is incident to exactly one edge. **Match** denotes the problem of computing a matching of minimum weight. We refer to a matching as the set M of its edges.

The minimum spanning tree problem, denoted by **MST**, is the problem of computing a spanning tree of minimum weight. We refer to a tree as the set T of its edges.

A **cycle cover** of a graph $G = (V, E)$ is a subgraph (V, C) that consists solely of cycles such that every vertex $v \in V$ is part of exactly one cycle. In most cases, we refer to a cycle cover as the set C of its edges. Hamiltonian cycles are cycle covers that consist of only a single cycle. Below we define two optimization problems concerning cycle covers.

Definition 2 (cycle cover problems). *The problem of computing cycle covers of minimum weight in undirected graphs is called* **SCC**. *The directed version of the problem is called* **ACC**.

Multi-criteria Optimization. Let us first formally define what a k-criteria optimization problem is. We assume in the following that the number k of criteria is fixed. The running-times of our algorithms are usually exponential in k. But since k is typically a small number, this does not cause any harm.

Definition 3 (k-criteria optimization problem). *A k-criteria optimization problem Π consists of a set I of instances, a set $\mathrm{sol}(x)$ of feasible solutions for every instance $x \in I$, k objective functions $w_1, \ldots, w_k$, each mapping pairs of $x \in I$ and $y \in \mathrm{sol}(x)$ to $\mathbb{N}$, and k types indicating whether w_i should be minimized or maximized.*

We refer to Ehrgott and Gandibleux [12,13] for surveys on multi-criteria optimization problems. Throughout this paper, we restrict ourselves to problems where all objective functions should be minimized. The optimization problems defined in Section 1.1 are generalized to their multi-criteria counterparts in the obvious way: We have k objective functions $w_1, \ldots, w_k$, each induced by edge weight functions (to which we also refer as $w_1, \ldots, w_k$) as described. If we have additional restrictions on the edge weights, like triangle inequality, every edge weight function is assumed to fulfill them. In general, the different objective functions are in conflict with each other, i. e., it is impossible to minimize all of them simultaneously. Therefore, the notion of Pareto curves has been introduced.

For the following definitions, let Π be a k-criteria optimization problem as defined above.

Definition 4 (Pareto curve). *A set $\mathcal{P}(x) \subseteq \mathrm{sol}(x)$ is called a Pareto curve of x if for all solutions $z \in \mathrm{sol}(x)$, there exists a solution $y \in \mathcal{P}(x)$ with $w_i(x, y) \leq w_i(x, z)$ for all $i \in [k]$.*

A Pareto curve contains all solutions that might be considered optimal. For completeness, let us mention that Pareto curves are not unique in general: In our definition, it is not forbidden to include dominated solutions in $\mathcal{P}(x)$ (a solution y is dominated if there exists a z with $w_i(x, z) \leq w_i(x, y)$ for all $i \in [k]$ and $w_i(x, z) < w_i(x, y)$ for some $i \in [k]$, i. e., z is strictly better than y). However, if we restrict ourselves to explicitly given sets of solutions, we can easily get rid of such dominated solutions.

For the majority of multi-criteria problems, computing Pareto curves is hard for two reasons: First, many two-criteria problems allow for a reduction from the knapsack problem. Second, Pareto curves are often of exponential size. Therefore, it is natural to consider the idea of an approximation to Pareto curves.

Definition 5 (β-approximate Pareto curve). *Let $\beta \geq 1$. Let $x \in I$ and $\mathcal{P}^{\mathrm{apx}}(x) \subseteq \mathrm{sol}(x)$. The set $\mathcal{P}^{\mathrm{apx}}(x)$ is called a β-approximate Pareto curve for x if, for every $z \in \mathrm{sol}(x)$, there exists a $y \in \mathcal{P}^{\mathrm{apx}}(x)$ with $w_i(x, y) \leq \beta \cdot w_i(x, z)$ for all $i \in [k]$.*

While Pareto curves itself are often of exponential size, it is known that $(1 + \varepsilon)$-approximate Pareto curves of size polynomial in the input size and $1/\varepsilon$ exist if the number k of criteria is fixed [19]. (The size of the approximate Pareto curve is in general exponential in k. The technical restriction is that the objective functions are restricted to assume values of at most $2^{p(|x|)}$ for $x \in I$ and some polynomial p. This is fulfilled for almost all natural optimization problems.).

The above definition leads immediately to the notion of an approximation algorithm for multi-criteria optimization problems.

Definition 6 (approximation algorithm). *Let $\beta \geq 1$. A β-approximation algorithm for Π is an algorithm that, for every input $x \in I$, computes a β-approximate Pareto curve for x in time polynomial in the size $|x|$ of x.*

A randomized β-approximation algorithm for Π is a polynomial-time algorithm that, for every input $x \in I$, computes a set $\mathcal{P}^{\mathrm{apx}}(x) \subseteq \mathrm{sol}(x)$ such that $\mathcal{P}^{\mathrm{apx}}(x)$ is a β-approximate Pareto curve for x with a probability of at least $1/2$.

By executing a randomized approximation algorithm ℓ times, we obtain a β-approximate Pareto curve with a probability of at least $1 - 2^{-\ell}$, i. e., the failure probability tends exponentially to zero: We take the union of all sets of solutions computed in the ℓ iterations and throw away all solutions that are dominated by solutions in the union.

Definition 7 (FPTAS, FPRAS). *An algorithm is a fully polynomial-time approximation scheme (FPTAS) for Π if, on input $x \in I$ and $\varepsilon > 0$, it computes a $(1 + \varepsilon)$-approximate Pareto curve in time polynomial in the size of x and $1/\varepsilon$.*

A fully polynomial-time randomized approximation scheme (FPRAS) for Π is a randomized approximation algorithm that, on input $x \in I$ and $\varepsilon > 0$, computes a $(1 + \varepsilon)$-approximate Pareto curve in time polynomial in the size of x and $1/\varepsilon$.

Definition 8 (randomized exact algorithm). *A randomized exact algorithm for Π is an algorithm that, on input x, computes a Pareto curve of x in time polynomial in the size of x with a probability of at least $1/2$.*

An optimization problem Π is said to be polynomially bounded if there exists a polynomial p such that the following holds for every objective function w_i of Π: For every instance x and every feasible solution y for x, $w_i(x, y) \leq p(|x|)$ for all $i \in [k]$. An exact algorithm can be obtained from an FPTAS for a polynomially bounded optimization problem.

Lemma 1. *Suppose that Π is polynomially bounded. If there exists an FPTAS for Π, then Π can be solved exactly in polynomial time. If there exists an FPRAS for Π, then there exists a randomized exact algorithm for Π.*

1.2   Previous Results

The approximability of single-criterion TSP has been studied intensively in the past. The currently best approximation ratios for the variants of single-criterion TSP considered in this paper are $3/2$ for Δ-STSP [10], $8/7$ for STSP$(1, 2)$ [5], $\min\{\frac{3\gamma^2}{3\gamma^2 - 2\gamma + 1}, \frac{2 - \gamma}{3 - 3\gamma}\}$ for $\Delta(\gamma)$-STSP [8], $0.842 \cdot \log n$ for Δ-ATSP [15], $5/4$ for ATSP$(1, 2)$ [6], and $\min\{\frac{1 + \gamma}{2 - \gamma - \gamma^3}, \frac{\gamma}{1 - \gamma}\}$ for $\Delta(\gamma)$-ATSP [7,9].

While single-criterion optimization problems and their approximation properties have been the subject of a considerable amount of research (cf. Ausiello et al. [3] for a survey), not much is known about the approximability of multi-criteria optimization problems.

Papadimitriou and Yannakakis [19], by applying results of Barahona and Pulleyblank [4], Mulmuley et al. [17], and themselves [18], showed that there exist FPTASs for multi-criteria MST and the multi-criteria shortest path problem and an FPRAS for multi-criteria Match. (More precisely, a fully polynomial RNC scheme.) The results were established by showing that a multi-criteria problem admits an FPTAS if the exact version of the single-criterion problem can be solved in pseudo-polynomial time. The exact version of a single-criterion optimization problem Π is the following decision problem: Given an instance $x \in I$ and a number $W \in \mathbb{N}$, does there exist a solution $y \in \mathrm{sol}(x)$ with $w(x, y) = W$? The exact versions of many single-criterion optimization problems are NP-hard since knapsack can be reduced to them easily. But this does not rule out the possibility of pseudo-polynomial-time algorithms for them.

Multi-criteria TSP has been investigated by Ehrgott [11] and Angel et al. [1,2]. Ehrgott [11] analyzed a generalization of Christofides' algorithm for Δ-STSP. Instead of considering Pareto curves, he measured the quality of a solution y for an instance x as a norm of the vector $(w_1(x, y), \ldots, w_k(x, y))$. Thus, he encoded the different objective functions into a single one, which reduces the problem to a single-criterion problem. The approximation ratio achieved is between $3/2$ and 2, depending on the norm used to combine the different criteria. However, by encoding all objective functions into a single one, we lose the special properties of multi-criteria optimization problems.

Angel et al. [1] considered two-criteria STSP(1, 2). They presented a $3/2$-approximation algorithm for this problem by using a local search heuristic. Finally, Angel et al. [2] generalized these results to k-criteria STSP(1, 2) by presenting a $2 - \frac{2}{k+1}$-approximation for $k \geq 3$. Although for every fixed k, the approximation ratio is below 2, it converges to 2 as k increases. Thus, the ratio tends to the trivial ratio of 2, which can be achieved by selecting any Hamiltonian cycle. These two are the only papers about the approximability of Pareto curves of multi-criteria TSP we are aware of.

1.3   Our Results

All our results hold for an arbitrary but fixed number of objective functions.

We present a deterministic polynomial-time algorithm that computes $(2 + \varepsilon)$-approximate Pareto curves for Δ-STSP (Section 2.1). This is the first efficient algorithm for computing approximate Pareto curves for this problem. In fact, we show the following more general result: If the edge weights satisfy γ-strengthened triangle inequality for $\gamma \in [\frac{1}{2}, 1]$, then the algorithm computes a $(1 + \gamma + \varepsilon)$-approximate Pareto curve for arbitrarily small $\varepsilon > 0$ in polynomial time.

We generalize Christofides' algorithm [10] (cf. Vazirani [22]) to obtain a randomized approximation algorithm for multi-criteria $\Delta(\gamma)$-STSP (Section 2.2). For $\gamma \in [\frac{1}{2}, 1]$, our algorithm achieves an approximation ratio of $\frac{2\gamma^3 + \gamma^2 + 2\gamma - 1}{2\gamma^2} + \varepsilon$. For $\gamma = 1$, this yields a ratio of $2 + \varepsilon$.

We consider cycle covers in Section 3. Cycle covers play an important role in the design of approximation algorithms for the TSP. We prove that there exists an FPRAS for computing approximate Pareto curves of multi-criteria cycle covers. Subsequently, we extend this result and show that the multi-criteria variant of the problem of finding graph factors of minimum weight admits an FPRAS, too.

Finally, we analyze a randomized cycle-cover-based algorithm for multi-criteria TSP (Section 4): We start by computing an approximate Pareto curve of cycle covers. Then, for every cycle cover in the set computed, we remove one edge of every cycle and join the paths thus obtained to a Hamiltonian cycle. We analyze the approximation ratio of this algorithm for $\Delta(\gamma)$-STSP (Section 4.1, approximation ratio $\frac{1+\gamma}{1+3\gamma-4\gamma^2} + \varepsilon$ for $\gamma < 1$), $\Delta(\gamma)$-ATSP (Section 4.2, ratio

---

Algorithm 1. The tree doubling algorithm for multi-criteria Δ-STSP

Input: undirected complete graph $G = (V, E)$; edge weights $w_i : E \to \mathbb{N}$ $(i \in [k])$;
     $\varepsilon > 0$
Output: an approximate Pareto curve $\mathcal{P}_{\text{TSP}}^{\text{apx}}$ to the multi-criteria STSP
 1: compute a $(1 + \frac{\varepsilon}{2})$-approximate Pareto curve $\mathcal{P}_{\text{MST}}^{\text{apx}}$ for MST on G
 2: **for all** trees $T \in \mathcal{P}_{\text{MST}}^{\text{apx}}$ **do**
 3:     duplicate all edges in T to obtain an Eulerian graph $\tilde{T}$
 4:     obtain a Hamiltonian cycle S from $\tilde{T}$ by taking shortcuts; put S into $\mathcal{P}_{\text{TSP}}^{\text{apx}}$
 5: **end for**

$\frac{1}{2} + \frac{\gamma^3}{1-3\gamma^2} + \varepsilon$ for $\gamma < 1/\sqrt{3}$), STSP$(1, 2)$, and ATSP$(1, 2)$ (Section 4.3, ratios $4/3$ and $3/2$, respectively).

As far as we know, our algorithms are the first approximation algorithms for Pareto curves for Δ-STSP, $\Delta(\gamma)$-STSP, $\Delta(\gamma)$-ATSP, and ATSP$(1, 2)$. Furthermore, we achieve a better approximation ratio for STSP$(1, 2)$ than the algorithms by Angel et al. [1,2] for all k.

2   Metric TSP

In this section, we present two algorithms for Δ-STSP and $\Delta(\gamma)$-STSP. Another approximation algorithm that can be used for approximating $\Delta(\gamma)$-STSP, which is based on computing cycle covers, will be presented in Section 4.

2.1   The Generalized Tree Doubling Algorithm

Consider the following approximation algorithm for single-criterion Δ-STSP, which was first analyzed by Rosenkrantz et al. [20] (cf. Vazirani [22]): First, we compute a minimum spanning tree. Then we duplicate each edge. The result is an Eulerian graph. We obtain a Hamiltonian cycle from this graph by walking along an Eulerian cycle. If we come back to a vertex that we have already visited, we omit it and take a short-cut to the next vertex in the Eulerian cycle. In this way, we obtain an approximation ratio of 2 for single-criterion Δ-STSP. Algorithm 1 is an adaptation of this algorithm to multi-criteria STSP.

Theorem 1. *For all $\gamma \in [\frac{1}{2}, 1]$, Algorithm 1 computes a $(1 + \gamma + \varepsilon)$-approximate Pareto curve for multi-criteria $\Delta(\gamma)$-STSP in time polynomial in the input size and $1/\varepsilon$.*

Corollary 1. *Algorithm 1 is a $(2+\varepsilon)$-approximation algorithm for multi-criteria Δ-STSP. Its running-time is polynomial in the input size and $1/\varepsilon$.*

2.2   A Generalization of Christofides' Algorithm

In this section, we generalize Christofides' algorithm to multi-criteria Δ-STSP, which is the best approximation algorithm for single-criterion Δ-STSP known

---

Algorithm 2. A generalization of Christofides' algorithm for multi-criteria Δ-STSP

Input: undirected complete graph $G = (V, E)$; edge weights $w_i : E \to \mathbb{N}$ ($i \in [k]$); $\varepsilon > 0$

Output: an approximate Pareto curve $\mathcal{P}_{\mathrm{TSP}}^{\mathrm{apx}}$ to the multi-criteria STSP (with a probability of at least $1/2$)

1: compute a $(1 + \frac{\varepsilon}{2})$-approximate Pareto curve $\mathcal{P}_{\mathrm{MST}}^{\mathrm{apx}}$ for MST on G
2: let p be the number of trees in $\mathcal{P}_{\mathrm{MST}}^{\mathrm{apx}}$
3: **for all** trees $T \in \mathcal{P}_{\mathrm{MST}}^{\mathrm{apx}}$ **do**
4:    let $V_{\mathrm{odd}} \subseteq V$ be the set of vertices of odd degree in T
5:    compute $\mathcal{P}_{\mathrm{Match}}^{\mathrm{apx}}(T)$ such that $\mathcal{P}_{\mathrm{Match}}^{\mathrm{apx}}(T)$ is a $\left(1 + \frac{\varepsilon}{2}\right)$-approximate Pareto curve for Match on the graph induced by V_{odd} with a probability of at least $1 - \frac{1}{2p}$
6:    **for all** matchings $M \in \mathcal{P}_{\mathrm{Match}}^{\mathrm{apx}}(T)$ **do**
7:       obtain a Hamiltonian cycle S from $T \cup M$ by taking shortcuts; put S into $\mathcal{P}_{\mathrm{TSP}}^{\mathrm{apx}}$
8:    **end for**
9: **end for**

---

so far. This algorithm computes Pareto curves of matchings. In case of single-criterion Δ-STSP, we can always find a matching with a weight of at most half of the weight of the optimal Hamiltonian cycle. This is in contrast to multi-criteria Δ-STSP, where the weights of the matchings can be arbitrarily close to the weight of the optimal Hamiltonian cycle. The reason is that we cannot choose the lighter of two different matchings since multiple objective functions are involved; the term "'lighter"' is not well defined. Therefore, we only get an approximation ratio of roughly two in this case. But for $\Delta(\gamma)$-STSP, we can show a better upper bound. The analysis of the algorithm exploits the following result due to Böckenhauer et al. [8].

Lemma 2 (Böckenhauer et al. [8]). *Let $G = (V, E)$ be an undirected complete graph with an edge weight function w satisfying γ-strengthened triangle inequality for some $\gamma \in [\frac{1}{2}, 1)$.*

Let $w_{\mathrm{max}} = \max_{e \in E}(w(e))$ and $w_{\mathrm{min}} = \min_{e \in E}(w(e))$ be the weights of a heaviest and lightest edge, respectively. Then $\frac{w_{\mathrm{max}}}{w_{\mathrm{min}}} \leq \frac{2\gamma^2}{1-\gamma}$.

Let e and e' be two edges with a common endpoint. Then $\frac{w(e)}{w(e')} \leq \frac{\gamma}{1-\gamma}$.

Theorem 2. *For $\gamma \in [\frac{1}{2}, 1]$, Algorithm 2 is a randomized $\left(\frac{2\gamma^3 + \gamma^2 + 2\gamma - 1}{2\gamma^2} + \varepsilon\right)$-approximation algorithm for multi-criteria $\Delta(\gamma)$-STSP. Its running time is polynomial in the input size and $1/\varepsilon$.*

We compare the ratios obtained by the two algorithms of this sections and the cycle cover algorithm of Section 4 in Section 5.1.

3 Matchings and Cycle Covers

A cycle cover of a graph is a spanning subgraph that consists solely of cycles such that every vertex is part of exactly one cycle. Many approximation algorithms for the single-criterion TSP are based on cycle covers. These approximation algorithms usually start by computing an initial cycle cover and then join the cycles to obtain a Hamiltonian cycle. This technique is called *subtour patching* [16].

3.1 Multi-criteria ACC

ACC, the cycle cover problem in directed graphs, is equivalent to finding matchings of minimum weight in bipartite graphs (assignment problem). An FPRAS for multi-criteria Match is also an FPRAS for the multi-criteria matching problem in bipartite graphs. Hence, multi-criteria ACC also admits an FPRAS.

Theorem 3. *There exists an FPRAS for the multi-criteria ACC.*

3.2 Multi-criteria SCC and f-Factors

To show that multi-criteria SCC admits an FPRAS, we reduce SCC using Tutte's reduction [21] to the matching problem in general graphs. Cycle covers in undirected graphs are also known as two-factors since in a cycle cover, every vertex is incident to exactly two edges. We obtain the following result from the fact that the matching problem admits an FPRAS [19].

Theorem 4. *Multi-criteria SCC admits an FPRAS.*

We can generalize the FPRAS for the undirected cycle cover problem to arbitrary f-factors by exploiting Tutte's reduction again. The proof goes along the same lines as the proof of Theorem 4. Let $G = (V, E)$ be an undirected graph and $f : V \to \mathbb{N}$ be any function. A subset $F \subseteq E$ is called an f-factor of G if all $v \in V$ are incident to exactly $f(v)$ edges in F. If $f(v) = 2$ for all $v \in V$, an f-factor is a cycle cover.

Definition 9 (GFP). *The graph factor problem GFP is the following minimization problem: An instance is an undirected graph $G = (V, E)$ with a function $f : V \to \mathbb{N}$ and an edge weight function $w : E \to \mathbb{N}$. The aim is to find an f-factor of minimum weight.*

Theorem 5. *Multi-criteria GFP admits an FPRAS.*

4 Approximations Based on Cycle Covers

The generic outline of a cycle-cover-based algorithm is the following: Start by computing a cycle cover. Then remove one edge of every cycle. Finally, join the paths thus obtained to form a Hamiltonian cycle. Algorithm 3 is our generalization of this algorithm to multi-criteria TSP. It achieves a constant approximation ratio if the quotient of the weight of the heaviest edge and the weight of the

Algorithm 3. An approximation algorithm based on cycle covers for multi-criteria TSP

Input: complete graph $G = (V, E)$; edge weights w_i $(i \in [k])$; $\varepsilon' > 0$
Output: approximate Pareto curve $\mathcal{P}_{\text{TSP}}^{\text{apx}}$ to multi-criteria TSP (with a probability of at least $1/2$)
1: compute a $(1 + \varepsilon')$-approximate Pareto curve $\mathcal{P}_{\text{CC}}$ to the multi-criteria cycle cover problem
2: **for all** cycle covers $C \in \mathcal{P}_{\text{CC}}$ **do**
3:     **for all** cycles c of C **do**
4:         remove one edge of c
5:     **end for**
6:     join the paths to form a Hamiltonian cycle S and add S to $\mathcal{P}_{\text{TSP}}^{\text{apx}}$
7: **end for**

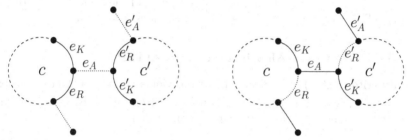

(a) Cycle cover, before the patching.   (b) Hamiltonian cycle, after the patching.

Fig. 1. Two cycles c and c' before and after joining the cycles to a Hamiltonian cycle. The edges e_R, e_K, and e_A belong to c while e'_R, e'_K, and e'_A belong to c'.

lightest edge is bounded. In this section, we present a general analysis of the approximation ratio of this algorithm. We will refine the analysis for multi-criteria $\Delta(\gamma)$-STSP (Section 4.1) to get an improved approximation ratio. Furthermore, we apply the analysis to get approximation results for multi-criteria $\Delta(\gamma)$-ATSP (Section 4.2) and STSP$(1, 2)$ and ATSP$(1, 2)$ (Section 4.3). We analyze Algorithm 3 in terms of the number αn of edges that have to be removed and the quotient $\beta = w_{\max}/w_{\min}$.

Lemma 3. *Assume that at most αn edges have to be removed from each cycle cover and that $\frac{\max_{e \in E} w_i(e)}{\min_{e \in E} w_i(e)} \leq \beta$ for all $i \in [k]$. Then Algorithm 3 is a randomized $(1 + \alpha(\beta - 1) + \varepsilon)$ approximation for every $\varepsilon > 0$. Its running-time is polynomial in the input size and $1/\varepsilon$.*

4.1   Refined Analysis for $\Delta(\gamma)$-STSP

From the general analysis, we obtain an approximation ratio of $\frac{2}{3} + \frac{1}{3} \cdot \frac{2\gamma^2}{1-\gamma} + \varepsilon$ for $\Delta(\gamma)$-STSP. In this section, we present a refined analysis that yields a better approximation ratio.

Consider any cycle c of a cycle cover of $\mathcal{P}_{CC}$. There will be an edge e_R of c that will be removed and an edge e_A adjacent to e_R that will be added during the joining process. Finally, there exists an edge e_K of c that is adjacent to both e_R and e_A (Figure 1 shows an example). Note that while e_R is unique, once the edges have been removed and added, the edge e_A is not since there are two edges that connect c to other cycles. However, once we have fixed e_A for one cycle c, the corresponding e_K is uniquely determined, and the e'_A and e'_K of all other cycles c' are also determined. By Lemma 2, we have $w_i(e_R) \geq \frac{1-\gamma}{\gamma} \cdot w_i(e_A)$ and $w_i(e_K) \geq \frac{1-\gamma}{\gamma} \cdot w_i(e_A)$. Exploiting these inequalities, we obtain the following result.

Theorem 6. *Algorithm 3 is a randomized $\left(\frac{1+\gamma}{1+3\gamma-4\gamma^2} + \varepsilon\right)$-approximation algorithm for all $\varepsilon > 0$. Its running-time is polynomial in the input size and $1/\varepsilon$.*

In Section 5.1, we compare the approximation ratios of the cycle cover algorithm for $\Delta(\gamma)$-STSP to the tree doubling and Christofides' algorithm.

4.2   The Cycle Cover Algorithm for $\Delta(\gamma)$-ATSP

For multi-criteria $\Delta(\gamma)$-ATSP, our algorithm yields a constant factor approximation if $\gamma < 1/\sqrt{3}$ since $w_{\max}/w_{\min}$ is bounded from above by $\frac{2\gamma^3}{1-3\gamma^2}$ for such γ. For larger values of γ, this ratio can be unbounded.

Lemma 4 (Chandran and Ram [9]). *Let $\gamma \in [1/2, 1)$. Let $G = (V, E)$ be a directed complete graph, and let $w : E \to \mathbb{N}$ be an edge weight function satisfying γ-triangle inequality. Let $w_{\min} = \min_{e \in E} w(e)$ and $w_{\max} = \max_{e \in E} w(e)$. If $\gamma < 1/\sqrt{3}$, then $\frac{w_{\max}}{w_{\min}} \leq \frac{2\gamma^3}{1-3\gamma^2}$. If $\gamma \geq 1/\sqrt{3}$, then $\frac{w_{\max}}{w_{\min}}$ can be unbounded.*

By combining Lemmas 3 and 4, we obtain the following result.

Theorem 7. *For $\gamma < 1/\sqrt{3}$, Algorithm 3 is a randomized $\left(\frac{1}{2} + \frac{\gamma^3}{1-3\gamma^2} + \varepsilon\right)$-approximation algorithm for $\Delta(\gamma)$-ATSP.*

We leave as an open problem to generalize the analysis to larger values of γ. However, it seems to be hard to find a constant factor approximation for $\gamma = 1$ since this would immediately yield a constant factor approximation for single-criterion Δ-ATSP

4.3   TSP with Weights One and Two

For both STSP(1, 2) and ATSP(1, 2), we have $\beta = 2$, i.e., $w_{\max}/w_{\min} = 2$. For STSP(1, 2), we have $\alpha \leq 1/3$, while we only have $\alpha \leq 1/2$ in case of ATSP(1, 2). The approximation ratio follows by exploiting Lemma 3. The edge weights and thus the objective functions are polynomially bounded for STSP(1, 2) and ATSP(1, 2). Thus, by Lemma 1, we can compute Pareto curves of cycle covers instead of only $(1 + \varepsilon)$-approximate Pareto curves. Hence, we can get rid of the additional ε in the approximation ratios.

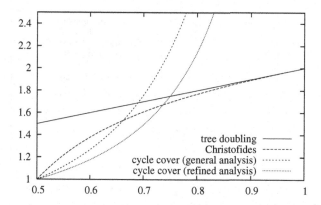

Fig. 2. Approximation ratios subject to γ achieved by tree doubling (Algorithm 1), Christofides' algorithm (Algorithm 2), and the cycle cover algorithm (Algorithm 3, Section 4), for which both the ratio obtained from the general and the refined analysis (Section 4.1) are shown

Theorem 8. *Algorithm 3 is a randomized 4/3 approximation algorithm for multi-criteria* STSP$(1, 2)$.

Theorem 9. *Algorithm 3 is a randomized 3/2 approximation algorithm for multi-criteria* ATSP$(1, 2)$.

5   Concluding Remarks

5.1   Comparing the Approximation Ratios for $\Delta(\gamma)$-STSP

Let us compare the approximation ratios achieved by the tree doubling algorithm (Section 2.1), Christofides' algorithm (Section 2.2), and the cycle cover algorithm (Section 4). Figure 2 shows the approximation ratios achieved by these algorithms subject to γ. Figure 3 shows the approximation ratios achieved deterministically (by the tree doubling algorithm) and randomized (by a combination of Christofides' and the cycle cover algorithm). The ratios are compared to the trivial ratio of $w_{\max}/w_{\min}$ and to the currently best known approximation ratio for single-criterion $\Delta(\gamma)$-STSP. Note that in particular for small values of γ, our algorithms for multi-criteria $\Delta(\gamma)$-STSP come close to achieving the ratio of the best algorithms for single-criterion $\Delta(\gamma)$-STSP.

5.2   Open Problems

Our approximation algorithm for multi-criteria $\Delta(\gamma)$-ATSP works only for $\gamma < 1/\sqrt{3}$. Thus, we are interested in finding constant factor approximation algorithms also for $\gamma \geq 1/\sqrt{3}$, which exist for all $\gamma < 1$ for single-criterion $\Delta(\gamma)$-ATSP [7,9].

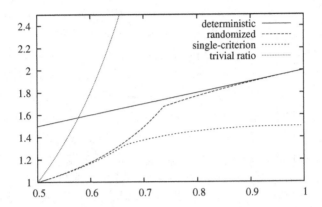

Fig. 3. Approximation ratios subject to γ. The deterministic ratio is achieved by tree doubling. Combining Christofides' and the cycle cover algorithm yields the randomized ratio. For comparison, the ratio for single-criterion $\Delta(\gamma)$-STSP and the trivial ratio $\frac{w_{max}}{w_{min}}$ are also shown.

The cycle-cover-based algorithm for Max-TSP, where Hamiltonian cycles of maximum weight are sought, does not seem to perform well for multi-criteria Max-TSP. The reason for this is that the approximation algorithms for Max-TSP that base on cycle covers usually contain a statement like "remove the lightest edge of every cycle". While this works for single-criterion TSP, the term "lightest edge" is not well-defined for multi-criteria traveling salesman problems. We are particularly curious about the approximability of multi-criteria Max-TSP.

References

1. Eric Angel, Evripidis Bampis, and Laurent Gourvés. Approximating the Pareto curve with local search for the bicriteria TSP(1,2) problem. *Theoretical Computer Science*, 310(1–3):135–146, 2004.
2. Eric Angel, Evripidis Bampis, Laurent Gourvès, and Jérôme Monnot. (Non-)approximability for the multi-criteria TSP(1,2). In Maciej Liśkiewicz and Rüdiger Reischuk, editors, *Proc. of the 15th Int. Symp. on Fundamentals of Computation Theory (FCT)*, volume 3623 of *Lecture Notes in Computer Science*, pages 329–340. Springer, 2005.
3. Giorgio Ausiello, Pierluigi Crescenzi, Giorgio Gambosi, Viggo Kann, Alberto Marchetti-Spaccamela, and Marco Protasi. *Complexity and Approximation: Combinatorial Optimization Problems and Their Approximability Properties*. Springer, 1999.
4. Francisco Barahona and William R. Pulleyblank. Exact arborescences, matchings and cycles. *Discrete Applied Mathematics*, 16(2):91–99, 1987.
5. Piotr Berman and Marek Karpinski. 8/7-approximation algorithm for (1,2)-TSP. In *Proc. of the 17th Ann. ACM-SIAM Symp. on Discrete Algorithms (SODA)*, pages 641–648. SIAM, 2006.

6. Markus Bläser. A 3/4-approximation algorithm for maximum ATSP with weights zero and one. In Klaus Jansen, Sanjeev Khanna, José D. P. Rolim, and Dana Ron, editors, *Proc. of the 7th Int. Workshop on Approximation Algorithms for Combinatorial Optimization Problems (APPROX)*, volume 3122 of *Lecture Notes in Computer Science*, pages 61–71. Springer, 2004.

7. Markus Bläser, Bodo Manthey, and Jiří Sgall. An improved approximation algorithm for the asymmetric TSP with strengthened triangle inequality. *Journal of Discrete Algorithms*, to appear.

8. Hans-Joachim Böckenhauer, Juraj Hromkovič, Ralf Klasing, Sebastian Seibert, and Walter Unger. Approximation algorithms for the TSP with sharpened triangle inequality. *Information Processing Letters*, 75(3):133–138, 2000.

9. L. Sunil Chandran and L. Shankar Ram. Approximations for ATSP with parameterized triangle inequality. In Helmut Alt and Afonso Ferreira, editors, *Proc. of the 19th Int. Symp. on Theoretical Aspects of Computer Science (STACS)*, volume 2285 of *Lecture Notes in Computer Science*, pages 227–237. Springer, 2002.

10. Nicos Christofides. Worst-case analysis of a new heuristic for the traveling salesman problem. Technical Report 388, Graduate School of Industrial Administration, Carnegie Mellon University, Pittsburgh, Pennsylvania, USA, 1976.

11. Matthias Ehrgott. Approximation algorithms for combinatorial multicriteria optimization problems. *International Transactions in Operational Research*, 7(1):5–31, 2000.

12. Matthias Ehrgott. *Multicriteria Optimization*. Springer, 2005.

13. Matthias Ehrgott and Xavier Gandibleux. A survey and annotated bibliography of multiobjective combinatorial optimization. *OR Spectrum*, 22(4):425–460, 2000.

14. Michael R. Garey and David S. Johnson. *Computers and Intractability: A Guide to the Theory of NP-Completeness*. W. H. Freeman and Company, 1979.

15. Haim Kaplan, Moshe Lewenstein, Nira Shafrir, and Maxim Sviridenko. Approximation algorithms for asymmetric TSP by decomposing directed regular multigraphs. *Journal of the ACM*, 52(4):602–626, 2005.

16. Eugene L. Lawler, Jan Karel Lenstra, Alexander H. G. Rinnooy Kan, and David B. Shmoys, editors. *The Traveling Salesman Problem: A Guided Tour of Combinatorial Optimization*. John Wiley & Sons, 1985.

17. Ketan Mulmuley, Umesh V. Vazirani, and Vijay V. Vazirani. Matching is as easy as matrix inversion. *Combinatorica*, 7(1):105–113, 1987.

18. Christos H. Papadimitriou. The complexity of restricted spanning tree problems. *Journal of the ACM*, 29(2):285–309, 1982.

19. Christos H. Papadimitriou and Mihalis Yannakakis. On the approximability of trade-offs and optimal access of web sources. In *Proc. of the 41st Ann. IEEE Symp. on Foundations of Computer Science (FOCS)*, pages 86–92. IEEE Computer Society, 2000.

20. Daniel J. Rosenkrantz, Richard E. Stearns, and Philip M. Lewis II. An analysis of several heuristics for the traveling salesman problem. *SIAM Journal on Computing*, 6(3):563–581, 1977.

21. William T. Tutte. A short proof of the factor theorem for finite graphs. *Canadian Journal of Mathematics*, 6:347–352, 1954.

22. Vijay V. Vazirani. *Approximation Algorithms*. Springer, 2001.

The Survival of the Weakest in Networks[*]

S. Nikoletseas[1,2], C. Raptopoulos[1,2], and P. Spirakis[1,2]

[1] Computer Technology Institute, P.O. Box 1122, 26110 Patras, Greece
nikole@cti.gr, raptopox@ceid.upatras.gr, spirakis@cti.gr
[2] University of Patras, 26500 Patras, Greece

Abstract. We study here dynamic antagonism in a fixed network, represented as a graph G of n vertices. In particular, we consider the case of $k \leq n$ particles walking randomly independently around the network. Each particle belongs to exactly one of two antagonistic species, none of which can give birth to children. When two particles meet, they are engaged in a (sometimes mortal) local fight. The outcome of the fight depends on the species to which the particles belong. Our problem is *to predict* (i.e. to compute) the eventual chances of species survival. We prove here that this can indeed be done in *expected polynomial time on the size of the network*, provided that the network is *undirected*.

1 Introduction and Our Results

In biological systems, successful genes and traits are the ones for which there is a *good chance* of a continuously fitness-increasing path leading from the current phenotype and genotype to the target ones. Similarly, in the context of the Internet, computational artifacts (like viruses and anti-virus programs, routing schemes etc) usually antagonize locally and the eventual prevalence of an artifact depends on this antagonism.

Evolutionary game theory has established ways of analysing multi-species competition. See e.g. Weibull [10] and the excellent work of Maynard Smith [4] for the interplay between biology and game theory. In all studies like [10], the "animals" of the competing species are randomly paired (as if they are in a "bag") and, depending on the local game played, some game participants may die or give birth, thus changing the population mixture. However, when individual members of each species "move" into a *finite* network (e.g. among neighbour nodes of a graph) then the basic "random mating" assumption of classical evolutionary game theory collapses. Only those individuals that happen to meet currently are involved in local fights.

[*] This work was partially supported by the IST Programme of the European Union under contact number IST-2005-15964 (AEOLUS) and by the Programme PENED under contact number 03ED568, co-funded 75% by European Union – European Social Fund (ESF), 25% by Greek Government – Ministry of Development – General Secretariat of Research and Technology (GSRT), and by Private Sector, under Measure 8.3 of O.P. Competitiveness – 3rd Community Support Framework (CSF).

T. Erlebach and C. Kaklamanis (Eds.): WAOA 2006, LNCS 4368, pp. 316–329, 2006.
© Springer-Verlag Berlin Heidelberg 2006

Given this new situation (which abstracts reality in networks) can we predict efficiently the "eventual" population mixture, e.g. the chance of survival of one (the weakest) of the species? Moreover, is our prediction method better than a simulation of the evolution of the population mixture?

We embark here in the study of this question. We first define the simplest, yet nontrivial, model of two-species antagonism: Our population members (called *particles* also) are either very malicious "hawks" or peaceful "doves". Hawks kill doves when they meet. Also, when two hawks meet they kill each other. Each particle performs an independent random walk in the graph. Doves do not harm each other when they meet.

We then concentrate on the simplest possible question: Can we calculate efficiently the chance of eventual survival of the weakest species? We assume that the graph (of motions) and the initial particles positions are given.

As a worst case scenario, we examine the *weakest possible case of the weak species* (just one dove). Note that the chance of eventual survival of a single dove *is a lower bound* for the case of many doves, since doves do not reproduce. We first prove the following result: When G is directed, then the probability of eventual survival of the dove can be exponentially small. We believe that in this case the problem is hard, thus, we do not expect to be able to efficiently predict the final outcome.

It turns out however that, when G *is undirected*, then:

1. we can decide in polynomial time (in the number n of the graph's vertices) when the probability of eventual survival of the dove is non-zero.
2. We prove that the probability of the dove's survival (when non-zero) is lower bounded by the inverse of a polynomial in n.
3. We can approximate the exact value of the probability of dove survival to any degree of accuracy in expected polynomial time (on n and the accuracy of the approximation). This result is a consequence of our main result in this paper, namely that the probability of dove's survival (when non-zero) is bounded below by the inverse of a polynomial on n.

We see our work as one first step of a new field, namely that of computational complexity of discrete evolutionary dynamics.

1.1 Previous Work and Comparison

An introduction to evolutionary game theory, stable strategies and replicator dynamics appears in [10]. A more stochastic approach is given in [8]. In both texts, the replicators dynamics used assume a population that contains individuals who play some incumbent strategy of some game Γ and at every step random matches between them take place. These kinds of dynamics either converge or not to an evolutionary stable strategy. A stochastic version of the classical replicator dynamics is considered by Imhof in [2]. In particular, he studies the long-run behaviour of the model when noise is added. Also, the authors in [3] study the case of stochastic replicators dynamics in 2×2 symmetric games. They add an element of randomness by making at each step every player to change his strategy with some probability ϵ, independently from the replicator principle. They show that this works quite well, as it favors paretto efficient Nash equilibria. None of these works considers interacting species when population members move inside a *network*. Our work is the first work that examines *species antagonism discrete dynamics of a stochastic nature*, over a *finite* graph.

In our model we consider a hawks-and-doves game and assume that the individuals perform random walks on the vertices of some graph. Only players who meet on a vertex play the game. Their payoffs are interpreted as birth or death rates. The evolution dynamics in our model can be described with Markov Chains. A nice introduction to Markov Chains is [6] and also [7] where other stochastic processes are described as well. An upper bound on the number of steps needed for all the players to meet can be derived by the result in [9] where the number of individuals is 2. A more general result for the case of k individuals is given in [1]. However, [9] and [1] do not consider the phenomenon of species antagonism, since they consider only one species. In fact, they do not offer any result about the chance of survival of the weakest species.

2   The Model

Suppose that k individuals, each being either a Hawk (H) or a Dove (D), perform independent random walks on the vertices of a graph G. When a hawk meets one (or more) dove, it kills them all. When two hawks meet then they are both killed. When two doves meet then nothing happens. More formally, when 2 individuals meet, they play the symmetric 2 players game Γ (of two pure strategies H, D) of payoff matrix

$$A_\Gamma = \begin{bmatrix} (0,0) & (1,0) \\ (0,1) & (1,1). \end{bmatrix}$$

Here, row 1 is strategy H and similar for column 1.

Given the initial position of the Doves and the Hawks on the graph G, we are interested in the probability that Doves survive the above process. We then say that *the Doves win the game*.

For simplicity of analysis we present the following more specific model.

Definition 1 (The Distinct Hawks-and-Doves Model)
We define the distinct Hawks-and-Doves model *in the following way:*

1. *Each individual starts on a different vertex of a connected graph G.*
2. *At every step a single individual is chosen uniformly at random and moves equiprobably to a neighboring vertex.*
3. *When a Hawk meets several Doves on a vertex, he plays the game Γ with every one of them, so that at the end of the step, no Doves remain on this vertex.* [1]

---

[1] An interesting variation of this is to assume that the Hawk plays the game with exactly one Dove on the vertex.

4. The process stops (and "the game ends") when only one type of individuals remains on the graph.

Representation of the Evolution Dynamics. It is easily seen that the evolution in the Distinct Hawks-and-Doves model can be represented by a Markov Chain. This is true, because all we need to know in order to specify the next configuration of the model at some discrete time t is the number and position of the individuals left on the graph. This can be a little tricky. The state space of the Markov Chain will need a way to describe the individuals that have not yet disappeared, as well as their exact location on G. The latter can be done perhaps with a k-tuple each element of which describes the position of a single individual. But the issue of which individuals have still remain on the graph remains. As we will see below, things are considerably simpler when we have exactly 1 Dove.

Single Dove Distinct Model. We will derive a Markov Chain that captures the evolution of the distinct Hawks-and-Doves model in the case where we have only 1 Dove. The state space of the Markov Chain is

$$I = \{(d, h_1, h_2, \ldots, h_{k-1}) \in V(G)^k\}$$

i.e. the set of all k-tuples whose first position denotes the vertex of the dove and whose i-th position denotes the vertex of the $(i-1)$-th Hawk. Note that if a vertex v appears an even number of times in some state $s \in I$, then all the individuals on this vertex have been killed.

In order to describe the transition matrix P of the Markov Chain we need some additional notation. For some $s \in I$, let $v(s)$ be the number of hawks that have v as their corresponding position in s. Then, (assuming that we have an even number of Hawks) the absorbing states of the chain are $A = \{s \in I : v(s) \text{ is even } \forall v \in V\} \bigcup \{s \in I : d(s) \text{ is odd}\}$, where $d(s)$ is the number of hawks that are in vertex d (i.e. the vertex of the dove) in situation s. If the chain is absorbed in the first set, then the Dove wins, otherwise he loses. Let now, for some transient state $s \in I$

$$k(s) = 1 + |v \in V : v(s) \text{ is odd}|$$

i.e. $k(s)$ is the number of players left when the state is s. Then the representation matrix P has

$$P_{s_1, s_2} = \begin{cases} \frac{1}{k(s_1)} \frac{1}{degree(x)} & \text{if } s_1 \notin A, \text{ the individual chosen is on vertex } x \\ & \quad \text{and } \exists u \in V : s_2 = s_1[x/u] \\ 1 & \text{if } s_1 = s_2 \in A \\ 0 & \text{otherwise.} \end{cases}$$

The notation $s_1[x/u]$ means that all occurencies of x in s_1 have been replaced by u.

2.1   Some Interesting Problems

Let us now discuss some interesting open problems on this model.

Open Problem 1 (Absorbtion Probability). *Given the initial position of every individual, what is the probability that the Hawks are eliminated before all the Doves do?*

It is easy to see that if the number of Hawks is odd, then the probability that the Doves win is 0 (as long as our graph is connected). *The question becomes interesting only in the case where there is an even number of Hawks.*

As we will see, if the number of individuals k is constant, then the above probability can be computed in $O(n^{poly(k)})$ steps. Of course, using the same method for $k = k(n) \rightarrow \infty$ we can get an exponential time algorithm. Is this the best we can do? This gives rise to the following obvious question:

Open Problem 2 (Hardness). *Given the initial position of every individual, how hard is it to compute the probability that the Hawks are eliminated before all the Doves do, in the case where $k = k(n) \rightarrow \infty$?*

Even if it is not likely that we are able to compute the above probability in polynomial time, we could possibly estimate it by running the model many times and looking at the outcome after a large period of time. In order for this technique to be usefull, the following should hold:

1. The outcome should be reached in a polynomial (expected) number of steps. Of course, using the results of [9], we can see that an upper bound on the expected time needed so that only Hawks or only Doves remain is at most $O(kn^3)$, which is polynomial considering that $k \leq n$. A slight improvement to this bound can be found in [1]. However, both these techniques do not take into consideration the strategy that each individual plays. This fact may lower considerably the upper bound on the time needed to come to a conclusion on whether the Doves or the Hawks finally win. For example, consider the extreme case where there are exactly two Hawks. Then, using the result of [9], the mean time needed until the time ends is just $O(n^3)$. But this does not tell us anything on the possibility of the dove winning the game.

2. The probability of dove survival should not be exponentially small (otherwize we should have to simulate the game for an exponential number of times to get a good result). In this paper we will show that when G is undirected then the probability of dove survival in the distinct hawks-and-doves model is lower bounded by the inverse of a polynomial in the number of vertices n of G. However, this is not the case for directed graphs, as is shown below.

Theorem 1. *There is a directed graph G and suitable initial positions of the individuals for which $0 < Pr(D \text{ survives}) \leq 1/2^{n/2}$ in the single dove distinct hawks-and-doves model.*

Proof. Let G and the initial positions of the $k-1$ hawks and the dove be as in the following figure. Clearly, the probability that the dove survives is equal to the probability that it reaches vertex Y. But the probability that this happens is exactly $\frac{1}{2^{k-1}} = \frac{1}{2^{n/2}}$.

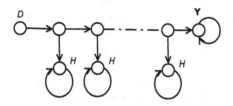

Fig. 1. An instance of a directed graph case with exponentially small probability of survival

$\square$

We conjecture that the exact calculation of $Pr(D$ survives$)$ is a #P-complete problem for directed graphs.

3   Probability of Absorbtion

As we have discussed, the evolution dynamics of the model can be described by a Markov Chain with state space I and transition matrix P. Without loss of generality, we may assume that I consists only of two absorbing states s_D and s_H, the first one being reached when only Doves remain (i.e. the Doves win) and the second one being reached when only Hawks remain. Also we will denote by f_s the probability that the Doves win the game, given that the chain starts at state $s \in I$. Let now $\mathcal{F} = [f_1, f_2, \ldots, f_{|I|}]^T$, for some ordering of the states in I. Clearly, $\mathcal{F}$ is the unique nonnegative solution that has $f_{s_D} = 1$ and $f_{s_H} = 0$, of the following (matrix) equation

$$\mathcal{F} = P \cdot \mathcal{F}. \tag{1}$$

By classical methods used for the transient analysis of Markov Chains we can get an algorithm that calculates the exact probability of absorbtion given any graph (directed or undirected) and any initial positions of the k individuals in $O(n^{3k})$ time. For details, see [5].

In the sequel we consider *undirected* graphs.

Note: The state space of the Markov Chain describing the evolution of the distinct model in the case of 1 Dove seems in a way easier to work with. However, the open problems 1 and 2 remain interesting even in this case.

3.1   The Exact Probability of Absorbtion (Some Cases)

Let G denote the graph on which the individuals of the distinct hawks-and-doves model move. Let also the number of Doves be 1 and the number of Hawks be

$k - 1$ (even). Assume for a moment that every individual chooses his starting vertex with some probability distribution on $V(G)$. For individual i, let us call this initial distribution F_i. Then, by symmetry we can see that the following holds:

Lemma 1. *If $F_i = F_j$ for all individuals $i \neq j$, then the probability that the Dove wins (given that all individuals start at a different vertex) is*

$$P(D \ wins) = \frac{1}{k}.$$

As we will see, when G is the complete graph, then Lemma 1 holds even in the case where every individual starts on a separate vertex.

Complete Graph. We here assume that the graph on which the individuals move is the complete graph K_n. Also, assume that we work with the distinct hawks-and-doves model where the number of Hawks is $k - 1$ and we have only one Dove (the general case with s Doves can be easily reduced to the case with 1 Dove and is discussed in the end of the section). Finally we assume that $k - 1$ is even, because if it is odd, the probability that the Dove wins is 0 (G is connected and there will always remain at least one Hawk). We can prove the following

Theorem 2. *In the distinct hawks-and-doves model, when $G = K_n$, the number of Hawks is $k - 1$ (even number) and we have only one Dove, the probability that the Dove wins the game is $\frac{1}{k}$.*

Proof. See [5].

$\square$

Notice that the above result is actually a special case of Lemma 1. Suppose now that we have s Doves instead of one and also, when a Hawk meets with several Doves on a vertex, it "eats" them all. Then the number of Doves that survive the process is actually a Binomial random variable $X \sim \mathcal{B}(s, \frac{1}{k})$. Hence

$$P(D \ wins) = 1 - P(X = 0) = 1 - \left(1 - \frac{1}{k}\right)^s.$$

Cycle. We now assume that the graph G on which the individuals move is the cycle C_n. We consider the distinct hawks-and-doves model where the number of Hawks is 2 and we have only one Dove. This choice of parameters is made because of the following reasons:

1. The probability that the Doves win in the case of one Dove is a lower bound on the probability that the Doves win in the case of more than one Doves. So, a lower bound for the first case applies also for the second (more general) case. Notice that this is true irrespectively of the graph G.
2. In the case where the graph is C_n and we have $k-1$ hawks and 1 Dove, a lower bound on the probability that the Dove wins is actually the probability that the Dove wins when it starts from a vertex with two Hawks on its adjacent

vertices. This is true because (a) only the first Hawk to the Dove's right and the first Hawk to the Dove's left can end the process in favor of the Hawks, i.e. "eat" the Dove and (b) all that the rest of the Hawks do is to turn these two further away from the Dove.

We can now prove the following (see [5] for the proof).

Theorem 3. *In the distinct hawks-and-doves model with 2 Hawks and one Dove, the probability that the Dove wins when $G = C_n$ and the initial distances of the two Hawks from the Dove are i_0, j_0 is at least $\frac{1}{n^2}$.* □

4   Testing for Survival in Polynomial Time

In the distinct hawks and doves model with a single dove D we examine here the decision problem:

SURVIVAL: Given G and initial positions I of the k individuals (at most one per vertex, so $k \leq n$) and having only one Dove (i.e. $k - 1$ hawks), has the Dove a positive probability to survive?

Note that when $k - 1$ is odd then D cannot survive because eventually a single hawk will remain with certainty and then meet D and kill it. Thus the problem is interesting when the number of hawks is even. In this case, two cases are absorbing states: (1) no particle survives or and (2) D survives.

We will show here that we can decide $Pr(D$ survives$) > 0$ in polynomial time. Note that the two cases shown in figures 2 and 3 (call them *extinction graphs*) have $Pr(D$ survives$) = 0$.

Fig. 2. *Line extinction graph*: a line with the dove between two groups of odd hawks each

In fact, we can prove the following

Theorem 4. *If we assume that we have an even number of hawks and a single dove, then the Line and the Star extinction graphs are the only cases for which $Pr(D$ survives$) = 0$.*

Proof. See [5].

□

In the case of many doves and an even number of hawks we can similarly show that

Theorem 5. *The Line with all the doves between two hawks and the star with all the doves in the center (and hawks on all other vertices) are the only cases for which the probability of some dove surviving is zero.*

Using Theorem 4 it is easy to see that we can decide in polynomial time if there is a positive probability that the dove survives.

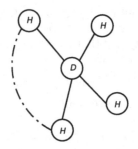

Fig. 3. *Star extinction graph*: a star with the single Dove in the middle and no free vertices

5   The Probability of Dove Survival Is at Least $1/poly(n)$

5.1   Introduction

We show here that when $Pr(D$ survives$)$ is non-zero, then it is at least $\frac{1}{poly(n)}$. We first analyse carefully two "hard" cases, that are produced from slight modifications of the extinction graphs. Then we use them to show the general case.

Our analysis methology is to bound the probability of dove's survival from below, by estimating the probability of particular sequences of moves of particles, in all of which the dove survives. When I, I' are two spontaneous descriptions of the positions of the particles (configurations) then the event "I moves to I' safely for D" means that there exists a sequence of moves of particles transforming I to I' without D being killed in the process. We denote this by $I \overset{s}{\Rightarrow} I'$ (s for "safely").

Clearly, if A is the event "Dove will eventually survive provided particles play the game starting from I" and B is the event "Dove will eventually survive provided particles play the game starting from I'", then

$$P(A) \geq P(I \overset{s}{\Rightarrow} I')P(B).$$

In the sequel, events like A are described (for economy) as "Dove survives given I".

5.2   The First "Hard" Case

Consider the case depicted in figure 4. We will denote this by (G_1, I_1).

Notice that this case is a slight modification (i.e. with one extra edge) of the Line extinction graph, in order to have non-zero probability of dove survival. Intuitively, we can see that this must be the worst case when we have only 2 hawks and 1 dove. We can prove that (see [5] for the proof).

Theorem 6
$$Pr(Dove\ survives\ given\ I) \geq c \cdot \frac{1}{n^5}.$$

where c is a constant. $\Diamond$

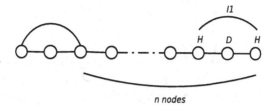

Fig. 4. Case (G_1, I_1)

5.3 The Second "Hard" Case

Consider the case depicted in figure 5. We will denote this by (G_2, I_2). $A_i, i = 1, \ldots, k$ are the initial positions of the hawks and 0 (root) the initial position of the Dove.

Notice that this case is a slight modification of the Star extinction graph, that has non-zero probability of dove survival. In order to lower bound the probability that the dove survives, we need the following definition

Definition 2 (Generalized Dove's Fortune). *Given an initial configuration I_2, the fortune of the dove is defined to be the k-tuple*

$$F(I_2) \overset{def}{=} \langle distance_{I_2}(H_1, D), \ldots distance_{I_2}(H_{k-1}, D) \rangle .$$

We will say that $F(I) \leq F(J)$ if for all hawks $distance_I(H_i, D) \leq distance_J(H_i, D), i = 1, \ldots, k - 1$. Although it is difficult to compare two configurations I, I' in the general case, it is easy to verify that when the Dove is in the root vertex of G_2 and $F(I) \leq F(J)$, then the probability that the dove survives is smaller for I. Indeed, the closer the hawks are to the dove, the less space does the latter have to move. Hence

Lemma 2. *When the dove is in the root vertex of G_2 we have*

$$F(I) \leq F(I') \quad \Rightarrow \quad Pr(D \text{ survives given } I) \leq Pr(D \text{ survives given } I').$$

So, the probability that the dove survives given I_2 is at least the probability that the dove survives when all the hawks are adjacent to it. Note also that in this situation, if we remove some edge from the end of a branch, then we "help" the hawk on this branch to move towards the dove, so the probability that the dove survives becomes smaller. Thus, the probability that the dove survives in the case (G_2, I_2) can be lower bounded by the probability that the dove survives in the case (G', I') that is depicted in figure 6.

To find the probability of dove survival in this case, it is easy to see that if the dove wants to survive we must first arrive at a situation where H' is on vertex 2 and D on vertex 1. Let us call this configuration I''. The easiest way to reach I'' from I' is to move H' first to 2 and then the D to 1. So we have

$$P(\text{we reach } I'' \text{ given } I') \geq \frac{1}{k}\frac{1}{2} \cdot \frac{1}{k}\frac{1}{k-1} \geq \frac{1}{2} \cdot \frac{1}{k^3}.$$

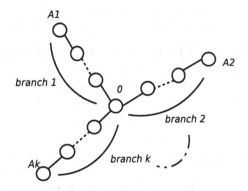

Fig. 5. Case (G_2, I_2): Dove is on the root initially

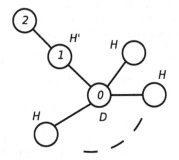

Fig. 6. Case (G', I')

Given I'', the probability that some two hawks, other than H' meet is obviously at least $\frac{k-2}{k} \cdot \frac{k-3}{k}$. In this new situation (call it I'''), let us condition on the event that the D moves before H' and it moves to vertex 0 (event E_1). This happens with probability $\frac{1}{4}$. Also, let us condition on the event that if the dove "escapes" from the center of the graph without being eaten and before H' has moved, then it does not return to vertex 1, but to one of the vertices of the two hawks that met before it moved (event E_2). This happens with probability $\frac{2}{3}$. But it is easy to see that given E_1, E_2 and I''', the probability that D survives is the same as some hawk's H other than H'. Also, it is clear that the probability that H' survives is strictly larger than the probability of survival of any other individual. Then, we must have that the probability that H' together with D and some $H \neq H'$ are the last three to survive is at least $\frac{k-4}{\binom{k-2}{3}} \geq \frac{6(k-4)}{k^3}$. Let us denote the situation where only H', D and some $H \neq H'$ have survived by I''''. Then we have seen that

$$P(\text{we reach } I'''' \text{ given } I') \geq \frac{1}{2}\frac{1}{k^3} \cdot \frac{k-2}{k}\frac{k-3}{k} \cdot \frac{1}{4}\frac{1}{3} \cdot \frac{6(k-4)}{k^3}$$

Given now that only these have survived, the probability that the dove survives is obviously constant. We have thus proved that the probability that the dove survives given I' is at least $c \cdot \frac{1}{k^5} \geq c \cdot \frac{1}{n^5}$. So

Theorem 7. *In the second hard case* (G_2, I_2) *the probability that the dove survives is at least* $c \cdot \frac{1}{n^5}$.

6   The Case of a General Graph G

Let I be the initial position of the particles in G. Since the probability that the Dove survives is non-zero, the number of hawks is even and also (if G is not a cycle which we handled earlier)

(a) either the tree of shortest paths from the Dove to the Hawks has at least 4 branches or
(b) the Dove is in a line with hawks left and right but the line connects (at least in one of its ends) to a subgraph G' with at least a node of degree 3 and a free position.

Set $P_S(D, I) \equiv Pr(D$ survives given $I)$. From the case (G_1, I_1) we get

$$P_S(D, I) \geq \frac{1}{poly(n)} \cdot P_S(D, I_1)$$

where the HDH block moves to the left end of the line and the left H and D enter G'. But then, if I is as in case (b) (call it I_b) we get

$$P_S(D, I_b) \geq \frac{1}{poly(n)} \cdot P_S(D, I_a)$$

where I_a is as in case (a). Note now that I_a is very similar to the case (G_2, I_2). The only difference is that there may be non-tree paths (with hawks) in the tree of shortest paths from D to the hawks positions. These hawks do not reduce the probability of dove's survival since they can only eat other hawks before eventually having a situation like (G_2, I_2). Eventually

$$P_S(D, I_a) \geq \frac{1}{q(n)} \cdot P_S(D, I_2)$$

but we know that $P_S(D, I_2) \geq \frac{1}{r(n)}$, where $q(n), r(n)$ are polynomials. So we get our main theorem.

Theorem 8. *For any undirected graph* $G = (V, E)$ *with an even number of hawks and a single dove and any initial positions* I *of the particles*

$$P_S(D, I) > 0 \Rightarrow P_S(D, I) > \frac{1}{\pi(n)}$$

where $\pi(n)$ *is a polynomial in* $|V| = n$.

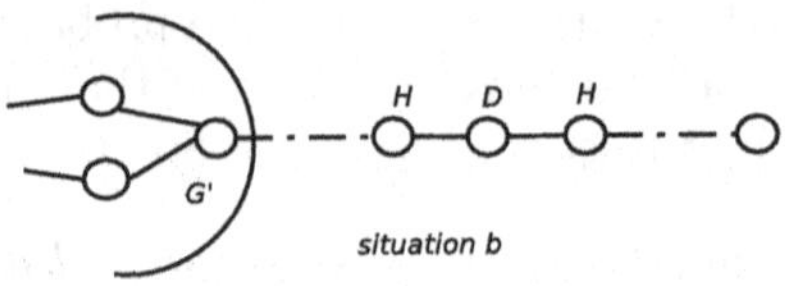

Fig. 7. The general case

Note that when we start with more doves, the probability that the doves win is even better. Hence

Corollary 1. *For any undirected graph* $G = (V, E)$ *with an even number of hawks, any number of doves and any initial positions* I *of the particles, if the probability of at least one dove surviving is non-zero, then it is at least* $1/\pi(n)$, *where* $\pi(n)$ *is a polynomial of the number of vertices* n *of* G.

From the above we can get the following

Corollary 2. *We can approximate the probability of only doves surviving, within any degree of accuracy in expected polynomial time, for any graph* G *and initial positions* I.

Our method just simulates the game (until it ends) polynomially many times and counts the frequency of cases where only doves survive at the end. For details, see the full version of this paper [5].

7   Conclusions and Future Work

We presented and analyzed here a simple but non-trivial model of antagonism in networks. We have proved that predicting the eventual survival chances of the antagonistic species can be done efficiently in the case of undirected graphs. We conjecture that the exact estimation of Dove's survival chances is a #P-complete problem in the case of directed graphs. We are currently working on extensions when Doves reproduce when they meet, e.g. let $\alpha \geq 2$ doves be born when 2 doves meet. Note that when dove reproduction is allowed, even the case of an odd number of hawks becomes non-trivial, as hawks and doves can co-exist. Thus, it is interesting to investigate the eventual population mixture as well as the existence (or not) of a threshold (for α) below which hawks win almost certainly at the end.

References

1. N. Bshouty, L. Higham and J. Warpechowska-Gruca, *"Meeting Times of Random Walks on Graphs"*, Information Processing Letters, 69:259-256, 1999.
2. L. A. Imhof, *"The long-run behaviour of the stochastic replicator dynamics"*, Annals of Applied Probability, 2005, Vol. 15, No. 1B, 1019-1045.
3. M. Kandori, G. J. Mailath and R. Rob, *"Learning, Mutation and Long Run Equilibria in Games"*, Econometrica, Vol. 61, No. 1 (Jan., 1993), 29-56.

4. J. Maynard Smith, *"Evolution and the Theory of Games"*, Cambridge University Press, 1982.
5. S. Nikoletseas, C. Raptopoulos and P. Spirakis, *"The Survival of the Weakest in Networks"*, `http://students.ceid.upatras.gr/~raptopox/HawksandDoves.ps`
6. J. Norris, *"Markov Chains"*, Cambridge University Press, 1997.
7. S. Ross, *"Probability Models for Computer Science"*, Harcourt Academic Press, 2000.
8. L. Samuelson, *"Evolutionary Games and Equilibrium Selection"*, MIT Press, Cambridge, Massachusetts, London, England.
9. P. Tetali and P. Winkler, *"On a random walk problem arising in self-stabilizing token management"*, in the Proceedings of the 10th Annual ACM Symposium on Principles of Distributed Computing, pages 273-280, 1991.
10. J. W. Weibull, *"Evolutionary Game Theory"*, MIT Press, Cambridge, Massachusetts, London, England.

Online Distributed Object Migration

David Scot Taylor

San José State University
taylor@cs.sjsu.edu

Abstract. We study Distributed Object Migration using competitive analysis. The problem is motivated by distributed object-oriented computing, for which intelligent dynamic migration of (Java or other object-oriented) objects during runtime is important for efficient implementation on multiprocessor systems. In the online version of the problem, k mobile objects reside at n nodes of a network and they respond to a sequence of requests. Each request specifies two objects which have to communicate, and the algorithm has to decide whether to bring the objects together or not. We focus on the case of uniform networks with relatively large communication costs and show tight upper and lower bounds of k, for any network size $n \geq 2$. Our algorithm TIMESTAMP uses a timestamp for each object, and we analyze it using an *implicit potential function argument*; the analysis is interesting in its own right, and may be applicable to a wider class of problems, but it doesn't seem to be widely used. This implicit potential function argument gives a simple and intuitive proof of the (suboptimal) competitive ratio of $2k - 1$, within a factor of 2 of the optimal deterministic competitive ratio. To show the optimal competitive ratio of k, we use an explicit, yet less intuitive, potential function.

Keywords: Online Algorithms, Competitive Analysis, Distributed Algorithms, Data Management, Object Oriented Programming.

1 Introduction

In distributed object-oriented computing on multiprocessors or on a computational grid, the time for communication between objects located on two separate processors (or nodes) may be very large compared to the time for communication within the same processor. To best exploit high processor speeds, objects may migrate to new processors, in an attempt to minimize inter-processor communication costs. Further complicating the issue is the fact that frequently, decisions about where the objects should reside must be made *online*, that is, without knowledge of future events.

We consider here the online version of Distributed Object Migration (DOM): a distributed program, consisting of a collection of self-contained objects, runs on an n processor network. At any given time, each object resides at some node (processor) of the network. Some objects are mobile and can move freely to every node of the network and some objects are immobile because they are associated

T. Erlebach and C. Kaklamanis (Eds.): WAOA 2006, LNCS 4368, pp. 330–344, 2006.
© Springer-Verlag Berlin Heidelberg 2006

with the particular node on which they run. When two objects on different machines need to communicate some amount of data c, and one of the objects is mobile, the system is presented with the following dilemma: should the objects be brought closer together or not? Further, if both objects are mobile, which object(s) should be moved? There are different costs associated with each option. The cost of moving objects from node u to node v depends in general on the type of the object as well as the nodes u and v. Also, the cost for communicating an amount of data c between nodes u and v is equal to $d(u,v) \cdot c$, where $d(u,v)$ is the distance in the network between nodes u and v. In this work, we mostly consider the uniform case, in which all distances are one, and all objects have size one. This case will help to illustrate key differences between this work and previously studied problems.

The problem can be simplified somewhat with respect to the immobile objects. Firstly, each node needs at most one immobile object, because for all purposes, multiple immobile objects at the same node are interchangeable. Next, in our uniform, complete network, a node that contains no immobile object can, with a fixed cost, be removed completely from consideration; mobile objects beginning at such a node can initially be moved to other nodes, after which the node will never be used again. In summary, for the uniform complete network, we can assume that each node of the network contains precisely one immobile object. Henceforth, we assume that there are n immobile objects, one on each node, and k mobile objects free to move between nodes. Also from the point of view of competitive analysis, it makes no sense to consider requests between two immobile objects. Both the online algorithm and the optimal offline algorithm are forced to service this type of request in the same way, with the same cost, and therefore they will only decrease the competitive ratio. If any such requests are made, any online algorithm could simply answer them while otherwise ignoring them, at no harm to its competitive ratio.

The problem belongs a general class of migration problems, including the intensively studied File Migration, File Allocation, and Distributed Paging Problems [2,3,4,8]. Of these, the problem closest to DOM is the *File Migration Problem* [8], in which files are stored on a distributed network, and different nodes access the files. The system attempts to minimize communication costs in case the files being accessed can be moved from one processor to another: within a distributed network, the cost of accessing a file is proportional to the distance from the requesting processor to the file. Alternatively, the file can move locations, at a cost of the distance moved multiplied by the size of the file.

DOM is a generalization of the File Migration Problem. While DOM allows requests between any two objects, the File Migration Problem only allows requests between an immobile and a mobile object.

We study DOM in an online setting, using competitive analysis [18] to measure the quality of the algorithm which determines object migration. Requests come one at a time, and after each request, the algorithm decides how each object should migrate before the requested objects communicate. Once the request is serviced, the next request is made. The algorithm has no knowledge of future requests. We measure the quality of an online algorithm by its competitive ratio, equal to the worst case ratio of the cost the algorithm over any sequence of requests, divided by the optimal cost required satisfy the same sequence. Oftentimes, the sequence is considered to be picked by an "adversary", which specifically chooses a sequence for which the online algorithm will have high costs, but the optimal (offline) solution costs will be small. The online algorithm is playing against the adversary, where the online algorithm tries to minimize competitive ratio, and the adversary tries to maximize it.

The abstract is organized as follows: in Section 2, we give a formal problem description, and introduce an important special case of the algorithm. This is followed by further discussion of related previous work. In Section 4, we prove a lower bound of k for the deterministic competitive ratio of every network metric space, even those with only two nodes. In Section 5, we describe a simple, natural algorithm, the TIMESTAMP algorithm. In Section 6, we prove a competitive ratio upper bound of $2k - 1$ for the basic problem variant using an intuitive *implicit* potential function, which is analyzed without explicitly writing any equation for the potential. In Section 7, a more complex potential function is given explicitly, and it is used to show that TIMESTAMP has the optimal deterministic competitive ratio of k for the variant. Although this section uses standard potential analysis techniques, we use intuition gained from the implicit potential analysis to explain *why* the explicit potential function works, and to break it into two intuitive parts. Section 8 summarizes contributions and outlines future research directions.

2   Problem Definition

Problem Setting: We are given a weighted graph of n nodes, where the nodes represent machines or processors, and the edge weights represent (metric) distances between the processors. There are also objects that reside on the nodes of the graph. There are k mobile objects and n immobile ones, one for every node of the graph. At every time step, each object is at some node of the graph; initial positions of mobile objects are given. In general, objects have sizes specified, but we mostly consider the case of same-size objects.

Online Input: Requests for object communication come online. Each request consists of a pair (x, y) of objects and a cost c representing the amount of data which needs to be exchanged between x and y. After each request, any of the k mobile objects can be moved from one processor to another. The cost of moving a mobile object z of size $s(z)$ from u to v is $s(z) \cdot d(u, v)$, equal to the product of its size and the distance moved. Then an additional cost of $d \cdot c$ is paid if the requested objects are still d distance apart.

Output: After each input request, the algorithm must decide any object movement before the next request.

For our upper bound results (Sections 6 and 7), we study a specific variant of the problem. First, all objects will have unit size, as might occur when the distributed program consists of many different instances of the same, fixed-size object type. Next, the network is uniform, with all nodes 1 unit apart. Finally, all communication costs will be relatively large, with $c \geq 2$. For this case, we can assume without loss of generality that any algorithm will answer queries between objects only after the objects are on the same machine. We will call this Basic Distributed Object Migration, or BDOM.

Although BDOM is somewhat restricted, it is an important DOM variant. Specifically, it helps to highlight DOM's differences from previously studied work. Given its similarities to the file migration problem, one might hope to see similarly competitive solutions for DOM, yet we show that this is impossible. In file migration, separate files can be considered independently; the system can achieve an $O(1)$ competitive ratio on one file, so the same algorithm will achieve an identical ratio for a multi-file system. The total number of files in the system is irrelevant. In contrast, for DOM, requests between two mobile objects allow for direct interaction between the objects, and so the number of objects in the system is an important ingredient for the competitive ratio. In DOM, if there is just one mobile object (and thus only requests between it and immobile objects, as in file migration), but all requests carry high communication costs, a trivial 1-competitive algorithm exists, but this proves nothing about performance with multiple mobile objects. We prove requests between two mobile objects add complexity to the system beyond that of file migration, even in simple systems: for BDOM, further restricted to just just two processors, there is a lower bound of k for the competitive ratio of any online algorithm of k mobile objects. This is quite different than the already known $O(1)$ upper bound for the file migration problem on general networks.

Requests between two mobile objects are fundamentally different than those previously considered in file migration, distributed paging, or caching problems. Beyond the previously considered questions of *if* an object should move and *where* an object should move to, we must also consider *which* object moves. BDOM, analyzed here, will concentrates study on this last facet, while the more general version of DOM also need to address previously considered issues.

3   Related Work

There is a long list of previous work on problems related to DOM, too long to be fully described here. Previous work related to the analysis technique introduced here (implicit potential analysis) is somewhat more limited.

DOM was motivated by a discussion with Miriam Busch about the distributing Java compiler Pangaea [9,20]. Panagaea acts as a front end for a Java compiler, and it is used to distribute objects which can dynamically migrate during runtime. Benchmark results show a marked improvement in performance over a static system. The benchmark test is based on the the the dining philosopher's problem, and the performances of various migration rules are compared.

DOM is a generalization of the file migration problem, introduced in [8], which spawned a family of online data management problems. For a survey on earlier systems, involving the (static) distributed file assignment problem, see Dowdy and Foster [15]. For a survey on early work on dynamic systems, see Bartal's 1996 survey in [16]. This includes many references to the file migration problem (in which one copy of each file is present on a network, and they migrate as they are called by users), the file allocation problem (in which files may replicate to help make *read* operations more efficient, but *write* operations must update all copies), and the distributed paging problem (in which each processor has a limited capacity). The survey covers work on both limited topologies and general networks. Although much work on these topics has occurred since 1996, most of it relates to adding more generality (network topology) or subtlety (such as direct mapping issues [2]) to the basic variant we will consider for DOM in this work. None of these systems need to address the question of which object should migrate, as only one object can satisfy a given request. Only in the paging problem do the moving objects (pages) interact at all, but there it is indirectly, through limited processor capacities.

The static offline case of object migration (which involves no migration) is equivalent to the MULTIWAY CUT problem [10,14]. The $n + k$ immobile and mobile objects represent the vertices of the graph to be cut, and the n immobile objects represent the specific vertices (terminals) which must be separated by the cut. Each pair of objects needs only one request between them, with cost equal to the weight between the matching vertices in the MULTIWAY CUT problem. This problem is known to be MAX SNP-Hard for 3 or more terminals since 1983. (Note, in the distributed Java objects case of [9], a static benchmark instance is used, but it is one which is easy to solve optimally.)

Of separate interest from DOM is the analysis technique used here, with a goal of making analysis simpler and more intuitive. The lazy potential, which we use in Section 6, is used (explicitly by equation, not implicitly) by Chrobak and others in [7,11,12]. Working towards the goal of making competitive analysis more intuitive, the potential function argument is completely eliminated by Koutsoupias in [17]. The author doesn't know of any previous specific attempts to perform implicit potential function analysis, but in the seminal online analysis papers by Sleator and Tarjan [18,19], potential functions are described in some detail rather than explicitly given by equations. This technique does not seem to be widely used, although it gives more intuition about the algorithms used, while reducing the complexity of equations analyzed.

4   Lower Bounds

Two helper lemmas will be useful to prove the problem lower bound (Lemma 3). Their proofs are omitted due to space constraints. Further, the proof for Lemma 3 has been greatly condensed. The proofs for these lemmas are either straightforward (for the helper lemmas) or follow well-established techniques (for Lemma 3). More difficult or novel proofs in following sections are given in full.

Lemma 1. *In* DOM, *for each request* (x, y), *regardless of cost* c, *the optimal online (or offline) algorithm need not move any object other than* x *or* y. *For* *the problem with two processors (n = 2), there is never a need to move both objects for any request.*

Lemma 2. *In* DOM, *for each request between objects x and y, if the size of the communication request c is at least twice the size of the smaller object, without loss of generality, the optimal online (and offline) algorithm can move x and y to be on the same machine before they communicate to satisfy the request.*

Lemma 2 shows that requests with large enough communication costs can always be answered by moving objects to the same node. Large requests make the problem simpler, eliminating "ski renter's dilemma" issues of whether or not an object should be moved. For related problems, see Karlin's chapter in [16].

Lemma 3. *On two (or more) processors with any k (positive weight) mobile objects, the deterministic competitive ratio of* DOM *is at least k.*

Proof. We use a standard lower bound argument, proving that for any online algorithm, the cost is equal to the combined cost of k specified offline algorithms, and thus must have cost at least k times larger than the smallest among them.

Consider just two processors. Suppose that initially, one processor contains only itself, called the (immobile) object 0, while the second processor contains itself (immobile object $k+1$) and regular (mobile) objects $1, 2, \ldots, k$. The adversary will always make large communication requests between consecutively numbered items not on the same server, starting with $(0, 1)$. Proof details are omitted due to space constraints, but the only configurations which need be considered (by Lemmas 1 and 2) are those with objects $0 \ldots i$ on one server and $i+1 \ldots k+1$ on the other, for $0 \leq i \leq k$. The online algorithm will be in one such configuration, and the adversary will have one offline algorithm for each of the other k such configurations.   □

This lower bound for the competitive ratio of DOM is much higher than that of the File Migration Problem (see Section 2), even with just $n = 2$ processors. This bound is proven tight for BDOM in Section 7, using algorithm TIMESTAMP.

5   The Timestamp Algorithm

The following algorithm is intuitively simple, and it achieves a competitive ratio of k for BDOM, the basic distributed object migration problem. It makes use of a unique "timestamp" for each online request, for which any strictly increasing function will suffice, including time, or sequential identification numbers. Given a request to objects (x, y), they can decide which should move without knowing information about any other objects. For simplicity in our proofs, we will assume that the timestamp is simply a globally known time.

Algorithm TIMESTAMP[1]:

1. Initialize all mobile objects with arbitrary timestamps, such that all are previous to the time of the first request, and no two objects residing on different processors have the same timestamp.
2. To satisfy a request between mobile objects (x, y), move the object with the least recent timestamp to the machine of the other object, and update the timestamp of the moving object to match the other object's timestamp. (That is, if x moves, give x the same timestamp that y has, ignoring the timestamp of the current request.) If the objects are already on the same machine, simply update the timestamp of the earlier object to match the object with the later timestamp.
3. When a mobile object is requested with an immobile object, the mobile object moves to the other's processor, and updates its timestamp to be the current time, that is, the time of the current request. (If it is already on the correct processor, it just updates its timestamp.)

Note that timestamp is a concept only used by the online algorithm, and thus, when we refer to the timestamp of an object, we refer to the timestamp assigned to that object by the online algorithm. Intuitively, when an object is called to a processor, we give it a new timestamp to indicate that the object *must* be at that processor at that time in any optimal algorithm. For requests between two mobile objects, the one with "less recent information" moves and "updates" its information to match the more recent object's timestamp.

In analyzing TIMESTAMP we will make use of the following simple lemma:

Lemma 4. *(a) If two mobile objects have the same timestamp, they reside on the same processor. (b) Other than the initial timestamps, all mobile objects will have the timestamp of a request involving the processor on which the reside. (c) Other than initial timestamps, new timestamps are only introduced to the system during requests involving processors (immobile objects).*

Proof. Part (c) is clear from the algorithm, where only the last rule introduces new timestamps. For parts (a) and (b), we use induction. (a) holds at the start by the initialization, and (b) is trivially true at the start. TIMESTAMP only changes the timestamp of an object when that object is moved, or when a request which can be answered "for free" is made (which will not violate the properties). If it is moved due to a request between two mobile objects, its timestamp is changed to match the timestamp of an object on the machine it moves to, which is (a) inductively different than any timestamp on any other machine, and also (b) inductively the same as a timestamp of a request of the processor to which the object is moving (or one of the initial timestamps). If it is in a request with a im-mobile object, the timestamp will be updated to one more recent than any other,

---

[1] TIMESTAMP is a natural name for this and many other algorithms for other prob-lems, especially those involving distributed systems, or "LRU" variants for paging algorithms (such as Albers [1]). The algorithm here is different, but the name is so natural, I have chosen to reuse it anyway.

that is, it will be updated to the current time, so (a) still holds. Its new timestamp matches the latest request, which also involves the processor to which the object is being moved, so (b) is also maintained.   □

In Section 6, using implicit potential function analysis, we prove that TIMES-TAMP has a competitive ratio of no more than $2k - 1$ on BDOM. This is followed in Section 7 by a tight, explicit potential function analysis which shows it to have the optimal competitive ratio k, matching the lower bound from Lemma 3.

6   The Lazy Potential and Implicit Analysis

For many online problems, it is difficult both to find and analyze a useful potential function. While finding potential functions, certain "tricks of the trade" are sometimes used, but they are not obvious to the reader of final paper versions, where only the successful potential functions are given. Instead, the potential function is often presented as a somewhat complex and seemingly arbitrary equation. While the potential and its analysis offer a convincing proof of algorithm performance, they offer little insight into how the algorithm works, or what the potential represents, especially for those new to online algorithm analysis.

Here, rather than pulling a complex equation out of a hat, we use a natural potential, the lazy potential[2], defined below. Next, we are able to analyze TIMESTAMP using the lazy potential implicitly: that is, *without ever giving an explicit equation for the lazy potential, we are able to prove that* TIMESTAMP *has competitive ratio at most* $2k - 1$. This proof should give the reader much more intuition than a standard, explicit potential argument, although an explicit potential (Section 7) can prove a tight bound of k on the competitive ratio of TIMESTAMP. There, we show how a more complex (and precise) potential is composed of two natural parts, one of which is the lazy potential.

Given configurations for both the online algorithm and an offline adversary, the *lazy potential* is defined as the maximum total amount of work that the adversary can force upon the online algorithm *while doing no work (and moving no objects) in its own configuration.* (Here, a configuration includes the position and the timestamp assigned by the online algorithm for each object.)

For BDOM, the lazy potential is simply the maximum total number of movements which the adversary can force upon the online algorithm while not moving any of its own objects, making requests it can answer for free. Note, we are not restricting the adversary to any subset of algorithms, it may make any movement at any time (even without requests). Nevertheless, at any point in time, the potential function between an online and offline configuration is equal to the maximal cost which the online algorithm can incur by a sequence of requests for which the adversary does not move. We will call any such request, which the adversary can answer without cost but which the online algorithm must move to answer, a *lazy request.*

---

[2] The term lazy potential, and its analysis, is explicitly used in (at least) [7,11,12], but the author believes that many more online algorithms make use of the lazy potential for all or part of their potential function, without using the name.

Here the lazy potential is defined for DOM, but it clearly generalizes to cover many online problems where the adversary can make requests which it can answer for free, but the online algorithm, in a different configuration, must pay. These requests force any competitive online algorithm to (eventually) converge to the offline algorithm's configuration.

Lemma 5. *For* BDOM *with k mobile objects on n machines,* TIMESTAMP *has a lazy potential of no more than $k(k+1)/2$.*

Proof. First, notice that in TIMESTAMP, a migrating object will never be assigned the same timestamp more than once. The algorithm only assigns timestamps when an object moves, and the new timestamp will be strictly larger (more recent) than the old timestamp.

Next, from given time t forward, during the course of any series of lazy requests, if TIMESTAMP ever assigns an object a timestamp of t or later, that object can never be forced to move again by a sequence of only lazy requests. In order to introduce timestamps larger than t, requests involving processors must be made, by Lemma 4.c. (Prior to time t, no object will have timestamp t or later.) Such a request between a mobile object and a processor freezes that mobile object onto that processor: it must be on that processor to answer that request in a lazy way, and thus the adversary cannot move it to any other processor without incurring movement costs, which is not possible with only lazy requests.

Putting these together, the object with the ith newest (largest) timestamp can only be assigned at most i new timestamps in any sequence which doesn't require any movement from the adversary. That is, $i-1$ more recent timestamps from other mobile objects may be possible, and one additional timestamp due to a request by a processor, for at most $\sum_{i=1}^{k} i$ total movements by lazy requests. $\square$

For any problem where there the minimum cost of a move for the offline algorithm is bounded away from 0, proving any finite bound on the lazy potential is enough to prove a finite bound on the competitive ratio of a problem:

Corollary 1. *For* BDOM *with k mobile objects on n processors,* TIMESTAMP *has a competitive ratio of $k(k+1)/2$ or less.*

Proof. Any movement by the adversary can at most increase the lazy potential of TIMESTAMP by $k(k+1)/2$, by Lemma 5. (The lazy potential is 0 here if the online and offline configurations match. It is never negative.) It costs the adversary one to move. $\square$

If we are somewhat more careful of how much a single move by the adversary can change the lazy potential, we can prove a stronger result, within a factor of 2 of the correct bound:

Lemma 6. *For* BDOM *with k mobile objects on n processors, if the offline algorithm moves an object from server i to server j, and server j has the set of objects S_j on it before the move, the increase to the lazy potential against the online configuration of* TIMESTAMP *is $2|S_j|+1$ or less.*

Proof. The lazy potential is finite, as shown above. Here we bound its change from one offline move.

Given an offline configuration, the only lazy requests possible are those between pairs of objects on the same offline machine, and those between objects and the offline processor on which they reside. Fix some sequence of lazy requests which would incur the lazy potential. We consider the lazy potential to be broken into n subsequences, based on which of the n offline processors the objects for that request reside. Each of these processors accounts for some part of the entire lazy potential, and the sum over all n processor subsequences will be equal to the lazy potential.

Consider the offline move of an object from processor i to processor j. The lazy potential due to processor i cannot increase: any sequence of lazy requests for objects on processor i without the moved object would have been valid before the object moved, so if the new potential due to objects on that processor is higher, the old potential from objects on that processor must not have been maximal. Further, for any processor besides i and j, the potential from objects on that processor does not change. How much can the potential due to the object moving to processor j?

We know that no object can be called to (or from) the same timestamp more than once. Let the set of objects originally on processor j is S_j, and suppose that there are m offline objects from the S_j with (online) timestamps older than that of the offline object being moved. The potential due to processor j is increased by at most $|S_j| + m + 1$: the m older objects might now have an additional lazy request possible in which they are assigned the timestamp of the moved object, and another where they have their timestamp updated from that timestamp, for $2m$ total. Also, the moved object might now be eligible for $|S_j| - m + 1$ lazy requests: one to each of the $|S_j| - m$ more recently timestamped objects on processor j, and one to the jth processor. This makes for $|S_j| + m + 1$ total, where $m \leq |S_j|$. Thus, the potential is increased by at most $2|S_j| + 1$. □

Corollary 2. *In* BDOM *on k mobile objects on n processors,* TIMESTAMP *has a competitive ratio no more than $2k - 1$.*

Proof. In Lemma 6, $|S_j| \leq k - 1$. □

Notice the simplicity of using the lazy potential. By separating the offline and online moves, above we analyze that the offline move (cost 1) raises the lazy potential by at most $2k - 1$. By definition, online moves will not lower the lazy potential below zero; it can always be used to pay for the online move, as it is defined to be the maximum amount of online work which can be forced without any offline work. Here, by not explicitly stating the potential function, almost all algebraic complexity is eliminated from the proof.

We do not expect the lazy potential to be sufficient to prove the optimal competitive ratio for most problems; here, it is sufficient to prove a ratio within a multiplicative factor of 2 from optimal, as we will see in the next section. Proving within an $O(1)$ factor with the lazy potential may the best we can hope for, because some configurations may be intrinsically easier for the online algorithm, or for the adversary, and this is not captured by the lazy potential. Next, we will see a potential for TIMESTAMP which is made of two parts: one part is the lazy potential, and the other very closely related, although it is independent of the online configuration.

7   Tight Bounds for Timestamp

Lemma 7. *For* BDOM *with k mobile objects on n processors,* TIMESTAMP *has a competitive ratio of at most k.*

Proof. To prove a tight upperbound, we use a standard potential argument. Let S_i, for $1 \leq i \leq n$ denote the n sets of objects located on each of the n *offline* processors. Given online and offline configurations *on* and *off*, let $x_{i,j}$, where $1 \leq j \leq |S_i|$, be the $|S_i|$ migrating objects from set S_i, where $j' > j$ implies that $timestamp(x_{i,j'}) \geq timestamp(x_{i,j})$. (That is, objects with higher indices have more recent timestamps, with ties broken arbitrarily.) Let $x_{i,|S_i|+1}$ be the non-migrating object on processor i. For object x, let $M(x)$ be the *online* machine which holds the object. (Note, for all $1 \leq i \leq k$, $M(x_{i,|S_i|+1} = i)$. Finally, let $G(i,j)$ be an indicator variable which has value 1 if $1 \leq j \leq |S_i|$ and $M(x_{i,j}) \neq M(x_{i,j+1})$. Otherwise, $G(i,j) = 0$. Then, we use potential:

$$\Phi(on, off) = \sum_{i=1}^{n} \sum_{j=1}^{|S_i|} \sum_{l=j}^{|S_i|} G(i,j) - \frac{1}{2} \sum_{i=1}^{n} (|S_i|^2).$$

As required, the initial potential value depends only on the configuration (and thus n and k) but not on the sequence of requests. (While this potential can have negative values, it is bounded below by $-k^2/2$. Should readers be more comfortable with non-negative potentials, consider adding $k^2/2$ to this one.)

Here, the first term (the triple summation) represents the lazy potential: over all n (offline) processors, over all objects on that (offline) processor, take the number of more recent objects which reside on a different (online) processor than the next oldest (by timestamp) object. Although this description relates the equation to to the number of times objects can be forced to move, the equation definition itself does not highlight this intuitively. We will return with intuition for the second term later.

Using Lemma 6, we can finish quickly: we know that any offline move of an object from processor i to j can only increase the lazy potential by at most $2|S_j| + 1$, where S_j is the set of objects on processor j from before the move. (After the move, S_j will have one more object, S_i will have one fewer object, and no other sets will change.) The change in the potential for second term will add $-|S_j| + |S_i| - 1$ (exactly) to the potential (where we use the set sizes from *before* the move). The sum of these two changes is $(2|S_j| + 1) + (-|S_j| + |S_i| - 1) = |S_i| + |S_j| \leq k$. Thus, any offline move can increase the potential by at most k.

For the online move, we have already argued that the lazy potential can be used to pay for any online moves. Here, the second term of the potential is independent of the online configuration, and is unchanged for any online move, so the lazy potential can still be used to pay for the online moves.   □

Of note is the intuitiveness of the preceding proof (paired with Lemma 6), compared to a proof in which the potential function above is simply given, and algebraically proved to be sufficient and correct, with little or no insight provided for the meaning of the potential.

As mentioned, the first half of the potential function above is the lazy potential. The second half acts as an "offline potential": given an offline configuration, with S_i objects on each processor, $k + \frac{1}{2} \sum_{i=1}^{n} (|S_i|^2)$ is the highest possible lazy potential for that configuration against *any* online configuration. (Which will be equal to the lazy potential term when all $G(i,j) = 1$.) Since we are concerned more with the change in potential instead of potential, dropping the leading k term makes no difference. To intuitively explain why the offline potential adds negatively to the total potential, if the adversary makes a move into a position with a "potentially" higher lazy potential, but the lazy potential does not in fact change, somehow the adversary has "lost ground" against its goal of getting a high lazy potential. Without knowledge of the online configuration, one would expect such a move to increase the lazy potential. Without a matching increase in the lazy potential, the adversary has not met "expectations", explaining the total drop in potential.

From this and Lemma 3, we trivially get:

Corollary 3. BDOM, *with k mobile objects on n processors, has deterministic competitive ratio k.* TIMESTAMP *achieves this ratio.*

8   Conclusions and Future Work

Our contributions include:

- The introduction of DOM, a generalization of the file migration problem
- A lower bound proof, demonstrating that a simple case of the object migration problem has a competitive ratio much higher than the known upper bound for the general file migration problem
- An algorithm TIMESTAMP and proof of its optimal competitive ratio for the BDOM variant
- Implicit potential analysis using the lazy potential, an intuitive potential function analysis, providing more insight and fewer calculations than explicit potential function analysis

In the future, we hope to explore the frequency with which online algorithms can be analyzed using natural potential functions, either by implicit potential analysis, or at least with intuitive potentials such as the lazy and offline potentials used here. The hope is that the more intuitive base for potential functions will allow them to be found more quickly.

Besides the analysis technique, DOM also merits further study, to find algorithms with good performance on more general versions of the problem, in both deterministic and randomized settings. A rich selection of variants are possible, many already considered for file migration problems. Some possibilities:

- Bounded Capacity Processors: Each processor in the network has a fixed capacity, and the total size of all objects on that processor must be no more than this size. This variant incorporates complexities which arise from the distributed paging algorithm ([2,6,13,21] and many more), a generalization of both the file migration problem and caching. In distributed paging, files do interact with each other, but only indirectly, through limited processor capacities, rather than through the direct requests of object migration.
- Distributed Control: In this work, we have implicitly used a centralized controller to manage movement, but we could require that decisions must be made locally at each node. Only information available at that processor (such as past requests to/by objects on the processor, or perhaps information carried with those requests) is available to make the decision. When a request for object y originates from object x, it may also be that the machine on which x resides has no way of knowing where y is currently located. Extra communication may be needed to track object movements. While we do not address this issue here, related issues for file migration problems are addressed in [5].
- Additional Customized Costs: It may be that even for two objects on the same machine, communication is not free, or perhaps it depends on how many objects are on the machine, or which specific machines. (This could be used to model different threads running in parallel for different objects, on non-identical machines.) Further, communication between two objects on different machines may depend on which two machines they are on. (These options can be used to add some of the generality of metrical task systems to the problem.)

Although we have not considered these more complex variants here, we hope that this work highlights the new difficulties that object migration poses. Requests between two mobile objects allow for extra chances to "make bad choices", thus increasing the competitive ratio well beyond what was seen, even for complex variants of file migration. When bounded capacity processors are added into the mix, distributed object migration becomes a generalization of distributed paging, but once again, requests between two mobile objects pose problems beyond those normally considered for distributed paging.

Acknowledgments

I would like to thank Miriam Busch, who introduced me to this problem in discussions work with her advisor [9,20]. I would also like to thank Peter Widmayer for helpful discussions on presentation, Elias Koutsoupias for extensive discussions about both research and the written presentation, and anonymous referees for their comments.

References

1. Susanne Albers. Improved randomized on-line algorithms for the list update problem. In *Proc. 6th Symp. on Discrete Algorithms (SODA)*, pages 412–419. ACM/SIAM, 1995.
2. Susanne Albers and Hisashi Koga. Page migration with limited local memory capacity. In *Proc. 4th Workshop on Algorithms and Data Structures (WADS)*, volume 955 of *Lecture Notes in Comput. Sci.*, pages 147–158. Springer, 1995.
3. Baruch Awerbuch, Yair Bartal, and Amos Fiat. Competitive distributed file allocation. In *Proc. 25th Symp. Theory of Computing (STOC)*, pages 164–173. ACM, 1993.
4. Baruch Awerbuch, Yair Bartal, and Amos Fiat. Competitive distributed file allocation. *Inform. and Comput.*, 185(1):1–40, 2003.
5. Baruch Awerbuch and David Peleg. Online tracking of mobile users. *J. ACM*, 42(5):1021–1058, 1995.
6. Yair Bartal. *Competitive analysis of distributed on-line problems - distributed paging*. PhD thesis, Tel-Aviv University, 1994.
7. Wolfgang Bein, Marek Chrobak, and Lawrence L. Larmore. The 3-server problem in the plane. *Theoret. Comput. Sci.*, 287:387–391, 2002.
8. David L. Black and Daniel D. Sleator. Competitive algorithms for replication and migration problems. Technical Report CMU-CS-89-201, Department of Computer Science, Carnegie-Mellon University, 1989.
9. Miriam Busch. Adding dynamic object migration to the distributing compiler Pangaea. Master's thesis, Institute of Computer Science, Freie Universität Berlin, September 2001.
10. Gruia Călinescuo, Howard Karloff, and Yuval Rabani. An improved approximation algorithm for MULTIWAY CUT. *Journal of Computer and System Sciences*, 60(3):564–574, 2000.
11. Marek Chrobak, Elias Koutsoupias, and John Noga. More on randomized on-line algorithms for caching. *Theoret. Comput. Sci.*, 290:1997–2008, 2003.
12. Marek Chrobak and Lawrence L. Larmore. HARMONIC is three-competitive for two servers. *Theoret. Comput. Sci.*, 98:339–346, 1992.
13. Marek Chrobak, Lawrence L. Larmore, Nick Reingold, and Jeffery Westbrook. Page migration algorithms using work functions. In *Proc. 4th International Symp. on Algorithms and Computation (ISAAC)*, volume 762 of *Lecture Notes in Comput. Sci.*, pages 406–415. Springer, 1993.
14. Elias Dahlhaus, David S. Johnson, Christos H. Papadimitriou, Paul D. Seymour, and Mihalis Yannakakis. The complexity of multiterminal cuts. *SIAM Journal on Computing*, 23(4):864–894, August 1994.
15. Lawrence W. Dowdy and Derrell V. Foster. Comparative models of the file assignment problem. *ACM Computing Surveys*, 14:287–313, 1982.
16. Amos Fiat and Gerhard J. Woeginger, editors. *Online Algorithms: The State of the Art*. Springer, 1998.
17. Elias Koutsoupias. Weak adversaries for the k-server problem. In *Proc. 40th Symp. Foundations of Computer Science (FOCS)*, pages 444–449. IEEE, 1999.
18. Daniel Sleator and Robert E. Tarjan. Amortized efficiency of list update and paging rules. *Commun. ACM*, 28:202–208, 1985.

19. Daniel D. Sleator and Robert Endre Tarjan. Self-adjusting binary search trees. *J. ACM*, 32:652–686, 1985.
20. André Spiegel. *Automatic Distribution of Object-Oriented Programs*. PhD thesis, FU Berlin, FB Mathematik und Informatik, 2002.
21. Jeffery Westbrook. Randomized algorithms for multiprocessor page migration. In Lyle A. McGeoch and Daniel D. Sleator, editors, *On-line Algorithms*, volume 7 of *DIMACS Series in Discrete Mathematics and Theoretical Computer Science*, pages 135–150. AMS/ACM, 1992.

Author Index

Lecture Notes in Computer Science

For information about Vols. 1–4271

please contact your bookseller or Springer

Vol. 4313: T. Margaria, B. Steffen (Eds.), Leveraging Applications of Formal Methods. IX, 197 pages. 2006.

Vol. 4312: S. Sugimoto, J. Hunter, A. Rauber, A. Morishima (Eds.), Digital Libraries: Achievements, Challenges and Opportunities. XVIII, 571 pages. 2006.

Vol. 4311: K. Cho, P. Jacquet (Eds.), Technologies for Advanced Heterogeneous Networks II. XI, 253 pages. 2006.

Vol. 4309: P. Inverardi, M. Jazayeri (Eds.), Software Engineering Education in the Modern Age. VIII, 207 pages. 2006.

Vol. 4308: S. Chaudhuri, S.R. Das, H.S. Paul, S. Tirthapura (Eds.), Distributed Computing and Networking. XIX, 608 pages. 2006.

Vol. 4307: P. Ning, S. Qing, N. Li (Eds.), Information and Communications Security. XIV, 558 pages. 2006.

Vol. 4306: Y. Avrithis, Y. Kompatsiaris, S. Staab, N.E. O'Connor (Eds.), Semantic Multimedia. XII, 241 pages. 2006.

Vol. 4305: A.A. Shvartsman (Ed.), Principles of Distributed Systems. XIII, 441 pages. 2006.

Vol. 4304: A. Sattar, B.-h. Kang (Eds.), AI 2006: Advances in Artificial Intelligence. XXVII, 1303 pages. 2006. (Sublibrary LNAI).

Vol. 4303: A. Hoffmann, B.-h. Kang, D. Richards, S. Tsumoto (Eds.), Advances in Knowledge Acquisition and Management. XI, 259 pages. 2006. (Sublibrary LNAI).

Vol. 4302: J. Domingo-Ferrer, L. Franconi (Eds.), Privacy in Statistical Databases. XI, 383 pages. 2006.

Vol. 4301: D. Pointcheval, Y. Mu, K. Chen (Eds.), Cryptology and Network Security. XIII, 381 pages. 2006.

Vol. 4300: Y.Q. Shi (Ed.), Transactions on Data Hiding and Multimedia Security I. IX, 139 pages. 2006.

Vol. 4299: S. Renals, S. Bengio, J.G. Fiscus (Eds.), Machine Learning for Multimodal Interaction. XII, 470 pages. 2006.

Vol. 4297: Y. Robert, M. Parashar, R. Badrinath, V.K. Prasanna (Eds.), High Performance Computing - HiPC 2006. XXIV, 642 pages. 2006.

Vol. 4296: M.S. Rhee, B. Lee (Eds.), Information Security and Cryptology – ICISC 2006. XIII, 358 pages. 2006.

Vol. 4295: J.D. Carswell, T. Tezuka (Eds.), Web and Wireless Geographical Information Systems. XI, 269 pages. 2006.

Vol. 4294: A. Dan, W. Lamersdorf (Eds.), Service-Oriented Computing – ICSOC 2006. XIX, 653 pages. 2006.

Vol. 4293: A. Gelbukh, C.A. Reyes-Garcia (Eds.), MICAI 2006: Advances in Artificial Intelligence. XXVIII, 1232 pages. 2006. (Sublibrary LNAI).

Vol. 4292: G. Bebis, R. Boyle, B. Parvin, D. Koracin, P. Remagnino, A. Nefian, G. Meenakshisundaram, V. Pascucci, J. Zara, J. Molineros, H. Theisel, T. Malzbender (Eds.), Advances in Visual Computing, Part II. XXXII, 906 pages. 2006.

Vol. 4291: G. Bebis, R. Boyle, B. Parvin, D. Koracin, P. Remagnino, A. Nefian, G. Meenakshisundaram, V. Pascucci, J. Zara, J. Molineros, H. Theisel, T. Malzbender (Eds.), Advances in Visual Computing, Part I. XXXI, 916 pages. 2006.

Vol. 4290: M. van Steen, M. Henning (Eds.), Middleware 2006. XIII, 425 pages. 2006.

Vol. 4289: M. Ackermann, B. Berendt, M. Grobelnik, A. Hotho, D. Mladenič, G. Semeraro, M. Spiliopoulou, G. Stumme, V. Svatek, M. van Someren (Eds.), Semantics, Web and Mining. X, 197 pages. 2006. (Sublibrary LNAI).

Vol. 4288: T. Asano (Ed.), Algorithms and Computation. XX, 766 pages. 2006.

Vol. 4287: C. Mao, T. Yokomori (Eds.), DNA Computing. XII, 440 pages. 2006.

Vol. 4286: P. Spirakis, M. Mavronicolas, S. Kontogiannis (Eds.), Internet and Network Economics. XI, 401 pages. 2006.

Vol. 4285: Y. Matsumoto, R. Sproat, K.-F. Wong, M. Zhang (Eds.), Computer Processing of Oriental Languages. XVII, 544 pages. 2006. (Sublibrary LNAI).

Vol. 4284: X. Lai, K. Chen (Eds.), Advances in Cryptology – ASIACRYPT 2006. XIV, 468 pages. 2006.

Vol. 4283: Y.Q. Shi, B. Jeon (Eds.), Digital Watermarking. XII, 474 pages. 2006.

Vol. 4282: Z. Pan, A. Cheok, M. Haller, R.W.H. Lau, H. Saito, R. Liang (Eds.), Advances in Artificial Reality and Tele-Existence. XXIII, 1347 pages. 2006.

Vol. 4281: K. Barkaoui, A. Cavalcanti, A. Cerone (Eds.), Theoretical Aspects of Computing - ICTAC 2006. XV, 371 pages. 2006.

Vol. 4280: A.K. Datta, M. Gradinariu (Eds.), Stabilization, Safety, and Security of Distributed Systems. XVII, 590 pages. 2006.

Vol. 4279: N. Kobayashi (Ed.), Programming Languages and Systems. XI, 423 pages. 2006.

Vol. 4278: R. Meersman, Z. Tari, P. Herrero (Eds.), On the Move to Meaningful Internet Systems 2006: OTM 2006 Workshops, Part II. XLV, 1004 pages. 2006.

Vol. 4277: R. Meersman, Z. Tari, P. Herrero (Eds.), On the Move to Meaningful Internet Systems 2006: OTM 2006 Workshops, Part I. XLV, 1009 pages. 2006.

Vol. 4276: R. Meersman, Z. Tari (Eds.), On the Move to Meaningful Internet Systems 2006: CoopIS, DOA, GADA, and ODBASE, Part II. XXXII, 752 pages. 2006.

Vol. 4275: R. Meersman, Z. Tari (Eds.), On the Move to Meaningful Internet Systems 2006: CoopIS, DOA, GADA, and ODBASE, Part I. XXXI, 1115 pages. 2006.

Vol. 4274: Q. Huo, B. Ma, E.-S. Chng, H. Li (Eds.), Chinese Spoken Language Processing. XXIV, 805 pages. 2006. (Sublibrary LNAI).

Vol. 4273: I. Cruz, S. Decker, D. Allemang, C. Preist, D. Schwabe, P. Mika, M. Uschold, L. Aroyo (Eds.), The Semantic Web - ISWC 2006. XXIV, 1001 pages. 2006.

Vol. 4272: P. Havinga, M. Lijding, N. Meratnia, M. Wegdam (Eds.), Smart Sensing and Context. XI, 267 pages. 2006.